Beetroot

Beetroot has been drawing increasing scientific and industrial interest due to its high nutritional value and range of potential applications. *Beetroot: Cultivation, Processing, and Food Applications* presents a roadmap for the entire lifecycle of beetroot, covering beetroot's nutritional value, health benefits, phytochemicals, as well as postharvest management. Featuring recent technological advancements, this book presents processing techniques and valorization strategies, expanding on the potential applications of beetroot in pharmacology and the packaging industry. This book will serve as a valuable resource for researchers, scientists, food technologists, health professionals, and industry stakeholders.

Beetroot

Cultivation, Processing, and Food Applications

Edited by
Zakir Showkat Khan, Sajad Ahmad Wani,
and Thameed Aijaz

CRC Press
Taylor & Francis Group
Boca Raton London New York

CRC Press is an imprint of the
Taylor & Francis Group, an **informa** business

Designed cover image: Shutterstock

First edition published 2026
by CRC Press
2385 NW Executive Center Drive, Suite 320, Boca Raton FL 33431

and by CRC Press
4 Park Square, Milton Park, Abingdon, Oxon, OX14 4RN

CRC Press is an imprint of Taylor & Francis Group, LLC

ISBN: 9781032857183 (hbk)
ISBN: 9781032863405 (pbk)
ISBN: 9781003527084 (ebk)

DOI: 10.1201/9781003527084

Typeset in Palatino
by codeMantra

Contents

Preface

Beetroot (*Beta vulgaris* L.), a vibrant red root vegetable, has emerged as a functional food of growing scientific and industrial interest. Its rich composition of bioactive constituents—including vitamins, minerals, phenolic compounds, carotenoids, nitrates, ascorbic acid, flavonoids, polyphenols, saponins, triterpenes, and betalains—contributes significantly to its therapeutic potential. These phytochemicals exhibit remarkable antibacterial, antioxidant, and pharmacological properties, making beetroot a promising candidate in the realm of nutrition and health promotion.

In recent years, beetroot has garnered considerable attention not only for its health benefits but also for its versatility in the food processing industry. Particularly, the betalain pigments, known for their non-toxic and non-carcinogenic nature, have found wide application as natural food colorants—enhancing the sensory appeal of tomato pastes, soups, sauces, and other products. Moreover, the incorporation of beetroot into various food matrices offers an excellent opportunity for developing innovative functional foods with added health benefits.

This book, comprising ten comprehensive chapters, provides an in-depth exploration of beetroot from multiple perspectives. It aims to offer a roadmap that encompasses postharvest management, nutritional profiling, extraction and characterization of phytochemicals, and therapeutic potential of beetroot and its derivatives. The volume further delves into its pharmacological applications, juice processing techniques, waste valorization strategies, and incorporation in the packaging sector. Special emphasis is placed on recent technological advancements, including both thermal and non-thermal processing techniques, positioning the book at the intersection of academic research and industrial innovation.

By compiling contemporary research findings and practical insights, this book aspires to serve as a valuable resource for researchers, scientists, food technologists, health professionals, and industry stakeholders. It not only highlights the multifaceted applications of beetroot but also discusses the opportunities and challenges associated with its processing and commercialization.

We hope this work stimulates further research and innovation in the domain of beetroot utilization and inspires novel approaches for functional food development and sustainable food processing.

About the Editors

Dr. Zakir Showkat Khan is currently working as an Assistant Professor, TEQIP III, in the Department of Food Technology, Islamic University of Science and Technology, Awantipora, J&K, India. He has completed his Master's degree in Food Technology from Sant Longowal Institute of Engineering and Technology, Punjab, India, and he is currently pursuing a Ph.D. from Guru Nanak Dev University, Amritsar, Punjab, India. Qualified Graduate Aptitude Test in Engineering (2011) was conducted by the Ministry of Human Resource Development (MHRD), India, in February 2011. He has published more than 18 research/review articles and two book chapters. He has attended ten national and five international conferences, seminars, and workshops throughout the world. He has participated in more than ten faculty development programs. He is a Potential Peer Reviewer of reputed international journals related to Food Science and Technology that belong to renowned publishing houses, such as Elsevier, Taylor Francis, Wiley, and Springer. He is also working as a member of various associations, both national and international. He is the recipient of prestigious research grants such as Collaborative Research Scheme TEQIP III funded by MHRD (2019–2021) and Entrepreneurship Development Scheme (2019) funded by the Government of India.

Dr. Sajad Ahmad Wani is currently working as a Lecturer in the Department of Food Science and Technology, Government Degree College Shopian, Kashmir, J&K, India. He has completed his Master's degree in Food Technology from IUST, Kashmir, India, and Ph.D. from Sant Longowal Institute of Engineering and Technology, Punjab, India. Qualified National Eligibility Test (2015) (NET) was conducted by the Indian Council of Agricultural Research (ICAR) in December 2015. He has published more than 40 research/review articles, ten book chapters, and ten books. He is also writing articles for *Food & Beverage News* (India's first magazine for Food & Beverage Industry). He has attended 50 national and 20 international conferences, seminars, and workshops throughout the world. He has participated in various faculty development programs. He served as a Honorary Associate Editor of the esteemed *International Journal of Food Science and Technology*, Wiley. He is a Potential Peer Reviewer of reputed international journals related to Food Science and Technology that belong to renowned publishing houses, including Elsevier, Taylor-Francis, Wiley, and Springer. He is the recipient of prestigious fellowships such as Dr. D. S. Kothari Post-doctoral Fellowship-2021 and Maulana Azad National Fellowship (MANF-2013-14) from University Grant Commission, India.

Dr. Thameed Aijaz is a distinguished figure in food engineering, renowned for his groundbreaking research and contributions. With a Ph.D. in Chemical Engineering from IIT Delhi, he currently serves as an Assistant Professor in Food Technology at the Islamic University of Science and Technology. His research focuses on solving complex problems in food engineering, from waste valorization to processing methodologies, integrating his chemical engineering background for innovative solutions. His work is widely published in reputed journals, and he has authored numerous chapters in academic volumes, influencing both academia and industry practices. He actively promotes collaboration and knowledge exchange through conferences and workshops, fostering advancements in food engineering. His journey from a passionate scholar to a trailblazing researcher and educator showcases his dedication to advancing the field, leaving a lasting impact on scientific innovation.

Contributors

Zara Arshad
School of Health Sciences
University of Management and Technology
Lahore, Pakistan

Sena Bakir
Department of Gastronomy and Culinary Arts, Tourism Faculty
Recep Tayyip Erdogan University
Rize, Türkiye
and
Blueberry Application and Research Center
Recep Tayyip Erdogan University
Rize, Türkiye

Moufida Chaari
Laboratory of Microbial Biotechnology and Enzymatic Engineering
Centre of Biotechnology of Sfax
University of Sfax
Sfax, Tunisia

Aishwary Dubey
Department of Food Science and Technology
Guru Nanak Dev University
Amritsar, India

Ceyda Durmaz
Faculty of Health Sciences
Nutrition and Dietetics Department
Istanbul Aydin University
Istanbul, Türkiye

Buşra Goz
Faculty of Health Sciences
Nutrition and Dietetics Department
Istanbul Aydin University
Istanbul, Türkiye

Zehra Gulsunoglu-Konuskan
Faculty of Health Sciences
Nutrition and Dietetics Department
Istanbul Aydin University
Istanbul, Türkiye

Elahesadat Hosseini
Department of Chemical Engineering
Payame Noor University
and
National Nutrition and Food Technology Research Institute
Faculty of Nutrition Sciences and Food Technology
Shahid Beheshti University of Medical Sciences
Tehran, Iran

Momina Javed
School of Health Sciences
University of Management and Technology
Lahore, Pakistan

Varsha Kanojia
Punjab Horticultural Postharvest Technology Centre
Punjab Agricultural University
Ludhiana, India

Amritpreet Kaur
Department of Botanical and Environmental Sciences
Guru Nanak Dev University
Amritsar, India

Sonali Khanal
School of Bioengineering and Food Technology
Shoolini University of Biotechnology and Management Sciences
Solan, India

Dinesh Kumar
School of Bioengineering and Food Technology
Shoolini University of Biotechnology and Management Sciences
Solan, India
and
Centre of Advanced Innovation Technologies
VSB – Technical University of Ostrava
Ostrava-Poruba Czech Republic

Supriya Kumari
Department of Food Science and Technology
Guru Nanak Dev University
Amritsar, India

Azime Irem Kuruhalil
Faculty of Health Sciences
Nutrition and Dietetics Department
Istanbul Aydin University
Istanbul, Türkiye

Abdul Mateen
Centre for Ayurveda Biology and Holistic Nutrition
The University of Trans-Disciplinary Health Sciences and Technology
Bengaluru, India

Swasti Mudgal
Centre for Ayurveda Biology and Holistic Nutrition
The University of Trans-Disciplinary Health Sciences and Technology
Bengaluru, India

Kritika Pandit
Department of Botanical and Environmental Sciences
Guru Nanak Dev University
Amritsar, India

Sadia Rafiq
University of Management and Technology
Lahore, Pakistan

Rishi Richa
College of Agricultural Engineering and Technology
Shere - Kashmir University of Agricultural Sciences and Technology
Shalimar, India

Anu Sharma
P.G. Department of Food Science & Technology
Khalsa College
Amritsar, India

Sakshi Sharma
Amity Institute of Biotechnology
Amity University
Noida, India

Shabnum Showkat
Division of Food Science and Technology
Shere - Kashmir University of Agricultural Sciences and Technology
Shalimar, Srinagar, India

Slim Smaoui
Laboratory of Microbial Biotechnology and Enzymatic Engineering
Centre of Biotechnology of Sfax
University of Sfax
Sfax, Tunisia

Kanchan Suri
P.G. Department of Agriculture
Khalsa College
Amritsar, India

Divyesh Suvedi
School of Biotechnology
Shoolini University of Biotechnology and Management Sciences
Solan, India

Zenebe Tadesse Tsegay
Department of Food Science and Post-Harvest Technology
College of Dryland Agriculture and Natural Resources
Mekelle University
Mekelle, Ethiopia

Hafsa Tahir
University of Management and Technology
Lahore, Pakistan

Theodoros Varzakas
Department of Food Science and Technology
University of the Peloponnese
Kalamata, Greece

Rachna Verma
School of Biological and Environmental Science
Shoolini University of Biotechnology and Management Sciences
Solan, India
and
Centre of Advanced Innovation Technologies
VSB – Technical University of Ostrava
Ostrava-Poruba Czech Republic

Sajad Ahmad Wani
Department of Food Science and Technology
Government Degree College Shopian
Kashmir, India

Neelesh Yadav
School of Biotechnology
Shoolini University of Biotechnology and Management Sciences
Solan, India

1

Beetroot: A Potent Food

Buşra Goz, Azime Irem Kuruhalil, Ceyda Durmaz, Sena Bakir, and Zehra Gulsunoglu-Konuskan

1.1 Introduction

In recent years, there has been a growing interest in functional foods. Among these, beetroot has garnered significant attention due to its rich nutritional profile and diverse range of bioactive compounds. Traditionally consumed as a root vegetable, beetroot is increasingly recognized as a functional food with therapeutic potential, offering antioxidant, anti-inflammatory, antidiabetic, and cardiovascular protective effects (Cervantes-Paz et al., 2016; Knez et al., 2022).

Belonging to the Chenopodiaceae family, beetroot (*Beta vulgaris* L.) is characterized by its deep red to purple color, primarily attributed to a unique group of phytonutrients known as betalains. These pigments, along with a wide array of vitamins, minerals, polyphenols, nitrates, and amino acids, contribute to beetroot's impressive pharmacological activity. Beyond the root, the leaves and stems of the plant are also nutrient-dense and hold potential for use in food and pharmaceutical applications (Mudgal et al., 2022; Vaitkevičienė et al., 2022).

The health-promoting properties of beetroot are being increasingly explored not only in traditional culinary practices but also in modern medicine and dietary interventions. It is natural nitrate (NO_3^-) content, for instance, has been linked to improved endothelial function and reduced blood pressure, making it a promising dietary option in the management of cardiovascular diseases. Furthermore, its antioxidant rich profile offers protection against oxidative stress, a key contributor to chronic diseases such as cancer and neurodegenerative disorders (Stoica et al. 2025). Beetroot's versatility and bioactive richness make it a valuable candidate in the development of functional foods and nutraceutical products. This chapter aims to provide an overview of the nutritional profile, utilization, and applications of beetroot as a potent food in promoting health and preventing disease.

1.2 Nutritional Composition of Beetroot

Beetroot contributes significantly to health maintenance and disease prevention due to its abundance in both macro- and micronutrients, as well as its diverse bioactive compounds. Its health effects, however, can vary depending on dosage and consumption method (Hord

DOI: 10.1201/9781003527084-1

et al., 2009). Red beetroot is considered one of the top ten antioxidant-rich vegetables due to its mineral and vitamin contents and the pharmacological efficacy of its phytochemicals (Masih et al., 2019). It is a moderate-calorie vegetable, rich in carbohydrates, dietary fiber, and proteins, with low fat content. Beetroot predominantly contains sucrose and minor amounts of glucose and fructose (Knez et al., 2022). However, the nutritional composition of beetroot varies with environmental and agricultural factors such as temperature, pH, irrigation, soil fertility, plant variety, and processing methods (Sawicki et al., 2016).

Beetroot contains an average of 1.89% protein in the tubers and 2.20% protein in the leaves per 100 g of fresh weight (Ceclu and Nistor, 2020). Additionally, the protein content of peel, pomace, stems, leaves, and juice was reported as 1.02%, 1.13%, 2.45%, 5.64%, and 1.21%, respectively (Abdo et al., 2020). The main amino acid found in beetroot was reported as glutamic acid, followed by aspartic acid and alanine (Moulick et al. 2023). The amino acid content between tubers and leaves varies significantly, and leaves, in particular, contain all essential amino acids. Especially with the drying processes applied to the leaves and tubers, they can provide approximately 50% of the daily requirement for histidine amino acid, 67% for threonine, 42% for tubers, and more than 50% of the daily requirement for isoleucine and leucine (Mella et al., 2024).

Regarding carbohydrates, beetroot contains a total of 9.56 g of carbohydrates per 100 g (USDA, 2019). A review study reported an average amount of carbohydrates as 7.23 g in the tuber part and 4.33 g in the leaf part (Ceclu and Nistor, 2020). In addition, it contains 6.76 g of sugar in the tuber part. Regarding dietary fiber, there are 3.25 g in the tuber and 3.7 g in the leaf part per 100 g of fresh weight (Ceclu and Nistor, 2020). Dietary fiber is crucial in promoting overall health by supporting digestive function, improving heart health, regulating blood sugar levels, aiding in weight management, and potentially reducing the risk of certain cancers. It helps maintain regular bowel movements, feeds beneficial gut bacteria, and slows the absorption of sugar and cholesterol, contributing to lower risks of type 2 diabetes and cardiovascular disease. Additionally, high-fiber foods promote satiety, helping to control appetite and support a healthy weight. Incorporating a variety of fiber-rich foods like fruits, vegetables, whole grains, and legumes in the diet is essential for long-term health and well-being (Anderson et al., 2009).

Beetroot has a structure that is poor in saturated fatty acids and rich in mono- and polyunsaturated fatty acids. Fatty acids are concentrated in the leaf parts. Especially, omega-3 fatty acids found in the leaf parts are among the main sources of fatty acids (Mella et al., 2024). The total oil amount is 0.15 g in the tuber part and 0.13 g in the leaf part per 100 g of fresh weight (Ceclu and Nistor, 2020).

Beetroots are rich in water-soluble vitamins like C and B group vitamins (thiamine, riboflavin, niacin, pantothenic acid, pyridoxine, folates, and cyanocobalamin) as well as fat-soluble vitamins (A, E, and K) (Bangar et al., 2022). B group vitamins found in beetroot help to reduce dementia and memory loss. Additionally, beetroot leaves can decrease blood pressure due to their high content of vitamin A and vitamin K (Yashwant, 2015). 100 g contains 6.326 IU of vitamin A, 30 mg of vitamin C, 0.106 μg of vitamin B6 in the leaves, 109 μg of B9, and 0.334 μg of vitamin B3 in the stem (USDA, 2019; Ceclu and Nistor, 2020). Minerals in beetroot, like manganese, iron, sodium, potassium, magnesium, calcium, and copper, are associated with the strengthening of bones, teeth, and tissues. Moreover, beetroot leaves contain a high amount of iron (2.57 mg/100 g), copper (0.191 mg/100 g), magnesium (70 mg/100 g), calcium (117 mg/100 g), and potassium (762 mg/100 g) (Bangar et al., 2022; Ceclu and Nistor, 2020). It is reported that green leaves and stems of beetroot can be effective in weight management due to their low energy content and can be effective in the production of proteins for bone health due to the vitamins A, K, and C, as well as calcium and iron (Suganya et al., 2017; Pavoković and Krsnik-Rasol, 2011). It was reported that when 1 g of beet leaves is consumed per day, it can meet the entire recommended daily intake for calcium and magnesium, while consuming the root parts can only meet the entire recommended daily intake of magnesium. However, it is reported that when 5–10 g is consumed per day, it can meet more than the entire recommended daily intake for all minerals (Mella et al., 2024).

Beetroot leaves are richer in carotenoids compared to the tuberous part. The most important carotenoids in red beet are β-carotene and lutein (Ceclu and Nistor, 2020). While α-carotene (22.0 μg/100 g) is found densely in the stem part, β-carotene (11.64 μg/100 g) and lutein+zeaxanthin (1.503 μg/100 g) are found in the leaf part. Beetroot is also a good source of NO_3^- (644–1,800 mg/kg), which exerts cardioprotective effects, helps modulate mitochondrial function, regulates blood pressure, and reduces cholesterol (Guldiken et al., 2016). Beetroot is also rich in some powerful antioxidants, such as triterpenes, sesquiterpenoids, carotenoids, coumarins, flavonoids (tiliroside, astragalin, rhamnositrin, rhamnetin, and kaempferol), betalains, and phenolic compounds. Other bioactive compounds found in beetroot include saponins, alkaloids (calystegin B1, calystegin B2, calystegin C1, calystegin B3, and ipomine), and tannins (Ceclu and Nistor, 2020; Mudgal et al., 2022).

Beetroot's distinctive color is due to betalains, which are subdivided into betacyanins (betanin, isobetanin, and neobetanin) and betaxanthins (vulgaxanthin I, II, and indicaxanthin) (Silva et al., 2019). It is especially dense in the tubers (129 μg/100 g) (Ceclu and Nistor, 2020). These compounds are mainly used for natural colorants in the food industry, while they have also received increasing attention due to their possible health effects, especially antioxidant and anti-inflammatory activities (Shaban et al., 2021; Ceclu and Nistor, 2020; Kavalcová et al., 2015; Mudgal et al., 2022). Betanin, in particular, has shown anti-tumorigenic effects by neutralizing free radicals and managing oxidative stress (Masih et al., 2019; Naseer et al., 2019). Scientific studies have demonstrated various health effects of beetroot and its bioactive compounds. Betalains exhibit strong antiradical activity, alone or in combination with other extracts (Spiegel et al., 2021). Compounds such as ferulic acid (stems 59.18 μg/100 g) aid in liver glycogenesis and reduce metabolic stress and inflammation (Lorizola et al., 2021; Al-Harbi et al., 2021; Paliwal et al., 2016). In addition, betaine demonstrates chemopreventive potential by inducing apoptosis and regulating oxidative stress (Lechner and Stoner, 2019). Thanks to its antioxidant properties and the bioactive compounds, it can be effective in preventing both infectious and non-infectious diseases by showing antitumorigenic, anti-inflammatory, antidiabetic, antilipidemic, antidepressant, antimicrobial, antifungal, hepatoprotective, and hypotensive effects (Mirmiran et al., 2020; Ceclu and Nistor, 2020; Ingle et al., 2017; Kale et al., 2018).

1.3 Toxic and Antinutritional Factors in Beetroot

It is known that vegetables are protective foods against many chronic diseases in terms of vitamins, minerals, and fiber (Slavin and Lloyd, 2012). However, in addition to these beneficial components, some vegetables, such as beetroot, contain substances like NO_3^- and nitrites (NO_2^-) that could pose health risks under certain conditions (Greer and Shannon, 2005). It has also been reported that beetroot contains antinutrient compounds such as oxalates and saponins (Siener et al., 2016; Spórna-Kucab et al., 2022). In addition, atmospheric and soil pollution causes agricultural products to be contaminated with heavy metals that are thought to have carcinogenic effects, and it has been stated that beetroot may also contain heavy metals such as cadmium and lead (Mohajer et al., 2014).

1.4 Nitrites and Nitrates in Beetroot

Beetroots contain compounds such as NO_3^- and NO_2^-, which are thought to pose a health risk (Greer and Shannon, 2005). Since excessive NO_3^- intake may pose a health risk, the maximum acceptable daily intake (ADI) has been determined to prevent harm. For NO_3^-, the ADI is 0–5 mg/kg body weight of NO_3^- ions (corresponding to 0–3.7 mg of $NaNO_3$), while for NO_2^-, it is 0–0.2 mg/kg body weight of NO_2^- ions (corresponding to 0–0.07 mg of $NaNO_2$) (Mortensen et al., 2017a, b). Vegetables are estimated to provide 80%–85% of the dietary NO_3^- consumed (van Velzen et al., 2008a).

Beetroot is consumed in many forms, such as fresh vegetables, pickles, fruit juices, and chips (Guldiken et al., 2016). It is thought that beet-derived products containing inorganic NO_2^- in appropriate concentrations may be a more effective ergogenic agent because they act faster than a product containing only NO_3^- salts (Broeder et al., 2021). Food supplements containing *Beta vulgaris* L., produced in the form of lozenges, tablets, capsules, and powders, are also used. It is thought that various potential risks may arise, such as excessive NO_3^- and/or NO_2^- intake and overloading the body with a product that does not have health-promoting properties due to the uncertainty about the effectiveness of such products compared to fresh vegetables and the content of bioactive substances in the products (Brzezińska-Rojek et al., 2023). Total NO_3^- intake from all sources should be monitored, particularly in children with low body weight, because of the risk of exceeding the maximum ADI. It is thought that NO_2^- poisoning may lead to the development of neoplasms or methemoglobinemia through the formation of N-nitrosamines, which are known to be carcinogenic (van Velzen et al., 2008b).

The conversion of vegetable-derived NO_3^- to NO_2^-, which produces methemoglobin, before and/or after ingestion, is considered potentially dangerous (Greer et al., 2005). It is thought that methemoglobin cannot bind to oxygen, causing a shift in the oxygen dissociation curve and causing hypoxemia (Chan, 1996). It is thought that avoiding excessive use of nitrogenous fertilizers to reduce NO_3^- accumulation in soil and vegetables can also reduce NO_3^- accumulation. It has been suggested that bacterial contamination can be reduced and the conversion of NO_3^- to NO_2^- can be prevented by proper storage and processing of vegetables. It has been reported that peeling beetroot can reduce NO_3^- content by 20%–62% (Chan, 2011).

1.5 Oxalates in Beetroot

Beetroot contains a significant amount of oxalic acid, known as an antinutrient compound. Oxalic acid is reported to prevent the absorption of calcium and iron and is also a strong metal chelator that triggers the formation of kidney stones (Lee et al., 2005). A salt formed from oxalic acid is called an oxalate: for example, calcium oxalate, which is found to be widely distributed in plants. Oxalate salts are formed by the formation of strong bonds between oxalic acid and minerals such as calcium, sodium, magnesium, and potassium (Gemede and Ratta, 2014).

It has been reported that beetroot juice contains an average of 800–1,000 mg oxalic acid/100 mL and that beets contain high levels of oxalate (European Commission, 2016). Some oxalate salts are soluble (e.g., sodium and potassium), but calcium oxalate salts are essentially insoluble. This insoluble calcium oxalate can precipitate in the kidney and/or urinary tract and form sharp-edged calcium oxalate crystals. It has been reported that these crystals, excreted in the urine, can cause kidney stones in the urinary tract (Fekadu Gemede, 2014). It has also been reported that oxalates can bind with minerals known to be important for health, such as calcium and magnesium, and their absorption can be prevented. It is thought that diets high in oxalates may increase the risk of rheumatoid arthritis and gout. Since beetroot juice is rich in oxalates, it is thought that regular consumption may pose a potential health risk in the long term (European Commission, 2016).

1.6 Saponins in Beetroot

Studies have shown that beetroot products contain saponin (Mikołajczyk-Bator et al., 2024; Stoica et al., 2025). Saponins are secondary compounds of plant origin (Kiranmayi, 2014). It has been reported that saponin content and types may vary depending on the plant cultivar and food processing (Da Silva et al., 2016). *In vivo* and *in vitro* studies have reported that saponins can exhibit antimutagenic, anti-inflammatory, antibacterial, antifungal, antiviral, and antitumor activities (Dong et al., 2020; Elekofehinti et al., 2021; Hu et al., 2025; Melek et al., 2015; Song et al., 2016). In addition, it is thought that extracts of plant materials containing triterpene saponins can be used in the production of nutraceuticals and functional foods (Mikołajczyk-Bator et al., 2024). On the other hand, it is reported that saponins can cause health problems related to digestion by inhibiting digestive enzymes such as glucosidase, amylase, lipase, chymotrypsin, and trypsin (Samtiya et al., 2020). Since digestive enzymes found in humans cannot easily hydrolyze saponins, they can therefore cause disorders in gastrointestinal digestion (Amin et al., 2011). In addition, it is thought that saponins can reduce vitamin absorption (Samtiya et al., 2020).

1.7 Heavy Metals in Beetroot

The level of heavy metals in food can be affected by the type of food, growing conditions, soil type, water, and agricultural practices (Figure 1.1). Additionally, the use of fertilizers containing heavy metals like cadmium and anthropogenic pollution can also influence the heavy metal content in foods (Schulte-Schrepping and Piscator, 2011). Lead, zinc, and cadmium can be released into the environment through both natural and anthropogenic processes such as volcanic eruptions, mining, and the extraction of metals from ores (Ćwieląg-Drabek et al., 2020). It is thought that beetroots tend to accumulate toxic elements known to be harmful, such as mercury, lead, and cadmium (Figure 1.1) (Ćwieląg-Drabek et al., 2020; Saletnik et al., 2016).

In a study conducted in Serbia, the amounts of heavy metals accumulated in the roots of beetroot were examined (Pajević et al., 2018), and the results showed that the average cadmium amount in beetroot was reported as approximately 0.1 μg/g. The maximum cadmium amount allowed by FAO/WHO was declared as 0.1 μg/g (FAO/WHO, 2001). While the limit value determined by FAO/WHO for the average lead amount was 0.3 μg/g

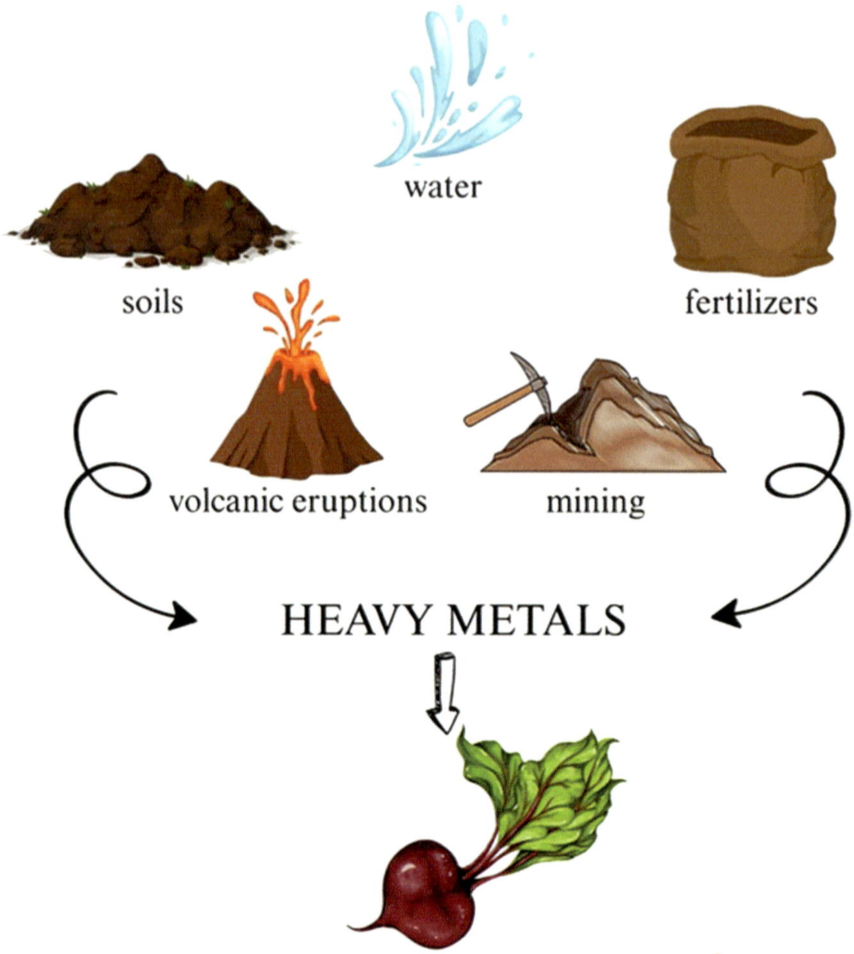

FIGURE 1.1
Reasons for the presence of heavy metals in vegetables.

(FAO/WHO, 2001), the average lead amount in beetroot was determined as 1.3 μg/g (Pajević et al., 2018). The average chromium amount in the beetroot samples was determined as 1.4 μg/g (Pajević et al., 2018), and it was reported that this value was below the limit value recommended by FAO/WHO (FAO/WHO, 2001). In addition, the average nickel amount in beetroot was determined as 0.7 μg/g, and it was seen that it had a lower nickel level than the other vegetables (Pajević et al., 2018).

There are studies showing that beetroots grown near mining or industrial areas can contain high amounts of manganese and zinc (Brzezińska-Rojek et al., 2024; Ćwieląg-Drabek et al., 2020). Toxic metals are thought to alter the pH value of the cytoplasm and the electrical potential of the cell membrane, thus disrupting plant metabolism (Brzezińska-Rojek et al., 2024). This disruption can lead to a decrease in the mineral content and bioactive composition of the vegetable (Brzezińska-Rojek et al., 2024). It has been reported that beetroots, along with carrots, are among the vegetables that accumulate the highest levels of heavy metals (Ćwieląg-Drabek et al., 2020; Dziubanek et al., 2017).

FIGURE 1.2
Examples of beetroot-based products across culinary and functional domains.

1.8 Utilization of Beetroot

Beetroot is used across various segments of processed food—including bakery, dairy, and dessert formulations—not only to enhance nutritional properties but also to deliver functional benefits. These beetroot-enriched products not only attract consumers through visual appeal but also contribute to overall product functionality and health-promoting effects. Figure 1.2 shows the example of beetroot-based products used for culinary and functional purposes in the food industry.

1.9 Traditional and Gastronomic Uses of Beetroot

In recent years, red beetroot and its derived functional products have attracted significant attention in the food industry due to their health-promoting properties and potential use as natural food additives (Akan et al., 2021). Rich in phytochemical compounds and natural pigments, beetroot continues to stand out as a valuable ingredient in both food and health-related applications (Chhikara et al., 2019; Tan et al., 2021). Interest in its use as a functional food additive has been steadily increasing, especially given its contributions to health (Punia Bangar et al., 2022).

Beetroot has been used for centuries due to its nutritional profile and health benefits. Historically, it has been utilized for medicinal purposes since the Roman era (Akan et al., 2021). Deep red varieties of beetroot are consumed both raw and cooked in human diets. In global cuisines, beetroot is commonly found in raw form as well as in salads, soups, vegetable dishes, baked recipes, and olive oil-based dishes (Akan et al., 2021). In particular, beetroot soup (such as borsch) is widespread in Eastern Europe, while in South America, pickled beetroot is recognized as a traditional food (Chhikara et al., 2019). Across various culinary cultures, beetroot is consumed in diverse forms such as juice, powder, bakery additives, gelled products, boiled or roasted roots, pickles, purées, and even jams (Punia Bangar et al., 2022). Despite its naturally high sugar content, beetroot remains a low-calorie and nutrient-dense vegetable, frequently used in salads and fruit-vegetable juices (Alshehry, 2019; Tan et al., 2021).

The morphological characteristics of beetroot significantly influence its processability and the ways it can be utilized in industrial applications. Belonging to the *Amaranthaceae*–Chenopodiaceae family, this plant is cultivated on a commercial scale. Often referred to as table beet or garden beet, this variety is primarily grown for its root (Tan et al., 2021). The edible part is predominantly the root, which may reach lengths of 1–2 m. Depending on the cultivar, the root can be spherical or cylindrical and varies in color, ranging from red-purple to golden yellow and even red-white striped. The leaves emerge from the hypocotyl and exhibit considerable variation in shape, size, and color (Chhikara et al., 2019). Its characteristic earthy aroma and mild flavor are largely attributed to geosmin, a volatile bicyclic alcohol compound present in the root (Tan et al., 2021).

1.10 Utilization of Beetroot in Functional and Processed Foods

Beetroot has gained recognition not only as a vegetable but also as a multifunctional ingredient in numerous value-added food products. Its rich color and nutritional profile make it highly versatile in processed food formulations. Examples of products incorporating beetroot include jams, confectionery, cupcakes, biscuits, cookies, noodles, pasta, jellies, mayonnaise, butter, and yogurt-based items. In such products, beetroot functions both as a natural colorant and a functional component, thus serving aesthetic and health-oriented purposes. In particular, its betalain content enhances visual appeal while contributing to antioxidant capacity and other functional properties (Narayanan et al., 2023).

In bakery applications, beetroot serves not only as a coloring agent but also as a functional additive. One study reported that the incorporation of beetroot in sponge cake formulations increased polyphenol content and antioxidant capacity, extended shelf life, and improved oxidative stability. These effects were further enhanced when combined with natural cocoa content. Notably, the addition of beetroot delayed staling during storage. However, it showed limited influence on lipid and protein oxidation in simulated digestion phases. These findings suggest that beetroot may contribute technologically to processed foods, especially those high in fat and protein (Ranawana et al., 2018).

Beetroot is not limited to solid products; it is also widely used in liquid-based functional formulations. For instance, one study compared the antioxidant capacities of beetroot-containing flavored milk and energy drinks, finding that the energy drink variant exhibited a higher functional potential. These beverages were also reported to possess balanced pH levels, low carbohydrate content, and a stable profile of proteins and phenolic compounds (Ashraf et al., 2022).

1.11 Beetroot as Natural Colorant

The red pigments of beetroot, known as betalains, are increasingly utilized as natural colorants in the food industry and are commonly incorporated into various product formulations (Akan et al., 2021). Due to this characteristic, betalains are positioned as promising natural alternatives to synthetic colorants in the market. While artificial dyes are known to cause allergic reactions and may lead to adverse health effects over time, beetroot-derived betalains—being water-soluble—can be easily incorporated into aqueous foods such as soups and jellies, enhancing both color and potential antioxidant activity. Growing consumer demand for "clean-label" or "green" products with fewer additives has also contributed to the widespread use of natural ingredients such as beetroot pigments. These pigments are considered safe, are generally non-allergenic, economically viable, and visually appealing (Chhikara et al., 2019).

Betalains are one of the four major classes of plant-based pigments available in the market, alongside anthocyanins, carotenoids, and chlorophylls. As natural red colorants, betalains are especially applied in dairy products, confections, gummy candies, and non-alcoholic beverages as alternatives to synthetic red dyes (Lombardelli et al., 2021). In addition, they are used in bakery goods, sweets, and ice cream. Their ability to provide natural red hues makes them suitable for dry mixes such as soup bases and powdered jellies (Alshehry, 2019).

Beetroot juice is also favored as a colorant in yogurt, processed cheeses, and confections. However, due to its thermal sensitivity, it is more frequently used in products that are processed at low temperatures or served cold. In some formulations, raw or freeze-dried beetroot components are used as substitutes for synthetic antioxidants (Chhikara et al., 2019).

For industrial applications, ensuring both high yield and structural stability of betalain pigments is essential. Studies show that traditional extraction methods often suffer from low pigment recovery and require significant energy and solvent input. In contrast, emerging technologies such as ultrasound-assisted extraction, microwave treatment, and pulsed electric field methods have demonstrated higher efficiency and better preservation of the betalain structure (Nirmal et al., 2021).

Nevertheless, betalain pigments remain highly sensitive to processing and storage conditions, posing a key challenge to their sustainable use in food manufacturing. Factors such as light, temperature, pH, and oxygen exposure can significantly affect pigment degradation (Fu et al., 2020).

1.12 Utilization of Beetroot By-Products in Functional Foods and Sustainability

Innovative studies have demonstrated the potential of beetroot by-products in the formulation of functional beverages. For instance, juice blends enriched with beet leaves and stalks have been shown to significantly enhance the levels of total phenolics, flavonoids, and betalains. These bioactive enrichments were associated with improvements in oxidative stress parameters, liver enzyme balance, and lipid profile regulation in high-fat diet animal models (Abdo et al., 2022).

Beetroot juice has also become increasingly popular in sports nutrition products due to its ability to boost nitric oxide (NO) production. Nevertheless, several controlled trials have reported inconsistent effects. For example, a post-marathon recovery study showed that athletes consuming beetroot juice did not experience significantly greater metabolic recovery compared to the placebo group within 48 hours. These findings suggest that individual variability may influence the efficacy of beetroot supplementation (Stander et al., 2021).

Sustainable utilization of beetroot waste materials has also been explored. One study formulated a protein- and antioxidant-rich functional beverage using whey protein, beet peel extract, and strawberry purée. Among the tested variations, the sample containing 2.5% beet peel extract demonstrated elevated levels of phenolic and flavonoid compounds, increased betalain content, and superior antioxidant capacity. Despite slight reductions during cold storage, the beverages maintained their functional properties over time and were well-received in sensory evaluations (Abdo et al., 2022). In another innovative formulation, marmalades enriched with red beetroot and sour cherry were developed to enhance functionality. These products showed notable increases in total phenolics, flavonoids, betalains, and anthocyanins. However, thermal treatment at 95°C for 30 minutes resulted in a 34% reduction in betalain content (Nistor et al., 2020).

Fermented beetroot juices have gained attention as functional beverages due to their high antioxidant capacity and potential probiotic benefits. A study evaluating six commercial variants—including plain, garlic-infused, radish, acerola, salt-free, and iodized versions—found that all had elevated levels of phenolics, flavonoids, and antioxidant capacity. Notably, the salt-free variant had a FRAP value of 5663.4 μM Fe(II)/L, while the iodized variant contained approximately 52 mg of vitamin C per 100 mL. These juices were positively evaluated in terms of taste, aroma, color, and overall acceptance (Jakubczyk et al., 2024).

Post-harvest beetroot residues, particularly rich in betalains, have attracted interest for their potential as sustainable sources of functional ingredients. In one study, beetroot peel extracts were encapsulated using a binary blend of gum Arabic and maltodextrin, resulting in powdered forms with high phenolic content, strong color quality, excellent solubility, and enhanced antioxidant capacity. These properties demonstrate the potential of valorizing food waste through circular bioeconomy approaches (Mkhari et al., 2023).

Another study explored the incorporation of beetroot peel powder (4%–10%) into meringue formulations as both a natural colorant and antioxidant agent. The enriched products showed increased phenolic concentrations, higher antioxidant capacity, and improved sensory properties such as color, aroma, and texture. These qualities remained stable over a 21-day storage period. The findings highlight the feasibility of repurposing beetroot by-products into value-added functional foods with both nutritional and visual appeal (Constantin et al., 2024).

The diversity of beetroot-derived products and their associated functional benefits has been extensively explored across multiple studies. To provide a comprehensive overview of these applications within food systems, selected examples are summarized in Table 1.1. These studies demonstrate how various beetroot matrices (e.g., juice, powder, and peel) contribute to enhanced phytochemical profiles, antioxidant activity, and improvements in sensory attributes and shelf-life stability across different product formulations.

TABLE 1.1

Beetroot Forms Used in Functional Products and Observed Effects

Product Type	Beetroot Form	Observed Effect	Additional Notes	Reference
Beetroot-enriched fruit juice (orange-based)	Leaves and stalks	↑Flavonoid, betalain, total phenolics	Reduced oxidative stress, liver enzyme regulation, improved lipid profile	Abdo et al. (2022)
Energy drink and flavored milk	Beetroot juice	↑Antioxidant capacity (higher in energy drink)	Balanced pH, low carbohydrate, stable phenolic-protein profile	Ashraf et al. (2022)
Biscuits	Beetroot powder (15%–25%)	↑Betalain and phenolic content, ↑Antioxidant properties	Bioactive retention during shelf life	Mitrevski et al. (2023)
Marmalade	Root beet+sour cherry	↑Total phenolics, flavonoids, betalains, anthocyanins	34% betalain loss after thermal treatment; strong gel structure	Nistor et al. (2020)
Fermented beetroot juice (various types)	Liquid beetroot	↑Flavonoids, phenolics, antioxidant capacity	Salt-free variant: FRAP 5663.4 µM Fe(II)/L; Iodized: 52 mg vitamin C/100 mL	Jakubczyk et al. (2024)
Meringue	Beetroot peel powder (4%–10%)	↑Phenolic compounds, ↑Antioxidant capacity	Improved color, aroma, texture; stable for 21 days	Constantin et al. (2024)

Note: ↑ indicates an increase. All information is derived from the reviewed literature and summarized accordingly.

1.13 Beetroot-Based Dietary Supplements

Red beetroot's biological activity and potential use as a functional food that promotes health and prevents disease have drawn increasing attention in recent years. Consuming beetroot, a NO_3^- source, offers a natural way to raise the availability of NO *in vivo* and has become a viable tactic to prevent and treat diseases linked to reduced NO bioavailability, including hypertension and endothelial dysfunction (Clifford et al., 2015). A strong vasodilator, NO, promotes blood flow and triggers a number of intracellular processes, including improved contractile and mitochondrial function (Zafeiridis, 2014). Direct ingestion of dietary NO_3^- and its subsequent conversion to NO_2^-, a NO-induced hypoxia signaling regulator, may enhance NO bioavailability. Consuming dietary NO_3^- improves endothelial dysfunction, lowers blood pressure, and guards against ischemia-reperfusion injury (Zafeiridis, 2014). Various clinical conditions linked to inflammation and oxidative stress are also being treated using beetroot, which is considered a viable therapeutic option. Its components, particularly the betalain pigments, exhibit strong anti-inflammatory, antioxidant, and chemopreventive properties both *in vitro* and *in vivo* (Clifford et al., 2015). It's also crucial to remember that a number of illnesses, such as type 2 diabetes, are characterized by chronic inflammation and an overabundance of reactive oxygen and nitrogen species (Cai and Harrison, 2000; Donath and Shoelson, 2011). Some studies related to beetroot-based dietary supplement consumption are listed in Table 1.2.

High blood pressure, often known as hypertension, is a prevalent condition that has been on the rise for a number of years. In order to effectively treat hypertension, the

TABLE 1.2

Studies About the Beetroot-Based Dietary Supplement Consumption

Participants	Dose	Application Conditions	Results	References
Fourteen (seven women and seven men)	70 mL of beetroot juice (~6.4 mmol NO_3^-) or placebo (~0.0055 mmol NO_3^- depleted)	Over a 10-minute baseline, blood pressure, continuous heart rate, and sympathetic nerve activity of muscle were recorded, along with discrete minute-to-minute brachial blood pressure readings.	• Reduction in central sympathetic output during rest and exercise.	Notay et al. (2017)
Fourteen moderately trained male swimmers	0.5 L/day organic beetroot juice (~5.5 mmol of NO_3^-)	Every swimmer executed the front crawl with a frequency of free movement, and athletes received food supplements for 6 days.	• Beneficial influence on skilled master swimmers' performance • Decrease in the cost of aerobic energy • Increased effort at the threshold for anaerobic	Pinna et al. (2014)
Ten healthy, recreationally active men	An acute dose of 70, 140, or 280 mL NO_3^- rich beetroot juice or 140 mL water (control)	Blood pressure measurement and obtainment of venous blood sample, 1, 2, 4, 8, 12, and 24 hours postingestion.	• Decrements in mean arterial pressure and systolic blood pressure that are dosage dependent	Wylie et al. (2013)
Ten male team-sport players	Intensity intermittent cycling tests using NO_3^- rich beetroot juice (8.2 mmol NO_3^-) and placebo (NO_3^- depleted beetroot juice; 0.08 mmol NO_3^-) throughout different 5-day supplementation periods	Twenty-four 6-second all-out sprints (24×6-second); seven 30-second all-out sprints (7×30-second); and six 60-second self-paced maximal efforts (6×60-second); on days 3, 4, and 5 of supplementation, respectively	• Improvement when engaging in specific forms of intermittent exercise	Wylie et al. (2016)
Older adults (~75 years)	A high NO_3^- diet (~12 mmol) including beetroot juice and a low NO_3^- diet as control	On day two, blood was taken before breakfast and one hour after breakfast. Following the blood draw one hour after breakfast, brain imaging was done right away	• Significant increments in the frontal cortical perfusion, a part of the brain involved in critical cognitive functions like executive function, working memory, and task-switching	Presley et al. (2011)
24 overweight participants	70 mL of concentrated beetroot juice for 21 days	Measurement of blood pressure: (i) resting clinic blood pressure, (ii) 24-hour ambulatory blood pressure monitoring, and (iii) daily resting blood pressure monitoring at home.	• Advantageous effects on daily systolic blood pressure	Jajja et al. (2014)
Male Sprague–Dawley rats (3–6 months)	Beetroot juice (1 mmol/kg NO_3^-, n=8) or tap water (control, n=11) for 5 days	Vascular conductance (radiolabeled microsphere infusions), mean arterial pressure, and blood flow to the hindlimb skeletal muscles were evaluated while jogging on a submaximal treadmill.	• Improvements in skeletal muscle oxygen transport during exercise, especially in fast-twitch type II muscles, and improvements in vascular regulation • Enhancements in metabolic regulation.	Ferguson et al. (2013)

American Heart Association has suggested the Mediterranean diet and dietary methods to reduce hypertension (Eckel et al., 2014). According to Kapil et al. (2010), males responded more strongly to raising their systolic blood pressure levels than women did. This is likely because men often have lower baseline blood pressure readings during premenopausal. Higher NO_2^- plasma levels may restrict the drop in blood pressure following food modification. Bonilla Ocampo et al. (2018) reviewed the use of beetroot juice's dietary NO_3^- for hypertension.

There is evidence that dietary supplementation with beetroot juice has a good effect on lowering blood pressure, primarily on systolic blood pressure compared to diastolic blood pressure, according to the results of 310 people who were represented across the analyzed trials. However, the characteristics of the investigated subjects (gender, age, nutritional status, and blood pressure) and the type of intervention carried out in the supplementation protocol (duration, juice volume, and NO_3^- concentration) are some factors that likely alter this response (Bonilla Ocampo et al., 2018). In a recent meta-analysis (Bahadoran et al., 2017), it was found that the impact of beetroot juice supplementation on systolic blood pressure is positively correlated with age. Consequently, further investigation is required to elucidate this age-dependent change following dietary NO_3^- supplementation.

Supplementing with beetroot juice and inorganic NO_3^- is linked to lower systolic blood pressure; however, these findings have mostly come from short-term studies involving young, healthy adults. Changes in systolic and diastolic blood pressure were measured both during and after supplementation. Supplementing with beetroot juice concentrate positively affected daily systolic blood pressure in overweight older adults (Jajja et al., 2014).

Recent research on humans looked at how dietary NO_3^- affected blood flow in the brain (Presley et al., 2011). In a group of older adults (~75 years old), magnetic resonance imaging was used to show that a high NO_3^- diet significantly and preferentially increased frontal cortex perfusion, a part of the brain involved in critical cognitive functions like working memory, executive function, and task-switching. In a study conducted with young persons (~21 years old), 500 mL of NO_3^--rich beetroot juice reduced cerebrovascular artery resistance, a sign of improved cerebral blood flow (Bond et al., 2013). It is crucial to remember that the study's participants were asymptomatic and seemingly disease-free, which restricts the applicability of the results in older and ill populations.

Muscle sympathetic nerve activity at rest was reduced by acute beetroot juice consumption, and this supplementation lessened handgrip exercise. Dietary NO_3^- is anticipated to provide cardio-vascular advantages due to reductions in sympathetic output (Notay et al., 2017). The idea that giving the exercising rat beetroot juice as a dietary NO_3^- supplement would lower mean arterial pressure and improve blood flow to the hindlimb muscles (Ferguson et al., 2013). This study offers a possible mechanism by which NO_3^- supplementation enhances metabolic control and supports the theory that it improves vascular control and increases skeletal muscle O_2 delivery during exercise, primarily in fast-twitch type II muscles (Ferguson et al., 2013). It is noted that dietary NO_3^- enhances vascular function; however, to determine whether NO_3/beetroot therapies may be useful tactics for the main and/or secondary prevention of endothelial dysfunction and atherosclerosis, long-term sustainability in vascular function needs to be assessed (Lara et al., 2016). To maximize the benefits of training and boost their athletic performance, athletes use nutritional supplements. NO, which is elevated in beetroot juice, has several benefits, including improved blood flow, gas exchange, mitochondrial biogenesis and efficiency, and muscle contraction strength (Domínguez et al., 2017). Nevertheless, some authors simplified that additional research is required to determine the ideal amount of beetroot to consume in order to improve athletes' performance during exercise, how dietary NO_3^- consumption affects training adaptations, how well beetroot supplements work to increase exercise tolerance in people with chronic illnesses, and whether long-term beetroot consumption is safe (Zafeiridis, 2014). Because of its high NO_3^- content, beetroot juice has been shown to improve performance (O_2 sparing and higher exercise tolerance), increasing plasma NO_2^- by 105% and lowering O_2 costs for constant work rate (Lansley et al., 2011). Beetroot juice, which contains roughly 5–8 mmol of inorganic NO_3^-, can be added to the diet to raise plasma NO_2^- content, lower blood pressure, and possibly improve the body's reactions to exercise. Young individuals' cardiovascular health and exercise performance are improved by beetroot juice drinking (Wylie et al., 2013). Supplementing with dietary NO_3^- lowers blood pressure and the oxygen cost of submaximal exercise in humans by reducing it to NO_2^- and then converting it to NO and other reactive nitrogen intermediates (Ferguson et al., 2013). Research has shown that supplementing with beetroot juice improves swimmers' performance by lowering their aerobic energy cost and raising their anaerobic threshold strain (Pinna et al., 2014). However, the circumstances of exercise should be established prior to supplementing with beetroot juice while engaging in high-intensity intermittent exercise (Wylie et al., 2016).

Furthermore, NO_3^- can alter calcium management following the translation of skeletal muscle contractile proteins. Beetroot can affect sports performance in a number of ways through NO. NO is an intracellular and extracellular messenger that causes blood vessels to dilate, increasing blood flow and controlling some cellular processes. Because it increases oxygen, glucose, and other nutrients for improved muscle fueling, NO appears to be useful in enhancing athletic performance (Arazi and Eghbali, 2021).

The profiles of 19 elements, including sodium, potassium, cadmium, and barium, were determined in order to evaluate the safety and health benefits of consuming the selected 37 beetroot-based dietary supplements (Brzezińska-Rojek et al., 2024). Factor analysis was used to distinguish between the pharmaceutical forms (capsules, tablets, and powders) and identify the source of the powder found in the capsules. In another research, the total mercury levels in 54 beetroot-based nutritional supplements and seven lots of conventional and organic beetroots (skinned, peeled, and unpeeled) were measured. The computed Target Hazard Quotient Index did not show the elevated non-carcinogenic risk brought on by consuming these products, and none of the examined beetroot samples or DSs exceeded the Provisional Tolerable Weekly Intake value. The difference in mercury content between skins and peeled and unpeeled beetroots was validated by statistical studies, with the latter having the lowest concentrations of this element. Since beetroot peels are the most polluted, peeling them can lower a consumer's exposure to mercury compounds. It is important to note that manufacturers do not disclose if peeled beetroots were utilized to produce their supplements (Brzezińska-Rojek et al., 2022). A different study evaluated commercially available 31 beetroot-based nutritional supplements in tablets and capsules based on specific physical parameters per Food and Drug Administration and United States Pharmacopoeia guidelines. The absence of particular standards in Europe may contribute to the improper grade of dietary supplements, posing a consumer risk (Brzezińska et al., 2021). In contrast to the United States of America, the European Union has not set any particular standards for the physicochemical characteristics of dietary supplements. Therefore, the European Union should consider criteria comparable to those in the United States. The study highlights the necessity of improving quality control over dietary supplements prior to their release onto the European market (Brzezińska et al., 2021). These findings emphasize how crucial it is to enforce stricter regulations and control mechanisms in the market for nutritional supplements containing beetroot in order to protect consumer safety (Brzezińska-Rojek et al., 2024).

1.14 Summary and Perspectives

Beetroot contains NO_3^-, which is responsible for cardioprotective, physiological, and metabolic benefits. Additionally, betalains and other phenolics are responsible for the anti-oxidative and anti-inflammatory properties. Consequently, foods rich in betalains and NO_3^- have a lot of potential as functional foods, as evidenced by their pharmacological effects. Due to its richness in nutritional and bioactive substances, red beetroot has emerged as a noteworthy ingredient in both the food industry and the health sciences, owing to its rich content of phytochemicals, natural pigments, and bioactive compounds. It has a broad potential for use in both traditional culinary applications and modern functional food innovations. Developing novel extraction techniques has played a critical role in facilitating the efficient recovery of bioactive components from beetroot. These technological advancements have enabled more sustainable integration of beetroot compounds into food formulations, thereby enhancing product functionality and overall health impact. In conclusion, the processing and incorporation of red beetroot into various functional and health-oriented products demonstrate considerable potential for future applications. Given its unique sensory attributes, nutritional richness, and technological adaptability, beetroot is expected to play an increasingly important role in advancing food science and product innovation.

References

Abdo, E. M., Allam, M. G., Gomaa, M. A. E., Shaltout, O. E., & Mansour, H. M. M. (2022). Valorization of whey proteins and beetroot peels to develop a functional beverage high in proteins and antioxidants. *Frontiers in Nutrition*, 9, 984891. https://doi.org/10.3389/fnut.2022.984891

Abdo, E. M., Shaltout, O. E., Ali, S., & Mansour, H. M. M. (2022). A functional orange juice fortified with beetroot by-products attenuates hyperlipidemia and obesity induced by a high-fat diet. *Antioxidants*, 11(3), 457. https://doi.org/10.3390/antiox11030457

Abdo, E., El-Sohaimy, S., Shaltout, O., Abdalla, A., & Zeitoun, A. (2020). Nutritional evaluation of beetroots (*Beta vulgaris* L.) and its potential application in a functional beverage. *Plants*, 9(12), 1752.

Akan, S., Tuna Gunes, N., & Erkan, M. (2021). Red beetroot: health benefits, production techniques, and quality maintaining for food industry. *Journal of Food Processing and Preservation*, 45(10), e15781.

Al Harbi, L. N., Pandurangan, S. B., Al-Dossari, A. M., Shamlan, G., Salamatullah, A. M., Alshatwi, A. A., & Alotiby, A. A. (2021). *Beta vulgaris rubra* L. (beetroot) Peel methanol extract reduces oxidative stress and stimulates cell proliferation via increasing VEGF expression in H_2O_2 induced oxidative stressed human umbilical vein endothelial cells. *Genes*, 12(9), 1380.

Alshehry, G. A. (2019). Utilization of beetroot as a natural antioxidant, pigment and antimicrobial in cupcake during the storage period. *International Journal of Engineering Research & Technology*, 8(10), 652–659.

Amin, H. A. S., Hanna, A. G., & Mohamed, S. S. (2011). Comparative studies of acidic and enzymatic hydrolysis for production of soyasapogenols from soybean saponin. *Biocatalysis and Biotransformation*, 29(6), 311–319. https://doi.org/10.3109/10242422.2011.632479

Anderson, J. W., Baird, P., Davis Jr, R. H., Ferreri, S., Knudtson, M., Koraym, A., & Williams, C. L. (2009). Health benefits of dietary fiber. *Nutrition Reviews*, 67(4), 188–205.

Arazi, H., & Eghbali, E. (2021). Possible effects of beetroot supplementation on physical performance through metabolic, neuroendocrine, and antioxidant mechanisms: a narrative review of the literature. *Frontiers in Nutrition*, 8. https://doi.org/10.3389/fnut.2021.660150

Ashraf, S., Sayeed, S. A., Ali, R., Vohra, F., Ahmed, N., & Alam, M. K. (2022). Assessment of potential benefits of functional food characteristics of beetroot energy drink and flavored milk. *BioMed Research International, 2022*, 1971018. https://doi.org/10.1155/2022/1971018

Bahadoran, Z., Mirmiran, P., Kabir, A., Azizi, F., & Ghasemi, A. (2017). The nitrate-independent blood pressure–lowering effect of beetroot juice: a systematic review and meta-analysis. *Advances in Nutrition*, 8(6), 830–838. https://doi.org/10.3945/an.117.016717

Bangar, S. P., Sharma, N., Sanwal, N., Lorenzo, J. M., & Sahu, J. K. (2022). Bioactive potential of beetroot (*Beta vulgaris*). *Food Research International*, 158, 111556.

Bond, V., Curry, B. H., Adams, R. G., Asadi, M. S., Millis, R. M., & Haddad, G. E. (2013). Effects of Dietary Nitrates on Systemic and Cerebrovascular Hemodynamics. *Cardiology Research and Practice, 2013*, 1–9. https://doi.org/10.1155/2013/435629

Bonilla Ocampo, D., Paipilla, A., Marín, E., Vargas-Molina, S., Petro, J., & Pérez-Idárraga, A. (2018). Dietary nitrate from beetroot juice for hypertension: a systematic review. *Biomolecules*, 8(4), 134. https://doi.org/10.3390/biom8040134

Borjan, D., Šeregelj, V., Cör Andrejč, D., Pezo, L., Tumbas Šaponjac, V., Knez, Ž., et al. (2022). Green techniques for preparation of red beetroot extracts with enhanced biological potential. *Antioxidants*, 11(5), 805. https://doi.org/10.3390/antiox11050805

Broeder, C. E., Flores, V., Julian, B., Wojan, F., Tauber, R., Schubert, L., Salacinski, A., & Ivy, J. L. (2021). Nitric oxide enhancement supplement containing beet nitrite and nitrate benefits high intensity cycle interval training. *Current Research in Physiology*, 4, 183–191. https://doi.org/10.1016/J.CRPHYS.2021.07.002

Brzezińska-Rojek, J., Rutkowska, M., Ośko, J., Konieczka, P., Prokopowicz, M., & Grembecka, M. (2022). Mercury content in beetroot and beetroot-based dietary supplements. *Journal of Food Composition and Analysis*, 114, 104828. https://doi.org/10.1016/j.jfca.2022.104828

Brzezińska-Rojek, J., Rutkowska, M., Ośko, J., Konieczka, P., Prokopowicz, M., & Grembecka, M. (2024). Evaluation of the safety and potential benefits of beetroot-based dietary supplements according to their elemental composition. *Biological Trace Element Research*, 202(7), 3318–3332. https://doi.org/10.1007/s12011-023-03902-x

Brzezińska-Rojek, J., Sagatovych, S., Malinowska, P., Gadaj, K., Prokopowicz, M., & Grembecka, M. (2023). Antioxidant capacity, nitrite and nitrate content in beetroot-based dietary supplements. *Foods, 12*(5), 1017. https://doi.org/10.3390/FOODS12051017/S1

Brzezińska, J., Szewczyk, A., Brzezicha, J., Prokopowicz, M., & Grembecka, M. (2021). Evaluation of physicochemical properties of beetroot-based dietary supplements. *Foods, 10*(8), 1693. https://doi.org/10.3390/foods10081693

Cai, H., & Harrison, D. G. (2000). Endothelial dysfunction in cardiovascular diseases: the role of oxidant stress. *Circulation Research*, 87(10), 840–844. https://doi.org/10.1161/01.RES.87.10.840

Ceclu, L., & Nistor, O. V. (2020). Red beetroot: composition and health effects – a review. *Journal of Nutritional Medicine and Diet Care*, 6(1), 1–9.

Cervantes-Paz, B., Victoria-Campos, C. I., & Ornelas-Paz, J. D. J. (2016). Absorption of carotenoids and mechanisms involved in their health-related properties. *Sub-Cellular Biochemistry, 79*, 415–454.

Chan, T. Y. K. (1996). Food-borne nitrates and nitrites as a cause of methemoglobinemia. *Southeast Asian Journal of Tropical Medicine and Public Health*, 27(1), 189–192.

Chan, T. Y. K. (2011). Vegetable-borne nitrate and nitrite and the risk of methaemoglobinaemia. *Toxicology Letters*, 200(1–2), 107–108. https://doi.org/10.1016/J.TOXLET.2010.11.002

Chhikara, N., Kushwaha, K., Sharma, P., Gat, Y., & Panghal, A. (2019). Bioactive compounds of beetroot and utilization in food processing industry: a critical review. *Food Chemistry*, 272, 192–200. https://doi.org/10.1016/j.foodchem.2018.08.022

Clifford, T., Howatson, G., West, D., & Stevenson, E. (2015). The potential benefits of red beetroot supplementation in health and disease. *Nutrients*, 7(4), 2801–2822. https://doi.org/10.3390/nu7042801

Constantin, O. E., Stoica, F., Mistrianu, S. L., Andronoiu, D. G., Turturică, M., Stănciuc, N., et al. (2024). A sustainable approach: repurposing red beetroot peels for innovative meringue products. *Foods*, 14(2), 317. https://doi.org/10.3390/foods14020317

Ćwieląg-Drabek, M., Piekut, A., Gut, K., & Grabowski, M. (2020). Risk of cadmium, lead and zinc exposure from consumption of vegetables produced in areas with mining and smelting past. *Scientific Reports*, 10(1), 3363. https://doi.org/10.1038/S41598-020-60386-8

Da Silva, D. V. T., De Oliveira Silva, F., Perrone, D., Pierucci, A. P. T. R., Conte-Junior, C. A., Da Silveira Alvares, T., Del Aguila, E. M., & Paschoalin, V. M. F. (2016). Physicochemical, nutritional, and sensory analyses of a nitrate-enriched beetroot gel and its effects on plasmatic nitric oxide and blood pressure. *Food & Nutrition Research*, *60*. https://doi.org/10.3402/FNR.V60.29909

Domínguez, R., Cuenca, E., Maté-Muñoz, J., García-Fernández, P., Serra-Paya, N., Estevan, M., Herreros, P., & Garnacho-Castaño, M. (2017). Effects of beetroot juice supplementation on cardiorespiratory endurance in athletes. a systematic review. *Nutrients*, 9(1), 43. https://doi.org/10.3390/nu9010043

Donath, M. Y., & Shoelson, S. E. (2011). Type 2 diabetes as an inflammatory disease. *Nature Reviews Immunology*, 11(2), 98–107. https://doi.org/10.1038/nri2925

Dong, S., Yang, X., Zhao, L., Zhang, F., Hou, Z., & Xue, P. (2020). Antibacterial activity and mechanism of action saponins from *Chenopodium quinoa* Willd. husks against foodborne pathogenic bacteria. *Industrial Crops and Products*, 149, 112350. https://doi.org/10.1016/J.INDCROP.2020.112350

Dziubanek, G., Baranowska, R., Ćwieląg-Drabek, M., Spychała, A., Piekut, A., Rusin, M., & Hajok, I. (2017). Cadmium in edible plants from Silesia, Poland, and its implications for health risk in populations. *Ecotoxicology and Environmental Safety*, 142, 8–13. https://doi.org/10.1016/J.ECOENV.2017.03.048

Eckel, R. H., Jakicic, J. M., Ard, J. D., de Jesus, J. M., Houston Miller, N., Hubbard, V. S., Lee, I.-M., Lichtenstein, A. H., Loria, C. M., Millen, B. E., Nonas, C. A., Sacks, F. M., Smith, S. C., Svetkey, L. P., Wadden, T. A., & Yanovski, S. Z. (2014). 2013 AHA/ACC Guideline on Lifestyle Management to Reduce Cardiovascular Risk. *Journal of the American College of Cardiology*, 63(25), 2960–2984. https://doi.org/10.1016/j.jacc.2013.11.003

Elekofehinti, O. O., Iwaloye, O., Olawale, F., & Ariyo, E. O. (2021). Saponins in cancer treatment: current progress and future prospects. *Pathophysiology*, 28(2), 250–272. https://doi.org/10.3390/PATHOPHYSIOLOGY28020017

European Commission. (2016). *Sugar and Oxalate Free Beetroot Juice: Production and Assessment of its Human Health Benefits*.

FAO/WHO. (2001). *Food Additives and Contaminants*. Joint Codex Alimentarius Commission, FAO/WHO Food standards Programme, ALINORM 01/12A. Rome, Italy.

Fekadu Gemede, H. (2014). Antinutritional factors in plant foods: potential health benefits and adverse effects. *International Journal of Nutrition and Food Sciences*, 3(4), 284. https://doi.org/10.11648/J.IJNFS.20140304.18

Ferguson, S. K., Hirai, D. M., Copp, S. W., Holdsworth, C. T., Allen, J. D., Jones, A. M., Musch, T. I., & Poole, D. C. (2013). Impact of dietary nitrate supplementation via beetroot juice on exercising muscle vascular control in rats. *The Journal of Physiology*, 591(2), 547–557. https://doi.org/10.1113/jphysiol.2012.243121

Fu, Y., Shi, J., Xie, S. Y., Zhang, T. Y., Soladoye, O. P., & Aluko, R. E. (2020). Red beetroot betalains: perspectives on extraction, processing, and potential health benefits. *Journal of Agricultural and Food Chemistry*, 68(42), 11595–11611. https://doi.org/10.1021/acs.jafc.0c04241

Gemede, H. F., & Ratta, N. (2014). Habtamu Fekadu Gemede, Negussie Ratta. Antinutritional factors in plant foods: potential health benefits and adverse effects. *International Journal of Nutrition and Food Sciences*, 3(4), 284–289. https://doi.org/10.11648/j.ijnfs.20140304.18

Greer, F. R., & Shannon, M. (2005). Infant methemoglobinemia: the role of dietary nitrate in food and water. *Pediatrics*, 116(3), 784–786. https://doi.org/10.1542/peds.2005-1497

Guldiken, B., Toydemir, G., Nur Memis, K., Okur, S., Boyacioglu, D., & Capanoglu, E. (2016). Home-processed red beetroot (*Beta vulgaris* L.) products: changes in antioxidant properties and bioaccessibility. *International Journal of Molecular Sciences*, 17(6), 858.

Hord, N. G., Tang, Y., & Bryan, N. S. (2009). Food sources of nitrates and nitrites: the physiologic context for potential health benefits. *The American Journal of Clinical Nutrition*, 90(1), 1–10.

Hu, Y., Li, Y., Zhu, H., Wang, D., Zhou, J., & Li, B. (2025). Saponin components exhibit antiviral properties against porcine epidemic diarrhea virus in vitro. *RSC Medicinal Chemistry*, 16(3), 1199–1207. https://doi.org/10.1039/D4MD00894D
Ingle, M., Thorat, S. S., Kotecha, P. M., & Nimbalkar, C. A. (2017). Nutritional assessment of beetroot (*Beta vulgaris* L.) powder cookies. *Asian Journal of Dairy & Food Research*, 36(3), 222–228.
Jajja, A., Sutyarjoko, A., Lara, J., Rennie, K., Brandt, K., Qadir, O., & Siervo, M. (2014). Beetroot supplementation lowers daily systolic blood pressure in older, overweight subjects. *Nutrition Research*, 34(10), 868–875. https://doi.org/10.1016/j.nutres.2014.09.007
Jakubczyk, K., Melkis, K., Janda-Milczarek, K., & Skonieczna-Żydecka, K. (2024). Phenolic compounds and antioxidant properties of fermented beetroot juices enriched with different additives. *Foods*, 13(1), 102. https://doi.org/10.3390/foods13010102
Kale, R. G., Sawate, A. R., Kshirsagar, R. B., Patil, B. M., & Mane, R. (2018). Studies on evaluation of physical and chemical composition of beetroot (*Beta vulgaris* L.). *International Journal of Chemical Studies*, 6(2), 2977–2979.
Kapadia, G. J., Tokuda, H., Konoshima, T., & Nishino, H. (1996). Chemoprevention of lung and skin cancer by *Beta vulgaris* (beet) root extract. *Cancer Letters*, *100*(1–2), 211–214.
Kapil, V., Milsom, A. B., Okorie, M., Maleki-Toyserkani, S., Akram, F., Rehman, F., Arghandawi, S., Pearl, V., Benjamin, N., Loukogeorgakis, S., MacAllister, R., Hobbs, A. J., Webb, A. J., & Ahluwalia, A. (2010). Inorganic nitrate supplementation lowers blood pressure in humans. *Hypertension*, 56(2), 274–281. https://doi.org/10.1161/HYPERTENSIONAHA.110.153536
Kavalcová, P., Bystrická, J., Tomáš, J., Karovičová, J., Kovarovič, J., & Lenková, M. (2015). The content of total polyphenols and antioxidant activity in red beetroot. *Slovak Journal of Food Sciences/Potravinarstvo*, 9(1), 77–83.
Kiranmayi, P. (2014). *Is Bio Active Compounds Inplantsacts as Anti Nutritonal Factors*. https://www.academia.edu/108266687/Is_Bio_Active_Compounds_Inplantsacts_as_Anti_Nutritonal_Factors.
Knez, E., Kadac-Czapska, K., Dmochowska-Ślęzak, K., & Grembecka, M. (2022). Root vegetables-composition, health effects, and contaminants. *International Journal of Environmental Research and Public Health*, 19(23), 15531.
Lansley, K. E., Winyard, P. G., Fulford, J., Vanhatalo, A., Bailey, S. J., Blackwell, J. R., DiMenna, F. J., Gilchrist, M., Benjamin, N., & Jones, A. M. (2011). Dietary nitrate supplementation reduces the O_2 cost of walking and running: a placebo-controlled study. *Journal of Applied Physiology*, 110(3), 591–600. https://doi.org/10.1152/japplphysiol.01070.2010
Lara, J., Ashor, A. W., Oggioni, C., Ahluwalia, A., Mathers, J. C., & Siervo, M. (2016). Effects of inorganic nitrate and beetroot supplementation on endothelial function: a systematic review and meta-analysis. *European Journal of Nutrition*, 55(2), 451–459. https://doi.org/10.1007/s00394-015-0872-7
Lechner, J. F., & Stoner, G. D. (2019). Red beetroot and betalains as cancer chemopreventative agents. *Molecules*, 24(8), 1602.
Lee, C. H., Wettasinghe, M., Bolling, B. W., Ji, L. L., & Parkin, K. L. (2005). Betalains, phase II enzyme-inducing components from red beetroot (*Beta vulgaris* L.) extracts. *Nutrition and Cancer*, 53(1), 91–103. https://doi.org/10.1207/S15327914NC5301_11
Lombardelli, C., Benucci, I., Mazzocchi, C., & Esti, M. (2021). Betalain extracts from beetroot as food colorants: effect of temperature and UV-light on storability. *Plant Foods for Human Nutrition*, 76(3), 347–353. https://doi.org/10.1007/s11130-021-00915-6.
Lorizola, I. M., Miyamoto, J. É., Vieira, A. L. F., Sumere, B. R., Bezerra, R. M. N., Torsoni, M. A., & Capitani, C. D. (2021). Beet (*Beta vulgaris* L.) stalk and leaf supplementation changes the glucose homeostasis and inflammatory markers in the liver of mice exposed to a high-fat diet. *Food Chemistry: Molecular Sciences*, 2, 100018.
Masih, D., Singh, N., & Singh, A. (2019). Red beetroot: a source of natural colourant and antioxidants: a review. *Journal of Pharmacognosy and Phytochemistry*, 8(4), 162–166.
Melek, F. R., Aly, F. A., Kassem, I. A. A., Abo-Zeid, M. A. M., Farghaly, A. A., & Hassan, Z. M. (2015). Three further triterpenoid saponins from *Gleditsia caspica* fruits and protective effect of the total saponin fraction on cyclophosphamide-induced genotoxicity in mice. *Zeitschrift Fur Naturforschung – Section C Journal of Biosciences*, 70(1–2), 31–37.

Mella, C., Rojas, N., Calderon-Bravo, H., & Muñoz, L. A. (2024). Evaluating biocompounds in discarded beetroot (Beta vulgaris) leaves and stems for sustainable food processing solutions. *Foods*, *13*(16), 2603.

Mikołajczyk-Bator, K., Błaszczyk, A., Czyżniejewski, M., Kachlicki, P., & Stochmal, A. (2024). Metabolite profiling of triterpene saponins from different red beetroot cultivars using ultra-high performance liquid chromatography high-resolution mass spectrometry. *Journal of Food Composition and Analysis*, 130, 106141. https://doi.org/10.1016/J.JFCA.2024.106141

Mirmiran, P., Houshialsadat, Z., Gaeini, Z., Bahadoran, Z., & Azizi, F. (2020). Functional properties of beetroot (*Beta vulgaris*) in management of cardio-metabolic diseases. *Nutrition & Metabolism*, 17, 1–15.

Mitrevski, J., Pantelić, N. Đ., Dodevska, M. S., Kojić, J. S., Vulić, J. J., Zlatanović, S., Gorjanović, S., Laličić-Petronijević, J., Marjanović, S., & Antić, V. V. (2023). Effect of beetroot powder incorporation on functional properties and shelf life of biscuits. *Foods*, 12(2), 322. https://doi.org/10.3390/foods12020322

Mkhari, T., Kaseke, T., & Fawole, O. A. (2023). Encapsulation of betalain-rich extract from beetroot postharvest waste using a binary blend of gum arabic and maltodextrin to promote a food circular bioeconomy. *Frontiers in Nutrition*, 10, 1235372. https://doi.org/10.3389/fnut.2023.1235372

Mohajer, R., Salehi, M., & Mohammadi, J. (2014). Lead and cadmium concentration in agricultural crops (Lettuce, cabbage, beetroot, and onion) of Isfahan Province, Iran. *Iranian Journal of Health and Environment*, 7(1), 1–10. https://ijhe.tums.ac.ir/article-1-5272-en.html

Mortensen, A., Aguilar, F., Crebelli, R., Di Domenico, A., Dusemund, B., Frutos, M. J., Galtier, P., Gott, D., Gundert-Remy, U., Lambré, C., Leblanc, J. C., Lindtner, O., Moldeus, P., Mosesso, P., Oskarsson, A., Parent-Massin, D., Stankovic, I., Waalkens-Berendsen, I., Woutersen, R. A., … Younes, M. (2017a). Re-evaluation of potassium nitrite (E 249) and sodium nitrite (E 250) as food additives. *EFSA Journal*, *15*(6), e04786. https://doi.org/10.2903/J.EFSA.2017.4786

Mortensen, A., Aguilar, F., Crebelli, R., Di Domenico, A., Dusemund, B., Frutos, M. J., Galtier, P., Gott, D., Gundert-Remy, U., Lambré, C., Leblanc, J. C., Lindtner, O., Moldeus, P., Mosesso, P., Oskarsson, A., Parent-Massin, D., Stankovic, I., Waalkens-Berendsen, I., Woutersen, R. A., … Younes, M. (2017b). Re-evaluation of sodium nitrate (E 251) and potassium nitrate (E 252) as food additives. *EFSA Journal*, *15*(6), e04787. https://doi.org/10.2903/J.EFSA.2017.4787

Moulick, S. P., Jahan, F., Islam, M. B., Al Bashera, M., Hasan, M. S., Islam, M. J., … & Bhuiyan, M. N. H. (2023). Nutritional characteristics and antiradical activity of turmeric (*Curcuma longa* L.), beetroot (*Beta vulgaris* L.), and carrot (*Daucus carota* L.) grown in Bangladesh. *Heliyon*, *9*(11), e21495.

Mudgal, D., Singh, S., & Singh, B. R. (2022). Nutritional composition and value added products of beetroot: a review. *Journal of Current Research in Food Science*, 3(1), 1–9.

Narayanan, M. V., Rasane, P., Singh, J., Kaur, S., Avinashe, H., Gunjal, M., Kaur, J., & Bhadariya, V. (2023). Beetroot bioactive and its associated health benefits: considerations for utilization of beetroot in value-added products. *Recent Advances in Food, Nutrition & Agriculture*, 14(3), 155–166. https://doi.org/10.2174/2772574X14666230725110541.

Naseer, S., Hussain, S., & Abid, A. (2019). Betalain as a food colorant: its sources, chemistry and health benefits: chemistry of betalain and its role as food colorant. *Proceedings of the Pakistan Academy of Sciences: B. Life and Environmental Sciences*, 56(2), 1–8.

Nirmal, N. P., Mereddy, R., & Maqsood, S. (2021). Recent developments in emerging technologies for beetroot pigment extraction and its food applications. *Food Chemistry*, 356, 129611. https://doi.org/10.1016/j.foodchem.2021.129611

Nistor, O. V., Șeremet Ceclu, L., Mocanu, G. D., Barbu, V., Andronoiu, D. G., & Stănciuc, N. (2020). Three types of red beetroot and sour cherry based marmalades with enhanced functional properties. *Molecules*, 25(21), 5090. https://doi.org/10.3390/molecules25215090

Notay, K., Incognito, A. V., & Millar, P. J. (2017). Acute beetroot juice supplementation on sympathetic nerve activity: a randomized, double-blind, placebo-controlled proof-of-concept study. *American Journal of Physiology-Heart and Circulatory Physiology*, 313(1), H59–H65. https://doi.org/10.1152/ajpheart.00163.2017

Pajević, S., Arsenov, D., Nikolić, N., Borišev, M., Orčić, D., Župunski, M., & Mimica-Dukić, N. (2018). Heavy metal accumulation in vegetable species and health risk assessment in Serbia. *Environmental Monitoring and Assessment*, 190(8). https://doi.org/10.1007/S10661-018-6743-Y

Paliwal, H., Goyal, S., Singla, S., & Daksh, S. (2016). Pigments from natural sources: an overview. *International Journal of Research in Pharmacy and Pharmaceutical Sciences (IJRPPS)*, 1(3), 1–12.

Pavoković, D., & Krsnik-Rasol, M. (2011). Complex biochemistry and biotechnological production of betalains. *Food technology and Biotechnology*, 49(2), 145–155.

Pinna, M., Roberto, S., Milia, R., Marongiu, E., Olla, S., Loi, A., Migliaccio, G., Padulo, J., Orlandi, C., Tocco, F., Concu, A., & Crisafulli, A. (2014). Effect of beetroot juice supplementation on aerobic response during swimming. *Nutrients*, 6(2), 605–615. https://doi.org/10.3390/nu6020605

Presley, T. D., Morgan, A. R., Bechtold, E., Clodfelter, W., Dove, R. W., Jennings, J. M., Kraft, R. A., Bruce King, S., Laurienti, P. J., Jack Rejeski, W., Burdette, J. H., Kim-Shapiro, D. B., & Miller, G. D. (2011). Acute effect of a high nitrate diet on brain perfusion in older adults. *Nitric Oxide*, 24(1), 34–42. https://doi.org/10.1016/j.niox.2010.10.002

Punia Bangar, S., Sharma, N., Sanwal, N., Lorenzo, J. M., & Sahu, J. K. (2022). Bioactive potential of beetroot (*Beta vulgaris*). *Food Research International*, 158, 111556. https://doi.org/10.1016/j.foodres.2022.111556

Ranawana, V., Moynihan, E., Campbell, F., Duthie, G., & Raikos, V. (2018). Beetroot improves oxidative stability and functional properties of processed foods: singular and combined effects with chocolate. *Journal of Food Science and Technology*, 55(7), 2401–2409. https://doi.org/10.1007/s13197-018-3157-3

Saletnik, B., Zaguła, G., Bajcar, M., & Puchalski, C. (2016). Accumulation of cadmium, lead and mercury in seedlings of selected sugar beet varieties as a result of simulated soil contamination. *Journal of Microbiology, Biotechnology and Food Sciences*, 5(4), 351–354. https://doi.org/10.15414/JMBFS.2016.5.4.351-354

Samtiya, M., Aluko, R. E., & Dhewa, T. (2020). Plant food anti-nutritional factors and their reduction strategies: an overview. *Food Production, Processing and Nutrition*, 2(1), 1–14. https://doi.org/10.1186/S43014-020-0020-5

Sawicki, T., Bączek, N., & Wiczkowski, W. (2016). Betalain profile, content and antioxidant capacity of red beetroot dependent on the genotype and root part. *Journal of Functional Foods*, 27, 249–261.

Schulte-Schrepping, K.-H., & Piscator, M. (2011). Cadmium and cadmium compounds. In IARC Monographs on the Evaluation of Carcinogenic Risks to Humans, 121–145. https://doi.org/10.1002/14356007.A04_499

Shaban, N. Z., Abd El-Kader, S. E., Mogahed, F. A., El-Kersh, M. A., & Habashy, N. H. (2021). Synergistic protective effect of Beta vulgaris with meso-2, 3-dimercaptosuccinic acid against lead-induced neurotoxicity in male rats. *Scientific Reports*, 11(1), 252.

Siener, R., Seidler, A., Voss, S., & Hesse, A. (2016). The oxalate content of fruit and vegetable juices, nectars and drinks. *Journal of Food Composition and Analysis*, 45, 108–112. https://doi.org/10.1016/J.JFCA.2015.10.004

Silva, D. V. T., dos Santos Baião, D., de Oliveira Silva, F., Alves, G., Perrone, D., Del Aguila, E. M., & Paschoalin, V. M. F. (2019). Betanin, a natural food additive: stability, bioavailability, antioxidant and preservative ability assessments. *Molecules*, 24(3):458.

Slavin, J. L., & Lloyd, B. (2012). Health benefits of fruits and vegetables. *Advances in Nutrition*, 3(4), 506. https://doi.org/10.3945/AN.112.002154

Song, G., Shen, X., Li, S., Li, Y., Si, H., Fan, J., Li, J., Gao, E., & Liu, S. (2016). Structure-activity relationships of 3-O-β-chacotriosyl oleanane-type triterpenoids as potential H5N1 entry inhibitors. *European Journal of Medicinal Chemistry*, 119, 109–121. https://doi.org/10.1016/J.EJMECH.2016.04.061

Spiegel, M., Gamian, A., & Sroka, Z. (2021). Antiradical activity of beetroot (*Beta vulgaris* L.) betalains. *Molecules*, 26(9), 2439.

Spórna-Kucab, A., Tekieli, A., Skalicka-Woźniak, K., Grzegorczyk, A., Świergosz, T., & Wybraniec, S. (2022). Characterization of Triterpene Saponin composition of white, yellow and red beetroot (*Beta vulgaris* L.). *Polish Journal of Food and Nutrition Sciences*, 72(2), 2083–6007. https://doi.org/10.31883/pjfns/149515

Stander, Z., Luies, L., van Reenen, M., Howatson, G., Keane, K. M., Clifford, T., Stevenson, E. J., & Loots, D. T. (2021). Beetroot juice – a suitable post-marathon metabolic recovery supplement? *Journal of the International Society of Sports Nutrition*, 18(1), 72. https://doi.org/10.1186/s12970-021-00468-8

Stoica, F., Râpeanu, G., Rațu, R. N., Stănciuc, N., Croitoru, C., Țopa, D., & Jităreanu, G. (2025). Red beetroot and its by-products: a comprehensive review of phytochemicals, extraction methods, health benefits, and applications. *Agriculture*, 15(3), 270. https://doi.org/10.3390/AGRICULTURE15030270

Suganya, M., Anitha, T., Lasyaja, A. B., Gayathri, P., Chithra, S., & Thomas, L. (2017). Biochemical and nutritive analysis of underutilized green leafy vegetable-*Beta Vulgaris*, L. *International Journal of Advanced Science and Research*, 2(3), 9–12.

Tan, M. L., & Hamid, S. B. S. (2021). Beetroot as a potential functional food for cancer chemoprevention, a narrative review. *Journal of Cancer Prevention*, 26(1), 1–17. https://doi.org/10.15430/JCP.2021.26.1.1

United States Department of Agriculture. (2019). *USDA FoodData Central* [Internet]. Available from: https://fdc.nal.usda.gov/

Vaitkevičienė, N., Sapronaitė, A., & Kulaitienė, J. (2022). Evaluation of proximate composition, mineral elements and bioactive compounds in skin and flesh of beetroot grown in Lithuania. *Agriculture*, *12*(11), 1833.

van Velzen, A. G., Sips, A. J. A. M., Schothorst, R. C., Lambers, A. C., & Meulenbelt, J. (2008a). The oral bioavailability of nitrate from nitrate-rich vegetables in humans. *Toxicology Letters*, 181(3), 177–181. https://doi.org/10.1016/J.TOXLET.2008.07.019

Yashwant, K. (2015). Beetroot: a super food. *IJESTA*, 1, 20–26.

Wylie, L. J., Bailey, S. J., Kelly, J., Blackwell, J. R., Vanhatalo, A., & Jones, A. M. (2016). Influence of beetroot juice supplementation on intermittent exercise performance. *European Journal of Applied Physiology*, 116(2), 415–425. https://doi.org/10.1007/s00421-015-3296-4

Wylie, L. J., Kelly, J., Bailey, S. J., Blackwell, J. R., Skiba, P. F., Winyard, P. G., Jeukendrup, A. E., Vanhatalo, A., & Jones, A. M. (2013). Beetroot juice and exercise: pharmacodynamic and dose-response relationships. *Journal of Applied Physiology*, 115(3), 325–336. https://doi.org/10.1152/japplphysiol.00372.2013

Zafeiridis, A. (2014). The effects of dietary nitrate (beetroot juice) supplementation on exercise performance: a review. *American Journal of Sports Science*, 2(4), 97. https://doi.org/10.11648/j.ajss.20140204.15

Zamani, H., de Joode, M. E. J. R., Hossein, I. J., Henckens, N. F. T., Guggeis, M. A., Berends, J. E., de Kok, T. M. C. M., & van Breda, S. G. J. (2021). The benefits and risks of beetroot juice consumption: a systematic review. *Critical Reviews in Food Science and Nutrition*, 61(5), 788–804. https://doi.org/10.1080/10408398.2020.1746629

2

Beetroot: Botany, Varieties, Cultivation, and Production

Kritika Pandit, Amritpreet Kaur, and Sakshi Sharma

2.1 Introduction

Plants are so crucial for human survival that we cannot imagine our lives without them. Many plant varieties have been grown for their agricultural, economic, and ethnomedicinal properties. One such plant with significant benefits is *Beta vulgaris*. It is commonly known as beetroot, a crop that belongs to the Chenopodiaceae family. Beetroot is a flowering plant that is predominantly biennial but can occasionally exhibit perennial growth. It is available in several varieties, with root colors ranging from vibrant yellow to deep red (Gokhale and Lele, 2014). It is a Middle Eastern species that has spread around the world, from the Americas to Europe and Asia (Mirmiran et al., 2020). Cultivated primarily in early spring for its thick, fleshy roots, beetroot belongs to the dicotyledonous family. This erect herb develops tuberous rootstocks and ranks among the top ten vegetables renowned for their potent antioxidant properties. The seeds of the crop are classified as multi-germ, meaning a single seed can produce multiple seedlings. The corky outer layer of the seed contains phenolic compounds, which act as a physical barrier and inhibit germination (Kezi and Sumathy, 2014).

Beetroot is a widely distributed vegetable cultivated across the globe. In 2014, the global production of beetroot reached approximately 269.7 million tons. France is the leading producer, followed by Russia. Other major producers include Germany, the United States, Ukraine, Egypt, and China (Food and Agriculture Organization of the United Nations, 2014). There are four main varieties of beetroot: Detroit Dark Red, Crimson Globe, Crosby Egyptian, and Early Wonder. (Chawla et al., 2016).

It makes an excellent dietary supplement by being abundant in carbohydrates, fats, proteins, micronutrients, and various functional components with significant health-enhancing properties (Yadav et al., 2016a). Beetroot is also a source of various vitamins, including vitamin A, thiamine, riboflavin, niacin, pantothenic acid, vitamin B6, ascorbic acid, and folate, as well as minerals such as sodium, calcium, iron, phosphorus, potassium, magnesium, and zinc (Yashwant, 2015). In addition, it also shows the presence of potent pigments such as betalains (Guldiken et al., 2016), ascorbic acid (Clifford et al., 2015), carotenoids (Ninfali and Angelino, 2013), polyphenols, flavonoids, saponins, and high levels of nitrate (644–1,800 mg/kg) (Lidder and Webb, 2013). Nitrate in beetroot juice is converted within the human body into nitrite and then into nitric oxide (NO),

 DOI: 10.1201/9781003527084-2

a compound known for its vasodilatory effects. This process helps lower blood pressure and enhances the delivery of oxygen and nutrients to active muscles (Jones et al., 2012). The nitrates in beetroot can lower blood pressure, protect against ischemia-reperfusion damage, and regulate mitochondrial function (Satyanand et al., 2014).

The deep red color of beetroots comes from high concentrations of betalains, which are used as natural colorants in the food industry. The primary betalains found in beetroot are betacyanins and betaxanthins (Gandia et al., 2010). Betalains have a wide range of therapeutic properties, including anticancer, liver-protective, and antitumor effects (Wootton-Beard et al., 2011). Beetroot offers a wide range of health benefits due to its rich content of bioactive compounds, which provide natural anti-anemic, anti-inflammatory, antihypertensive, antioxidant, anticarcinogenic, antipyretic, antibacterial, detoxifying, and diuretic properties. The phytochemicals found in beetroot also play a crucial role in reducing age-related oxidative stress and supporting cognitive functions such as perception, learning, communication, and decision-making.

This book chapter provides a comprehensive overview of beetroot botany, its diverse varieties, cultivation methodologies, and production trends. It highlights the crop's significance and outlines potential pathways for future innovations in beetroot cultivation and utilization.

2.1.1 Botany

Beetroot (*Beta vulgaris* L.) is a herbaceous crop belonging to the Chenopodiaceae family (Angiosperm polygamy group: Amaranthaceae) (Oyen, 2004). Linnaeus provided the first reliable description of the binomial name *Beta vulgaris* in Species Plantarum, Ed. I, (1753). Eight varieties with Latin names were listed in the description. The species was divided into cultivated and wild plants in Species Plantarum, Ed. 2, (1762). The cultivated plants were kept in *Beta vulgaris,* which was further divided into varieties with Latin names, while the wild beets were maintained in *Beta maritima*. Linnaeus, in the year 1767, further divided the cultivated species into two species, viz., *Beta vulgaris* and *Beta cicla*.

Taxonomic Position: The taxonomic position of the Beetroot is as follows (Table 2.1) (Lange et al., 1999; Yamane, 2023):

TABLE 2.1
Taxonomic Position of Beetroot

Kingdom	Plantae
Sub-Kingdom	Viridiplantae
Super Division	Embryophyta
Division	Tracheophyta
Subdivision	Spermatophytina
Class	Magnoliopsida
Superorder	Caryophyllanae
Order	Caryophyllales
Family	Amaranthaceae
Tribe	Betae
Genus	*Beta*
Species	*Beta vulgaris* L.

TABLE 2.2

The Botanical Description of the Crop Plant

Overview	All the cultivated forms of beetroots are herbaceous, dicotyledonous plants in the genus Beta
Root	Roots are thick and fleshy, which store a large amount of reserve food in the form of sucrose. The lower portion of the root develops from the taproot, from which the secondary roots arise, while the upper portion develops from the hypocotyl. During the development of fleshy roots, storage parenchyma tissues and vascular tissues are formed alternately, which result in the formation of the concentric rings and are visible in a median cross-section of mature beetroot. The interior color of the beetroot variety is influenced by the following factors such as temperature, nutrition, soil, and season. On the other hand, the exterior color of the root varies from orange-red to dark purple-red, depending on the cultivar. The color of the root is due to the red-violet pigments, known as β-cyanins and yellow pigment β-xanthins
Leaves	The leaves are dark green or reddish in color, cordate, and ovate. They develop in a rosette fashion and form a spiral with the oldest leaves on the outer side. Season, temperature, soil moisture, and spacing influence the color and shape of the leaves
Flower	Flowers are inconspicuous and small. Sepals are present and green in color, which become thicker toward the base as the fruit ripens. Petals are absent. Stamens are five in number. Stigma is present and has three segments. Inflorescence is of a large spike type, which usually develops in the second year. The flowers are sessile, arising in clusters of three to four in the axils of bracts of the inflorescence axis and its secondary branches. Cross-pollination is reported by means of wind
Fruit	Fruits are mostly aggregates made up of two or more fruits held together by a swollen perianth base and thus form irregular dry cork-like bodies known as a seed ball or seed
Seed	The seed is kidney-shaped and tiny and ranges in color from shiny to reddish-brown, measuring around 3 mm in length. There are multiple seeds in the seed ball, and under favorable storage conditions, they will all remain viable for five to six years
Common names	Beet, beetroot, common beet, Chard, fodder beet, garden beet, mangel, red beet, sea beet, Sugar beet, spinach beet

2.1.2 Botanical Description

The botanical description of the crop plant is given in a tabulated form (Table 2.2) (McGrath et al., 2007; Moore, 2017).

2.1.3 Varieties

The beetroot varieties based on various characteristics can be classified under multiple categories.

1. **Based on the Shape of the Roots:**

   Root shape plays a potent role in differentiating beetroot varieties. Some of the important beet varieties based on shape include (Sood and Gupta, 2017)

   a. **Flat-shaped**: Considered a modern variety that leads to the production of thick roots, as in the Flat Egyptian Beetroot variety

   b. **Short top-shaped**: Such varieties have flattened roots at the top and bottom, and sides are rounded, consisting of tapered ends. Examples of such kind of beetroot varieties include Crosby Egyptian, Strawberry Crosby, Ferry Crosby, and Vermilion Crosby.

   c. **Deep oblate**: These varieties have globular roots. Examples include Early Wonder, Green Top Early Wonder, Ohio Canner, and Perfected Detroit.

d. **Oval-shaped**: The roots of such varieties are oval. Examples include Detroit Dark Red, Crimson Globe, Green Top Bunching, Asgrow Canner, and Crimson King

e. **Long-shaped**: Such varieties have long roots and are commonly found in Europe. Examples include Long Dark Blood, Long Smooth Blood, and Winter Keeper.

2. **Based on the Maturation Time of Heirloom:**

a. **Detroit Dark Red:** Detroit Dark Red beetroots have smooth, uniform roots with dark red flesh. The maturation time of the heirloom of this variety is 58 days.

b. **Early Wonder:** Early Wonder beetroots have flattened roots with dark red flesh and some lighter red zones. Their tops feature heavy green leaves with red veins, and the globes have rounded shoulders with a smooth texture. This variety's heirloom maturation time is less than Detroit Dark Red's. In this case, the heirloom matures in 52 days.

c. **Sangria:** The heirloom maturation period is 56 days.

d. **Sweetheart:** The heirloom maturation takes place at 58 days.

3. **Based on Some Specialized Characters (Kumar and Brooks, 2018):**

a. **Cylindrica**: This variety is known for its long, cylindrical shape. Its characteristic feature is that it can be cut into slices of equal size.

b. **Touchstone Gold**: This new variety is known for its small yellow roots, which retain their color when cooked.

c. **Green Top Bunching**: This variety is known for its bright red-colored roots having superior tops.

d. **Golden**: The golden variety of beetroot is well known for its bright yellow color and gives a sweet, mild flavor.

e. **Di Chioggia**: This variety is considered to be special due to its Italian heirloom and has red and white interior. It has a mild taste and is also known for its early maturation.

4. **Hybrid Varieties of Beetroot:**

a. **Avenger:** This variety is a hybrid variety known for its globe-shaped red-colored roots.

b. **Big Red:** This hybrid variety has a maturity period of 55 days and is considered to be the best late-season producer.

c. **Gladiator**: The Gladiator variety matures rapidly at only 48 days and is an excellent variety for canning.

d. **Red Ace**: This hybrid variety is known for its vigorous growth as well as sweet roots. The maturity period of this variety is 53 days.

e. **Warrior**: This variety is known for its specialized uniform globe-shaped roots and rapid development. The maturity period of this variety is 57 days.

A study also illustrated the comparison between varieties Cylindra, Kahira, Chioggia, and Crosby Egyptian based on the amount and concentration of different pigments. The study showed that the total polyphenol content was maximum in the beetroot variety named

Cylindria (887.75±7.73 mg/kg), followed by Crosby Egyptian (882.4±11.90 mg/kg), Kahira (373.8±11.38 mg/kg), and Chioggia (368.75±5.14). This indicates that the varieties also differ from each other based on pigment content. The amount of anthocyanin also varied, and these four varieties followed the same trend: Cylindria (84.50±4.71 mg/kg), Crosby Egyptian (73.50±5.19 mg/kg), Kahira (50.36±4.02 mg/kg), and Chioggia (14.48±0.40 mg/kg) (Kovarovič et al., 2017).

Based on the color and the consumption of beetroot all over the world, various varieties can be categorized. Among all, the Red Beet is found to be the most popular and prominent for human consumption. In the present era, North America, Britain, and Central America produce Red Beet in enormous amounts. Red Beets are also considered the most important variety due to their extreme commercial importance in decorative art, paintings, medicinal products, cosmetics, and so forth (Kumar and Brooks, 2018). Recently, a few F1 hybrids of beetroot have been introduced from abroad by private Indian seed companies for commercial cultivation. The varieties include Solo (British-bred monogerm), Alto, Super King, Red Cloud, Majesty, Fancon, Cardinal, and Super Queen (OECD, 2006). Therefore, many varieties of beetroot are available, having tremendous significance in all aspects, be it commercial, medicinal, or food industry (Sood and Gupta, 2017).

2.2 Beetroot Cultivation

Beetroot is relatively easy to cultivate and ranks among the top ten vegetables in India (Arya, 2020). They have a long growth season and are considered a cool-season vegetable. They thrive optimally in cooler climates, although they can withstand moderate heat and brief periods of frost (Hole et al., 2023).

Basic Considerations for Beetroot Cultivation: These are the following considerations for the cultivation of beetroot:

a. **Temperature:** Beetroot exhibits resilience to lower temperatures and thrives in cooler climates. While it can grow in warmer conditions, optimal development of characteristics such as color, texture, and sugar content occurs in cooler environments. Elevated temperatures can lead to zoning, characterized by the formation of alternating light and dark red concentric rings within the root. Additionally, exposure to extremely low temperatures ranging from 4.5°C to 10.0°C for 15 days may trigger bolting. Based on Markov–Haen's law, the ideal temperature for beetroots is 19°C. Germination initiates at temperatures ranging from 5°C to 6°C; however, the germination rate is most rapid at temperatures between 25°C and 26°C. They often require a cold period (4°C–10°C or longer) for 2 weeks to initiate flowering. Colder temperature also promotes the synthesis of deep red pigments. Adequate sunlight is essential for the proper growth and development of storage roots (Jasmitha et al., 2018; Székely and Máté, 2023).

b. **Soil:** Beetroot cultivation is unfavorable in sandy or stony soils. Loamy, sandy loam, and humus-rich sandy soils yield the highest quantity as well as quality of beetroots. Heavy clayey soils cause a soil crust to form after irrigation or rain, which hinders crop germination and establishment. In such heavy soils, the roots may develop irregular shapes and fail to grow properly. Furthermore, beetroot is particularly sensitive to soil acidity, with the ideal pH range for its growth being between 6.5 and 7.5. It is advisable to avoid acidic soils with lower pH levels, as they can obstruct nutrient uptake and hinder root development (Balázs, 1994; Székely and Máté., 2023).

c. **Land Preparation**: Beetroot is considered a cool season crop, grown in the plains during winter and in hilly regions during spring and summer, usually by March or April, with sowing periods ranging from September to November in the plains, August to March in winter rainfall areas, and late August to mid-March in cooler summer regions. The ploughing process transforms the land into a fine tilth, resulting in a loose and friable texture. To achieve this, it is essential to plough to a depth of 15–20 cm, ensuring that all clods are thoroughly removed. The soil should be leveled adequately, possess a well-developed crumb structure to maintain sufficient moisture, and be devoid of any unrooted plant debris. Ridges and furrows are prepared to enhance the effective soil depth, facilitating improved drainage, concentrating topsoil in the root zone, and supplying more oxygen for optimal root growth (Jasmitha et al., 2018; Hole et al., 2023).

d. **Seed Sowing:** Beetroot is typically planted directly into the field at a depth ranging from 15 to 20 mm. The seeds should be spaced 25–50 mm apart in rows drawn 200–400 mm apart. It is advisable to lightly firm the soil after covering the seeds to promote uniform germination. If plant emergence is successful, any excess plants should be thinned out at an early stage to achieve a final in-row spacing of 50–70 mm. Generally, 6–10 kg of seed is needed for 1 ha of beetroot (Joubert, 1974).

e. **Soil Nutrients and Fertilizing Agents:** Beetroots, like all plants, have particular nutrient requirements that are crucial for their growth and development. Understanding these requirements, one can decide which fertilizer to apply and in what quantity to produce the highest-quality crop. A consistent supply of potassium, phosphate, and nitrogen is crucial throughout the growing season to achieve high quality and good yields. To produce one ton of beetroots of desired quality 6 kg of potassium; 2.4 kg of nitrogen, and 1.4 kg of phosphorus are required (Balázs, 1994). Based on soil analyses, the application of nitrogen fertilizers is crucial, with 300–400 kg/ha of ammonium sulfate, ammonium nitrate, or limestone being utilized in 2–3 applications (10, 20, or 30 days after sowing) throughout the growing season. Typically, around 150 kg/ha of nitrogen is administered at the time of planting, with the remaining amount applied when plants reach a height of 10–15 cm. It is important to monitor the nitrogen application, as an excessive amount can lead to a decline in the quality of the beet body, ultimately leading to lower nitrate storage; diminished color and dry matter content; and decreased disease resistance in the plant. Just before seeding, 200–300 kg/ha of potassium chloride and 500–600 kg/ha of superphosphate are applied. An alternative is to apply a fertilizer combination of 2:3:2 at a rate of 1,000–1,200 kg/ha. Beets exhibit sensitivity to boron levels and are susceptible to boron deficiency, which can lead to the formation of cankers or necrotic patches within the root, making them inedible (Joubert, 1974; Lang, 2023; Székely and Máté, 2023).

f. **Irrigation:** Beetroot has shallow roots, which require consistent moisture to yield high-quality roots. It is advisable to irrigate every 7–10 days, with increased frequency during the initial growth phases and in periods of dry or elevated temperatures. On average, beetroot needs approximately 25–30 mm of water weekly, although this requirement may fluctuate based on soil characteristics and seasonal variations; for example, sandy soils tend to drain more rapidly and thus require more regular watering. Drip irrigation is a highly effective method, as it provides uniform water distribution while reducing the risk of wet foliage, which can contribute to disease development. It is equally important to avoid excessive watering, as this can lead to root rot and oxygen deprivation. It is beneficial to utilize organic mulch to retain moisture, inhibit weed growth, and maintain consistent soil temperature around the plants. Finally, it is recommended to stop watering about two weeks before harvest to achieve the best quality of the roots (8,9).

g. **Disease Control:** Beetroot is prone to many pathogens, which significantly reduce its yield and quality. These pathogens can be bacterial, fungal, viral, nematodes, or pests; they can damage plants above or below the ground and cause seedling diseases, foliar, or root diseases. In Table 2.3, information about a few diseases, their causing agent, symptoms, and management is provided.

TABLE 2.3

Common Diseases of Beetroot

Disease Category	Name of Disease	Causing Agent	Symptoms	Management	References
Bacterial	Bacterial blight	*Pseudomonas syringae* pv. aptata	Irregular circular spots on infected leaves	Use disease-free and healthy seeds	Gaulke and Goldman (2022)
			Dry leaf margins	Crop rotation	
	Scab	*Streptomyces scabies*	Small brown round spots on roots	Avoid crop rotation with the potato crop	Munyaneza and Bizimungu (2022)
			Raised corky lesions on the root surface	Reduce soil pH	
				Plant in soil that drains well	
Viral	Beet curly top disease	Beet curly top virus	Foliar chlorosis and curling of leaves	Grow resistant cultivars	Strausbaugh et al. (2012)
		Beet mild curly top virus	Dwarf and crinkly leaves	Spray insecticide to manage leafhoppers (virus transmitter)	
		Beet severe curly top virus	Leaves rolled upward and inward	Regulate the planting time	
			Unevenly enlarged leaf veins and vascular tissue discoloration	Keep overwintered weeds out of the field	
			Dwarfed young roots	Cover the seedlings with a physical barrier	
	Beet Western yellows virus	Beet Western Yellows Virus	Mottling and yellowing of young leaves	Control aphids	Duffus (1973)
			Brittle and thick leaves	Grow the resistant variety	
			Stunted plant growth	Control weeds	

(*Continued*)

TABLE 2.3 (*Continued*)

Common Diseases of Beetroot

Disease Category	Name of Disease	Causing Agent	Symptoms	Management	References
Fungal	Cercospora leaf spot	*Cercospora beticola*	Small yellow round lesions on older leaves	Crop rotation	Pethybridge et al. (2017)
			Necrotic or black irregular spots	Apply fungicide	
				Bury crop debris post-harvest into the soil	
	Damping off	*Aphanomyces cochlioides*	Seedling collapse	Avoid overwatering	Lamichhane et al. (2017)
		Rhizoctonia solani	Absent or blackened roots	Treat seeds with fungicide before planting	
		Phoma betae	Constricted plant crown	Soil mulching	
		Pythium ultimum	Mushy stems	Avoid seed overcrowding	
			Seeds not able to germinate		
			Uneven or stunted seed growth		
	Downy mildew	*Peronospora farinosa*	Angular chlorotic leaf spots	Grow resistant plant varieties	Salcedo et al. (2021)
			Fragile, leathery, and thickened leaves	Remove and destroy the infected crop	
			Both leaf surfaces have flaky gray growth	Apply fungicide	
Fungal	Fusarium yellows and root rot	*Fusarium oxysporum*	Plant wilting	Grow resistant plant varieties	Husna et al. (2021)
			Dry, brittle, and clustered leaves	Crop rotation	
			Rotting of the taproot	Use fungicide	
			Interveinal chlorosis	Maintain soil pH	
Nematode	Beet cyst nematode	*Heterodera schachtii*	Cysts formed on beetroots	Grow nematode-resistant plant varieties	Kumar et al. (2020)
			Reduced and stunted leaf growth	Crop rotation	
			Premature plant wilting	Early planting	
			Abnormal lumps on the root	Deep summer plowing (exposing soil cysts to sunlight)	
	Root-knot nematode	*Meloidogyne* spp	Root galls	Soil solarization	Bahadur (2020)
			Yellowing of leaves	granular nematicides	
			Stunted growth	Field sanitation	

(*Continued*)

TABLE 2.3 (*Continued*)

Common Diseases of Beetroot

Disease Category	Name of Disease	Causing Agent	Symptoms	Management	References
Insects	Root Aphids	*Rhopalosiphum rufiabdominalis*	Yellowing and wilting of leaf	Insecticides Use resistant varieties	Singh et al. (2024)
	Web Worms	*Hymenia* spp	Egg clusters on leaf's lower side	Use insecticide	
		Loxostege spp	Restrict flowering and pod formation		
	Leaf miners	*Liriomyza* spp	White spots on leaves Dry leaf Trails and lesions on leaves Leaves fall prematurely	Immediately remove the plant post-harvest Use insecticide Field sanitization Avoid water stress	Liu et al. (2015)

2.3 Harvesting

Beetroot is harvested 75–90 days after seeding in the summer and 100–120 days in the winter. Before the cutting or pulling of the beets, the soil should be moist enough. Extremely dry soil may hinder the roots' cleaning process, and the risk of top breakage may increase significantly. Harvesting of beetroots is recommended when their roots have grown to a size of 1–3 inches, depending upon the cultivar. The harvesting of beetroots on a smaller scale may be carried out manually (pulling the upper part of the plant with hands) or with tools like a potato fork. In contrast, larger-scale harvesting requires specialized machinery like multiple-row diggers. Finally, to increase their perishability, four to five beets are bunched and tied from the top. To prevent bunched beetroot from decay and discoloration, they are stored below 4°C at a relative humidity of 90%–94% after harvest. The yield of the crop often varies from 25 to 30 tins/ha (Kumar, 2015).

2.4 Post-Harvesting

The following are a few post-harvesting handling stages (Jasmitha et al., 2018):

- **Washing**: Wash the beetroots to get rid of dirt after harvesting.
- **Grading**: Sort the beets based on size and remove the injured or diseased roots while sorting.
- **Packaging**: Beetroot can be packed in crates or cartons.
- **Storage**: Fresh beets should be kept in cool and dry conditions (refrigerated in an airtight plastic bag) to avoid fungal contamination.
- Transportation
- Marketing

TABLE 2.4

Top Beetroot-Producing States

State	Production (in tons)
Maharashtra	150,000
Haryana	120,000
Uttar Pradesh	100,000
Arunachal Pradesh	80,000
Karnataka	70,000
Gujarat	60,000
Punjab	50,000
West Bengal	40,000
Tamil Nadu	30,000
Rajasthan	20,000

2.5 Production

Beetroot is a significant vegetable crop worldwide, particularly in the Middle East, North America, Europe, and specific parts of Asia, notably India. According to 2023 data from the IndiaStat website, the following are the states producing beetroot in India (Table 2.4).

2.6 Summary and Future Perspectives

Beetroot is a predominantly biennial plant consisting of an enormous number of varieties, which are characterized based on the shape of roots, maturation time of heirloom, and some specialized characteristics. The plant is cultivated in early spring and is known for its thick fleshy roots. It is widely distributed all over the globe, having a global production of approximately 269.7 million tons. The manufacturing period ranges from 55 to 60 days and goes up to 90 days in winter. Beetroot is a source of various vitamins, including vitamin A, thiamine, riboflavin, niacin, folate, and so forth. It is also known for the presence of potent pigments such as betalains, which provide red color to the plant, ascorbic acid, saponins, and high levels of nitrate. An ambient temperature, generally warmer, loamy, humus-rich soil, and an adequate amount of soil nutrients and fertilizing agents lead to a good yield of crops. Beetroot is prone to many pathogens that reduce its yields. The pathogen includes Bacterial Blight, Beet Curly Top disease, Beet Western Yellow Virus, Cercospora Leaf Spot, Beet Cyst nematode, and so forth. The harvesting of the plant is done 75–90 days after seeding, which goes up to 100–120 days in the winters. The post-harvesting handling stages include washing, grading, packaging, transportation, and marketing.

Beetroot, being a model plant for its commercial and industrial application, should be grown in a well-defined manner following great agronomic practices, which help in improving its yield, thereby playing a significant role in the pharmaceuticals, food industry, and so forth. The research conducted on beetroot provides information regarding beetroot breeding programs. The growth of hybrid beetroot varieties plays a significant role in resolving pest and disease-related problems. Information formulated by beetroot research was not available to the breeders in the West. Various measures have to be taken into consideration to build a good collaboration between the beetroot crop breeders. Beetroot is exposed to fluctuating temperatures throughout its life cycle, but gives a good yield when exposed to a humid environment. Recent studies have carried out multiple omics approaches to outline the genes that are upregulated in winter. Such studies will help to build up frost-resistant varieties in the future.

Conflict of Interest

The authors have no conflict of interest.

Acknowledgment

The authors are grateful to the UGC-BSR (316/Bot and Env Sci) program of the University Grants Commission, New Delhi (India), Amity Research Fellowship (ARF), the Department of Botanical and Environmental Sciences, Guru Nanak Dev University, and Amity University for providing the necessary facilities and support.

References

Arya, P. J., 2020. *Performance of Beet Root (Beta vulgaris L) for Growth Yield and Quality* (Doctoral dissertation, Department of Vegetable Science, College of Agriculture, Vellayani).

Bahadur, A. (2020). Beet-root (Beta vulgaris) found infection with root-knot nematode (Meloidogyne incognita) in Tripura, India. *Indian Phytopathology*, 73(3), 561–562.

Balázs, S. (1994). *Zöldségtermesztők kézikönyve* (pp. 631–639). Budapest: Mezőgazda Kiadó.

BEETROOT: Cultivation Characteristics and Irrigation. (2024). https://palaplast.com/2024/10/17/beetroot-cultivation-characteristics-and-irrigation/.

Chawla, H., Parle, M., Sharma, K., & Yadav, M. (2016). Beetroot: a health promoting functional food. *Nutraceuticals*, 1, 976–3872.

Clifford, T., Howatson, G., West, D. J., & Stevenson, E. J. (2015). The potential benefits of red beetroot supplementation in health and disease. *Nutrients*, 7, 2801–2822.

Duffus, J. E. (1973). The yellowing virus diseases of beet. In: K. M. Smith & M. A. Lauffer (Eds.), *Advances in Virus Research* (Vol. 18, pp. 347–386). New York and London: Academic Press.

Food and Agriculture Organization of the United Nations. (2014). FAOSTAT Statistical Database [Internet]. Available from: https://www.fao.org/faostat/en/#data/

Gandia, F., Escribano, J., & Garcia, F. (2010). Structural implications on color, fluorescence, and anti-radical activity in betalains. *An International Journal of Plant Biology*, 232(2), 449–460.

Gaulke, E., & Goldman, I. L. (2022). Screening table beet and Swiss chard for resistance to Pseudomonas syringae pathovar aptata. *HortScience*, 57(11), 1436–1446.

Gokhale, S. V., & Lele, S. S. (2014). Betalain content and antioxidant activity of Beta vulgaris: Effect of hot air convective drying and storage. *Journal of Food Processing and Preservation*, 38, 585–590.

Guldiken, B., Toydemir, G., Nur Memis, K., Okur, S., Boyacioglu, D., & Capanoglu, E. (2016). Home-processed red beetroot (*Beta vulgaris* L.) Products: Changes in anti- oxidant properties and bioaccessibility. *International Journal of Molecular Sciences*, 17, 858.

Hole, B., Gaikwad, A., Giramkar, A., Deepak, G., & Umalkar (2023). A review on beta vulgaris (Beet Root). *PENSEE*, 51, 1–11.

Husna, A., Zakaria, L., & Mohamed Nor, N. M. (2021). Fusarium commune associated with wilt and root rot disease in rice. *Plant Pathology*, 70(1), 123–132.

Datanet India. 2025. Agriculture – Beetroot (Chukandar) Statistics and Growth Figures Year-wise of India. *Indiastat*. https://www.indiastat.com/data/agriculture/beetroot-chukandar [accessed 13 November 2025].

Jasmitha, S. K., Shenoy, A., & Hegde, K. (2018). A review on beta vulgaris (beet root). *International Journal of Pharma and Chemical Research*, 4, 136–140.

Jones, A. M., Bailey, S. J., & Vanhatalo, A. 2012. Dietary nitrate and O_2 consumption during exercise. *Medicine and Sport Science*, 59, 29–35. https://doi.org/10.1159/000342062

Joubert, G. 1974. *The Cultivation of Vegetables in South Africa, the Cultivation of Root Crops: The Cultivation of Beetroot. Root Crops Series No.A.1/1974*. Pretoria: Horticultural Research Institute.

Kezi, J., & Sumathy, J. H. (2014). Betalain a boon to the food industry. *Discovery*, 20, 51–58.

Kovarovič, J., Bystrická, J., Tomáš, J., & Lenková, M. (2017). The influence of variety on the content of bioactive compounds in beetroot (Beta vulgaris L.). *Potravinarstvo Slovak Journal of Food Sciences, 11*(1), 106–112.

Kumar, S., & Brooks, M. S. L., 2018. Use of red beet (Beta vulgaris L.) for antimicrobial applications – a critical review. *Food and Bioprocess Technology*, 11, 17–42.

Kumar, V., Kumar, V., & Yadav, P. (2020). Importance of summer ploughing for sustainable agriculture. *International Journal of Trend in Research and Development*, 7(3), 33–35.

Kumar, Y. (2015) Beetroot: a super food. *International Journal of Engineering Studies and Technical Approach*, 1(3):20–26.

Lamichhane, J. R., Dürr, C., Schwanck, A. A., Robin, M. H., Sarthou, J. P., Cellier, V., ... & Aubertot, J. N. (2017). Integrated management of damping-off diseases. A review. *Agronomy for Sustainable Development*, 37, 1–25.

Lang, K. (2023, March 30). *Table Beets: Harvest and Storage*. South Dakota State University Extension. Table Beets: Harvest and Storage. Brookings: South Dakota State University Extension.

Lange, W., Brandenburg, W. A., & De Bock, T. S. (1999). Taxonomy and cultonomy of beet (Beta vulgaris L.). *Botanical Journal of the Linnean Society*, 130(1), 81–96.

Lidder, S., & Webb, A. J. (2013). Vascular effects of dietary nitrate (as found in green leafy vegetables and beetroot) via the nitrate-nitrite-nitric oxide pathway. *British Journal of Clinical Pharmacology*, 75, 677–696.

Linnaeus, C. (1753). *Species Plantarum*, Vol. 1. Stockholm: Laurentii Salvii.

Linnaeus, C. (1762). *Species Plantarum*, 2nd ed., Vols. 1–2. Stockholm: Laurentii Salvii.

Linnaeus, C. (1767). *Systema Naturae*, 12th ed., Vol. 1, pp. 11–12. Stockholm: Laurentii Salvii.

Liu, W. H., Dai, X. H., & Xu, J. S. (2015). Influences of leaf-mining insects on their host plants: a review. *Collectanea Botanica, 34*, e005.

McGrath, J. M., Saccomani, M., Stevanato, P., & Biancardi, E. (2007). Beet. In C. Kole (Ed.), *Vegetables* (pp. 191–207). Berlin, Heidelberg: Springer.

Mirmiran, P., Houshialsadat, Z., Gaeini, Z. et al. (2020). Functional properties of beetroot (*Beta vulgaris*) in management of cardio metabolic diseases. *Nutrition & Metabolism, 17*, 3. https://doi.org/10.1186/s12986-019-0421-0

Moore, P. H. (2017). "Sugarcane and Sugarbeet." In *Encyclopedia of Applied Plant Sciences* (2nd ed., Vol. X, pp. 273–280). Amsterdam: Elsevier/Academic Press.

Munyaneza, J. E., & Bizimungu, B. (2022). Management of potato pests and diseases in Africa. In A. Alyokhin, S. I. Rondon, & Y. Gao (Eds.), *Insect Pests of Potato* (pp. 407–426). London: Academic Press.

Nintali, P., & Angelino, D. (2013). Nutritional and functional potential of Beta vulgaris cicla and rubra. *Fitoterapia*, 89, 188–199.

OECD, P., 2006. *Safety Assessment of Transgenic Organisms: OECD Consensus Documents.*
Oyen, L. P. A. (2004). Beta vulgaris L. In G. J. H. Grubben & O. A. Denton (Eds.), *Plant Resources of Tropical Africa 2: Vegetables* (pp. 110–113). Wageningen/Leiden: PROTA Foundation/Backhuys Publishers/CTA.
Pethybridge, S. J., Vaghefi, N., & Kikkert, J. R. (2017). Management of Cercospora leaf spot in conventional and organic table beet production. *Plant Disease*, 101(9), 1642–1651.
Pink, D. A. C. (1993). Beetroot: *Beta vulgaris* subsp. vulgaris. In G. Kalloo & B. O. Bergh (Eds.), *Genetic Improvement of Vegetable Crops* (pp. 473–477). Oxford/New York: Pergamon Press.
Salcedo, A. F., Purayannur, S., Standish, J. R., Miles, T., Thiessen, L., & Quesada-Ocampo, L. M. (2021). Fantastic downy mildew pathogens and how to find them: advances in detection and diagnostics. *Plants*, 10(3), 435.
Satyanand, V., Vali, S. M., & Krishna, B. P. (2014). A study of beet root derived dietary nitrate efficacy on the performance of runners. *Indian Journal of Basic and Applied Medical Research*, 3, 690–695.
Singh, A., Kaur, A., Srivastava, P. K., & Kumar, D. (2024). Integrated pest management in beetroot cultivation: a systematic review. *World Journal of Advanced Research and Reviews*, 21(1), 1687–1698.
Sood, S., & Gupta, N. (2017). Beetroot. In M. K. Rana (Ed.), *Vegetable Crop Science* (pp. 247–260). CRC Press, Taylor & Francis Group.
Strausbaugh, C. A., Wenninger, E. J., & Eujayl, I. A. (2012). Management of severe curly top in sugar beet with insecticides. *Plant Disease*, 96(8), 1159–1164. https://doi.org/10.1094/PDIS-01-12-0106-RE
Székely, D., & Máté, M. (2023). *Red Beetroot (Beta Vulgaris L.).* IntechOpen. https://doi.org/10.5772/intechopen.106692
Wootton-Beard, P. C., & Ryan, L. (2011). A beetroot juice shot is a significant and convenient source of bioaccessible antioxidants. *Journal of Functional Foods*, 3, 329–334.
Yadav, M., Chawla, H., Parle, M., & Sharma, K. (2016a). Beetroot: a health promoting functional food. *Inventi Rapid: Nutraceuticals, 2015*(1), 1–5.
Yamane, T. (2023). Sugar beet. In *Encyclopaedia Britannica.* Chicago, IL: Encyclopaedia Britannica. https://www.britannica.com/plant/sugar-beet
Yashwant, K. (2015). Beetroot: a super food. *International Journal of Engineering Studies and Technical Approach*, 1, 20–26.

3

Beetroot Harvesting and Post-Harvesting

Neelesh Yadav, Sonali Khanal, Rachna Verma, and Dinesh Kumar

3.1 Introduction

Beetroot (*Beta vulgaris*) is a versatile crop with nutritional, agricultural, and industrial importance. It is a root crop that contains important nutrients like vitamins (C and B-complex), minerals (potassium, magnesium, and iron), dietary fiber, and bioactive substances such as betalains and nitrates. These ingredients are responsible for its health-benefiting activities, such as antioxidant, anti-inflammatory, and cardiovascular effects (Singh et al., 2024). Beetroot's tolerance of variable climates and soils contributes to its being a valuable crop in ensuring food security as well as practising sustainable agriculture. Additionally, its uses include fresh consumption and processing to make products such as beetroot powder and juice, which are commonly used in food, beverages, and nutraceutical companies (Chen et al., 2021). Several economic, climatic, and technological aspects influence world consumption and production trends of beetroot. It is cultivated worldwide, with prominent producers including Russia, the United States, China, Ukraine, and other European countries such as Poland and Germany (Tayyab et al., 2023). The demand for beetroot is due to its rich nutritional value, such as high antioxidant content, dietary fiber, and essential minerals. In the past few years, consumption has been on the rise due to rising awareness about health benefits, especially in functional foods, drinks, and natural colorants. Growth in plant-based diets has further increased the demand for beetroot in the food processing market. Production trends, however, are impacted by climate change, soil health, and technological innovations in agricultural methods. Effective processing, storage, and supply chain management are key to addressing the increased global demand. Generally, the beetroot industry continues to grow, with innovation in product formulation and sustainable agriculture defining its future development (Rud et al., 2024). Harvest and post-harvest operations are essential for retaining the quality, shelf life, and marketability of beetroot. Avoiding physical injury to the crop during harvesting is important for the retention of its visual quality and the prevention of microbial contamination. Post-harvest activities, such as cleaning, packaging, and storage, mainly contribute to preserving the nutritional contents and freshness of beetroot. Cleaning eliminates dirt and microbes and minimizes spoilage potential, while packaging prevents physical damage and dehydration. Storage, in terms of temperature regulation, also plays a significant role; cold storage retards physiological activities and microbial growth, improving shelf life and consumer satisfaction (Paulauskienė et al., 2023). Moreover, drying methods such as osmotic dehydration can also increase shelf life by lowering weight and volume, which will ease transportation while maintaining nutritional value (Šuput et al., 2024).

DOI: 10.1201/9781003527084-3

3.2 Fruit Development

Beetroot, a versatile crop with utility in its leaves and roots, must be handled with care at harvesting and post-harvest operations to ensure optimal yield and quality. Beetroot fruit development is characterized by the formation of its storage root, which accumulates nutrients and pigments that contribute to its quality and nutritional value. Nitrogen application during cultivation significantly contributes to increasing leaf and root yields. Nitrogen concentrations of 90–120 kg·ha^{-1} substantially enhance root diameter, length, fresh weight, and mineral composition, such as magnesium and iron. Excessive nitrogen, however, decreases protein concentration in the roots. Leaf harvesting management also affects beetroot growth; moderate leaf removal (30%) does not adversely affect root yield but enhances protein content in the roots (Mampa et al., 2017). Post-harvest handling is pivotal for beetroot marketability and storage. Organic beetroot production is challenged by rots that develop in storage due to diseases caused by pathogens, including Fusarium, Phoma, Botrytis, Rhizoctonia, and Pythium. Research has demonstrated that seed, field, or post-harvest environments can be sources of pathogen infection. Prophylactic measures such as seed steam sterilization, biocontrol treatments, optimal cooling practices, and cultivar choice have been tried in an attempt to minimize losses. The measures improve the health of seeds, seedling establishment, and stored beetroot quality. Storage losses continue to exceed 50% in spring due to infection mechanisms being unknown, opening up areas of research into storage practices and pathogen control (Bernasconi et al., 2023). Proper post-harvest handling practices, such as temperature control, humidity management, and modified atmosphere storage, are necessary to preserve beetroot quality over long periods of storage. Sophisticated technologies like forced ventilation and computational modeling of storage environments have been found to be effective in minimizing moisture-related losses while increasing sucrose concentration in sugar beet roots. These methods could be used for beetroot to enhance its storability and economic value (Schouten & van Schaik, 1981).

3.2.1 Growth Stages of Beetroot

Beetroot goes through three growth phases: germination, first growth, and second growth with sucrose storage. In germination, water intake by the seed stimulates metabolic activities essential for sprouting. The root apical meristem triggers cell division to form the initial root structure. This process is essential for vascular tissue development, facilitating nutrient uptake and elongation. Hormonal control is important, with phytohormones like cytokinin and auxins controlling cell differentiation and the patterning of tissue. During secondary growth, the plant, supernumerary cambia form results in taproot thickening and sucrose storage (Jammer et al., 2020). The rapid expansion of shoots characterizes the vegetative growth phase and leaves through photosynthesis. At the same time, the taproot becomes thicker, as lateral roots develop to improve water and nutrient uptake. The phase then shifts to the root development phase, during which the beet's fleshy taproot develops. Secondary growth involves cambium cell division and sucrose accumulation in storage cells. Beetroot is simultaneously a source of metabolic activities and an energy storage sink. There are definable physiological changes, such as increased enzyme activity and sucrose partitioning, marking this phase that determines the final yield and quality of the crop (Jammer et al., 2020).

3.2.2 Factors Influencing Beetroot Development

Beetroot development depends on a number of factors, such as soil, climate, irrigation, nutrients, and fertilizer. Soil is a key factor in beetroot development since it has to develop in well-drained and loamy soil with a pH of 6.0–7.5. Excess moisture in the soil leads to root rot, while insufficient moisture results in poor development. Climate plays an important role in beetroot growth; temperatures must be moderate since excessive heat or cold will interfere with its growth cycle. Irrigation is also essential in maintaining consistent levels of moisture in the soil to facilitate healthy root growth and nutrient uptake (Paulauskienė et al., 2023; Batoqui et al., n.d.). The availability of nutrients and fertilizer application also affects beetroot growth. Nutrients such as nitrogen, phosphorus, and potassium are important in stimulating healthy leaf growth and root development. It has been demonstrated that potassium content rises during beetroot maturation processes, which maintains the metabolic processes of the plant (Batoqui et al., n.d.). Also, organic fertilizer use has been associated with enhanced nutrient quality and antioxidant potential in beetroot leaves and roots. For instance, it's betalains and nitrates are responsible for their biological activity, increasing their health benefits and resistance (Chen et al., 2021). Effective nutrient management provides maximum growth and enhances the functional qualities of beetroot as a food item.

3.2.3 Physiological Changes During Development

Beetroot undergoes extreme physiological changes during its growth, including the storage of sugars, colorants, and nutrients. The most extreme change is the development and deposition of betalains, primarily betanin, which are water-soluble pigments responsible for their color. Not only do these pigments add to the aesthetic appeal of the plant, but they also function as antioxidants that defend the plant against oxidative stress. Moreover, beetroot also stores large amounts of inorganic nitrate during its development. Nitrate is converted to nitric oxide in the human body, which is an important component in vasodilation and enhancing blood flow, making it a functional food of significance (Arazi & Eghbali, 2021; Chen et al., 2021).

Its production further involves the accumulation of sucrose and glucose, which give it sweetness and calories. Both sugars play vital roles in storing energy in the plant and, upon being consumed by humans, are utilized as an instantaneous source of energy. Beetroot also has dietary fiber, vitamins (such as vitamin C), minerals (potassium, magnesium), and polyphenolic substances. The synergy of these nutrients collectively supports beetroot's functional properties, including its role in cardiometabolic disease management and physical performance enhancement. Such synergy of these bioactive molecules makes beetroot a valuable crop for nutrition as well as pharmaceutical applications (Olsson et al., 2019; Arazi & Eghbali, 2021).

3.3 Maturity

Beetroot maturity needs to be determined to increase its nutritional and market value. It can be assessed based on visible characteristics such as shape, color, and size. Healthy, mature beetroot roots are uniformly deep red or purple, reflecting the presence of high levels of betalain pigments. Root shape and size are also involved; mature beetroots typically have a round or slightly oblate shape, depending on the variety. These visual characteristics are important to ensure the beetroot is of market quality and meets consumer requirements. Physiological markers offer a more accurate quantitative measure of beetroot maturity. Sugar content, for example, rises with the development of the root and is responsible for its typical sweetness (Stoica et al., 2025). Another important physiological factor is firmness; older beetroots are denser and firmer because structural carbohydrates such as cellulose and hemicellulose build up in the root tissues. Research has demonstrated that the polyphenolic composition also changes with increasing maturity. For instance, young shoots of beetroot have much greater concentrations of polyphenols than mature roots, such as gallic acid and chlorogenic acid, which contribute to antioxidant activity (Piasna-Słupecka et al., 2023). These physiological indicators not only confirm maturity but also express the nutritional and functional qualities of the beetroot in the various phases of growth.

3.3.1 Optimal Harvest Time

The maturity phase of beetroot plays an important role in determining its quality traits and storability. Beetroots taken 90 days after planting are found to possess higher antioxidant activity and stable betalain pigments than the beetroots taken at different phases. Ideal physical characteristics, including preferred size, specific gravity, and juice yield, are also exhibited by these beetroots. On the other hand, beetroots picked at 105 days or later have higher betalain content but a lower redness index, which means that there is some trade-off between color quality and pigment concentration (Abarna, 2010). Proper evaluation of beetroot maturity plays a pivotal role in maximizing harvest time. Conventional destructive techniques, including measurement of physical and chemical parameters, are informative but typically entail production loss. Further development of non-destructive methods, such as Fourier transform infrared spectroscopy integrated with principal component analysis, presents effective options. These methods enable immediate care to be taken for determining the amount of moisture and quality parameters without damaging the produce, thereby facilitating better decision-making regarding harvest and storage (Nesakumar et al., 2018).

3.4 Handling

3.4.1 Harvesting Techniques

Beetroot can be harvested using manual or mechanical means, both with their own set of advantages and disadvantages. Manual harvesting is done by pulling the roots manually or loosening the soil with tools such as garden forks, making it ideal for small-scale agriculture and home gardens because of minimal damage and selective picking according to maturity. It is necessary to break up hard soil in advance to avoid breakage, and watering the field a day in advance can make extraction easier. Handling carefully while pulling and cleaning after harvesting maintains root quality and reduces bruising. Conversely, mechanical harvesting is more effective for commercial production, using machines that lift and cut leaves at the same time. Calibration of the machines is important to avoid damage, as incorrect settings may cause root bruising or breakage. Mechanically harvested beetroots must be sorted to eliminate damaged roots to ensure uniformity and minimize post-harvest losses. Soil conditions have a great influence on the effectiveness of mechanical harvesting; wet or dry soils can result in more breakage or compaction, reducing overall yield and storability. Irrespective of the process, post-harvest handling best practices are essential for preserving the quality of beetroot. Sorting and grading help remove diseased or physically damaged roots to avoid spoilage. For storage, beetroots should not be washed directly after harvesting; rather, loose soil can be removed with a gentle brushing process before storing them in cool, dry environments. Storage of beets at 0°C–2°C with 98%–100% relative humidity greatly increases shelf life by avoiding moisture loss and microbial spoilage. Placing stored beets under leaves or rice straw additionally minimizes weight loss and maximizes preservation. These practices collectively ensure that beetroots remain in optimal condition from harvest to consumption (Singh et al., 2024; Abarna, 2010; Akan et al., 2021; Hoffmann et al., 2018; Seadh et al., 2021).

3.4.2 Post-Harvest Handling

Correct post-harvest handling of beetroot is essential to preserve its quality, reduce losses, and increase its shelf life. Cleaning is the initial step and eliminates soil, dirt, and microbially derived contaminants that could promote spoilage. Cleaning is normally accomplished by water sprays or dry brushing, subject to market demands and storage conditions (Kader, 2002). Upon washing, grading and sorting are performed according to size, shape, color, and external blemishes. Grading provides homogeneity in presentation to the market and facilitates optimum price realization. Proper grading of beetroot, according to studies, has better consumer acceptability and less wastage during transit and storage (Ahmad & Siddiqui, 2015). Mechanical damage like bruising and cutting seriously impacts the post-harvest quality of beetroot by making it more vulnerable to microbial infection and dehydration. Research has indicated that poor handling leads to enhanced weight loss and a reduction in textural and visual quality (Wills & Golding, 2016). In order to avoid mechanical damage, it is necessary to adopt soft handling practices, like padded conveyors and gentle manual handling during harvesting and packing. Furthermore, ideal storage conditions, like relative humidity of 90%–95% and 0°C–2°C temperature, assist in minimizing moisture loss and the firmness of the root being preserved (Yahia & Carrillo-Lopez, 2018). Harvesting and Post-harvesting technologies are showcased in Figure 3.1.

3.5 Packing and Distribution

Beetroots' shelf life and post-harvest quality are dramatically affected by pre-harvest agricultural management practices, such as irrigation water management and application of organic solutes. A test to analyze the interaction between the depth of irrigation and ascorbic acid dosage revealed that the application of ascorbic acid reduced the undesirable impacts of water stress, where the post-harvest qualities such as total soluble solids, titratable acidity, and retention of moisture increased. Particularly, under limited irrigation (40% evapotranspiration), the use of 2 mM ascorbic acid maximized beetroot quality, indicating that pre-harvest treatments can minimize post-harvest degradation during packing and storage (Carlos De Melo Gonçalves et al., 2019). Such a discovery emphasizes the critical need to consolidate agronomic management with post-harvest handling so that beetroot quality is kept intact, especially in areas prone to water deficits. Through structural stabilization of cell structures and relief from oxidative stresses, ascorbic acid

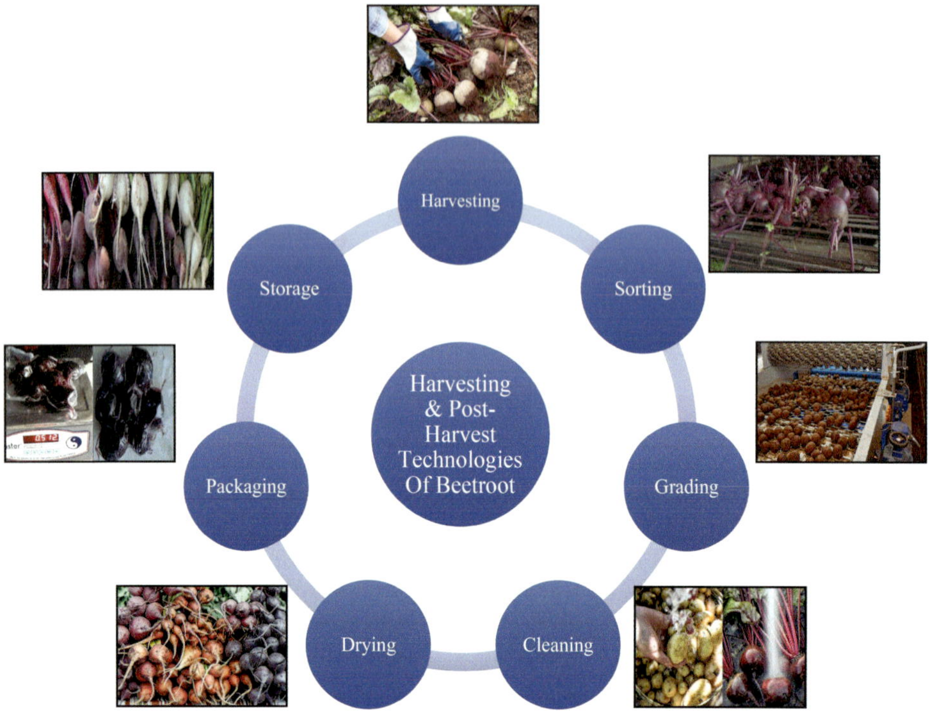

FIGURE 3.1
Harvesting and post-harvesting technologies of beetroot.

applications could extend the shelf life of beetroots for marketing during delivery, even in poor storage facilities. The packaging material selection is of significant importance in maintaining the nutritional and phytochemical integrity of stored and distributed processed beetroot products. A study comparing laminated aluminium pouches (LAP) and high-density polyethylene (HDPE) for storage of beetroot powder under accelerated storage conditions (40°C, 90% relative humidity) indicated that LAP packaging retained better betalain content, color stability, and moisture control during six months of storage. Betalains, which are light- and heat-sensitive pigments that give beetroot its bright color and antioxidant activity, broke down faster in HDPE because it is more permeable to oxygen and water (Singh & Hathan, 2017). These findings indicate the importance of using barrier-resistant packaging in order to avoid oxidative and hygroscopic damage, especially in value-added beetroot products with long-distance transportation. This study's proposal for LAP comes into line with business trends in new-generation packaging innovations that deliver increased shelf life as well as preservation of bioactive compounds for an effective circular bioeconomy by limiting post-harvest waste (Mkhari et al., 2023). Post-harvest handling procedures must also take into account the physiological effects of selective leaf harvesting on root quality throughout storage. Nitrogen fertilization (up to 120 kg·ha^{-1}) and moderate selective leaf harvesting (30%) did, however, improve root yield and mineral concentrations without affecting dry matter accumulation, but excessive leaf pulling (50%) lowered photosynthesis, resulting in lower root biomass and possible exposure during transit (Mampa et al., 2017). This would indicate that beetroots have controlled leaf harvest before packing and can reconcile their dual market as root and leaves while retaining structure firmness as well as resisting mechanical damage. Such practices, implemented together with optimum packaging systems, assure beetroots' quality status in the chain of supply, from farm through to consumer level.

3.5.1 Packaging Materials and Methods

Packaging is vital in facilitating the safe handling, storage, and transport of products. The packaging method and materials to be used rely on the product type, whether it is fragile, and where it is headed. The usual packaging types are crates, bags, and bulk containers.

1. **Crates:** Crates, most often made of wood or hard plastic, are often utilized due to their reusability and strength. They excel in protecting the commodities from mechanical shock and compression when they are being handled and transported. Plastic crates are particularly suited for perishable products like fruits and vegetables, as they are capable of being subjected to a high-humidity environment. They can also be sanitized using chlorinated water to prevent contamination with ease (Kitinoja & Kader, 2002).
2. **Bags:** Bags are an affordable choice for mass packaging but are not as protective as rigid boxes. To further their use, ventilation systems like woven bamboo tubes can be introduced to mitigate the heat generated through product respiration (Kitinoja & Kader, 2002). Bags, however, are less robust in general and could fail to offer protection from the outside forces associated with stacking and shipping.
3. **Bulk Containers**: Bulk containers, including large bins or pallets with secondary packaging layers, are ideal for industrial-scale storage and transportation. These containers provide robust protection while allowing for efficient handling in warehouses or distribution centres. Tertiary packaging, like shrink wraps or palletized unit loads, is often used to secure bulk containers during transit (Kitinoja & Kader, 2002).

For processed beetroot in the form of cubes or powder, dedicated packaging material such as HDPE or LAP is used. Research has indicated that LAP offers improved protection against moisture and color degradation than HDPE in long-term storage under accelerated conditions (Kumari & Divakar, 2017).

3.5.1.1 Importance of Ventilation and Protection

Ventilation is essential in beetroot packaging to ensure freshness and avoid spoilage. Fresh beetroot is very sensitive to water loss, causing it to shrink and lose market value. Ventilated packaging enables air circulation, ensuring the right relative humidity (95%–100%) and avoiding excessive moisture accumulation that may result in fungal growth or bacterial rot (FR2557071A1, n.d.; WO2008076075A1, n.d.; Kumari & Divakar, 2017).

Also, protective measures are necessary to reduce mechanical damage and physiological degradation. For example:

1. Perforated liners in bags or crates provide uniform ventilation while minimizing direct contact with air currents that may dehydrate the roots.
2. Controlled atmosphere packaging stabilizes oxygen and carbon dioxide levels, lowering respiration rates and prolonging storage life without substantial loss of quality.

3. For processed beetroot products such as cubes or powder, vacuum sealing or nitrogen flushing can be used to inhibit oxidation and microbial contamination while preserving visual appearance

By combining proper materials with proper ventilation approaches, the fresh and processed beetroot shelf life can be largely prolonged to its quality right from harvest time up to the delivery point.

3.5.2 Logistics and Transportation

3.5.2.1 The Use of Cold Storage in the Preservation of Nutritional Value

Plant phenolics, which are health-oriented compounds, are studied more and more for their post-harvest changes. Cold storage has a crucial function in this respect. Kujala et al. (2000) observed the influence of cold storage on red beetroot, a plant that contains high levels of phenolics and betacyanins. The analysis showed that total phenolic content, betanin, isobetanin, and a certain ferulic acid ester changed considerably during cold storage (5°C, 0–196 days).

3.5.2.2 Effect of Storage Time on Phenolic Compounds

The study revealed that the betanin content of red beetroot peels declined until 140 days of storage and then began to rise. The concentration of a certain ferulic acid derivative rose with storage time. Total phenol content fell progressively until 63 days and then changed little subsequently. This suggests the need for controlling storage time to maximize levels of useful compounds in crops (Yasaminshirazi et al., 2021).

3.5.2.3 Distribution of Phenolics within the Plant

The positioning of phenolic compounds in the plant is also important. It was discovered that the majority of the phenolics are present in the external regions of the red beetroot, with the highest in the peel, followed by the crown, and then the flesh. This can be useful in identifying what regions of the plant to use to get the most nutritional value and how processing could impact the final product's quality (Akan et al., 2022).

3.6 Post-Harvest Storage

3.6.1 Storage Conditions of Beetroot

Beetroot is a root crop that needs to be kept under strict post-harvest conditions in order to preserve its quality and shelf life. The storage conditions of optimal temperature, humidity, and atmospheric control are essential for the retention of the physical and biochemical characteristics of beetroot.

3.6.1.1 Temperature and Humidity Control

Beetroot should be stored at temperatures between 0°C and 1°C with relative humidity levels of 90%–95%. These conditions reduce water loss, sprouting, and microbial growth, which are the major factors responsible for post-harvest deterioration. Storage of beetroots above 2°C substantially shortens their storage life, and low freezing temperatures can damage the tissue. For beetroots with foliage, storage is only 7–10 days because of the quick yellowing and quality deterioration of the leaves. Leaf removal before storage can prolong the roots' lifespan (Viskelis et al., 2019).

3.6.1.2 Atmospheric Control

3.6.1.2.1 Pre-Cooling

Pre-cooling shortly after harvesting at high speed is necessary to prevent biochemical changes from bruising or mechanical injuries while handling. Commercial operations widely implement this procedure in order to retain freshness.

3.6.1.2.2 Edible Coatings and Dehydration

Edible biopolymer coatings were found to improve the stability of osmotically dehydrated beetroots under cold storage (4°C). Moisture loss and oxidation are inhibited by these coatings, thus improving the shelf life while maintaining nutritional quality. Furthermore, drying processes employing sugar beet molasses have been optimized for maintaining bioactive compounds such as phenols and flavonoids during storage (Šuput et al., 2024).

3.6.1.2.3 Pickling and Freezing

Pickling is an old preservation technique that acidifies beetroot to stabilize it. Alternatively, freezing cooked or peeled beets in tightly sealed containers can store them for up to 10 months without appreciable loss of quality.

3.6.1.2.4 Varietal Selection

Some beetroot cultivars, like "Kestrel H," "Pablo H," "Rhonda H," and "Joniai," have proved to store better as a result of their physical and biochemical characteristics. These cultivars have lower betacyanin degradation rates, less variation in nitrate content, and improved firmness retention under long-term storage (Viskelis et al., 2019).

By using these advanced storage methods and best environmental controls, beetroot farmers can considerably prolong the shelf life of the vegetable while maintaining its nutritional quality and marketability.

3.6.2 Common Storage Issues

3.6.2.1 Spoilage, Wilting, and Microbial Contamination

Spoilage, wilting, and microbial contamination are the most critical problems of food storage. Spoilage is caused by enzymatic activity, microbial growth, and chemical reactions that reduce the quality of food. For example, perishable items such as fruits, vegetables, milk products, and fish are most susceptible to spoilage when exposed to unsuitable temperature and humidity conditions. Enzymatic action can cause texture and flavor modification, whereas microbial contamination—for example, by bacteria such as Bacillus cereus or mould—is likely to cause food poisoning or apparent spoilage (Mafe et al., 2024). Wilting is a particular issue with fresh produce like leafy greens and flowers. Wilting commonly occurs as a result of water loss in storage or transport through poor humidity control. Wilting not only lowers the visual quality but also lowers the nutritive value of the produce. Microbial contamination also adds to storage problems; pathogens may enter the food chain at various stages, ranging from harvesting, packaging, or storing. For instance, psychrotolerant bacteria in milk can resist pasteurization and cause spoilage during cold storage (Ukuku et al., n.d.).

3.6.2.2 Strategies for Minimizing Storage Losses

To meet these challenges, various strategies have been devised. Temperature management is of the greatest importance; keeping refrigeration at optimal levels (below 40°F or 4°C) is effective in preventing bacterial growth within the "danger zone" of 41°F–135°F (5°C–57°C). Modified atmosphere packing (MAP) and cold storage transportation are increasingly also being deployed to extend shelf life by regulating oxygen and humidity levels. Fast detection kits and nano-filters, for example, could be employed early on in the supply chain for spoilage hazard detection. Hygiene is crucial in preventing microbial contamination. Sanitary storage facilities with clean racks, walls, floors, and handling equipment minimize contamination risks. Clean packaging materials also avoid cross-contamination by separating various food categories. Predictive modeling for microbial growth is an innovation that aids in designing efficient storage protocols specific to commodities. Traceability systems are kept open across the supply chain by monitoring batch numbers and expiration dates. This allows for timely intervention when spoilage risk is detected. Further, education and training of perishable handlers can minimize losses through human error or lack of knowledge of storage conditions (Rudra et al., 2022; Ukuku et al., n.d.). To summarize, the control of spoilage, wilting, and microbial contamination involves a mix of technological innovation, scrupulous hygiene, and effective management of logistics. These efforts not only improve the safety of food but also help prevent dehydration-related losses—something that is pivotal to ambitious sustainable development efforts.

3.7 Innovations and Latest Trends

Recent advances in beetroot harvesting and post-harvest handling have focused on optimizing yield, reducing waste, and enhancing the nutritional and economic quality of by-products. In mechanical harvesting, advances in harvester design have played a significant role in precision and efficiency. For instance, prolonged field tests of beet harvesters (1984–2012) demonstrated a reduction of mass losses because of root breakage (2.7% surface loss and 0.9% subsoil loss) and improved topping accuracy, with more modern machines offering better retention of the crown and less over-topping. This is attributed to improved sensor technology and adaptive adjustments based on soil conditions, particularly with loess soils, where the tare of the soil averages 9.8%. These advances guarantee beetroots of better quality with reduced physical damage, which has a direct effect on marketability and storage life (Lammers & Schmittmann, 2024). Agronomic management has also been developed to accommodate dual-purpose beetroot production, where leaves and roots are both harvested. Studies show that the use of 120 $kg{\cdot}ha^{-1}$ of nitrogen (N) optimizes fresh leaf and root yields and increases magnesium, iron, and zinc levels in leaves. In parallel, partial leaf harvests of up to 30% at 35 days post-transplanting do not affect root yield, allowing farmers to benefit from the use of leafy greens as a secondary crop without reducing root biomass. Such a method is in line with circular bioeconomy reasoning

TABLE 3.1

Recent Innovations and Trends in Post-Harvest Technology and Preservation

S. No.	Aspect	Innovations/Trends	References
1.	Technological Advancements	Incorporation of intelligent packaging systems with sensors and Internet of Things technologies to track and control storage conditions, thus prolonging shelf life and maintaining food safety	Du et al. (2025)
2.	Sustainable Practices	Use of biodegradable and recyclable packaging materials to reduce environmental impact and minimize waste related to post-harvest operations	Biji et al. (2015)
3.	Organic post-harvest treatments	Application of organic and natural procedures for post-harvest handling to preserve quality and shelf life without the use of synthetic chemicals	Biji et al. (2015)
4.	Genetic upgrades for enhanced post-harvest qualities	Studies aimed at improving the genetic characteristics of crops to enhance post-harvest quality, such as greater resistance to spoilage and longer shelf life	Du et al. (2025)
5.	Novel preservation methods (e.g., edible coatings)	Innovative preservation technique exploration and use of food commodities, such as edible coatings, for sustaining freshness and improving the shelf life of fruits and vegetables	Biji et al. (2015)

by using the whole plant, thus minimizing post-harvest waste. In addition, partial leaf harvesting induces regrowth, and photosynthetic ability and continued nutrient translocation to roots throughout senescence are maintained (Mampa et al., 2017). Post-harvest innovations place focus on the valorization of wastes using high-technology encapsulation processes. Process waste of beetroot, such as stems and deteriorated roots, consists of high proportions of betalains, phenolic acids, and flavonoids. Using gum Arabic (GA) or maltodextrin (MT) to encapsulate these extracts retains bioactive compounds, but GA-based powder has higher solubility (89.2%) and retention of betalains (84.5%) than MT mixes. Such powders are potential natural colorants or functional compounds in fortified food products, preventing 20%–40% of post-harvest vegetable waste from going to landfill sites (Mkhari et al., 2023). These technologies not only increase shelf life but also generate new revenue avenues, highlighting the transition toward sustainable resource use in the beetroot industry. Table 3.1 shows recent innovations and trends in post harvest technology and preservation.

3.8 Summary and Conclusion

Beetroot has gained substantial interest across the world due to its nutrient-rich nature, bioactive metabolites, and diverse industrial usage. Beetroot is particularly coveted for betalain pigments, fiber, vitamins, and antioxidants that help in many of its health-supportive features, as well as in its increased usage in food, pharmaceuticals, and cosmetic sectors. This chapter gave an elaborate account of the harvesting and post-harvesting of beetroot, going through traditional techniques, new technology, and challenges related to post-harvest management. Harvesting of beetroot is a very important process in maintaining its quality and nutritional content. Maturity indicators like root diameter, days after sowing, and intensity of color are important factors in determining the best time to harvest. Harvesting techniques, either manual or mechanical, must be executed with care to prevent mechanical damage, which can lead to post-harvest losses. The chapter stressed that timely harvesting and proper handling are necessary to ensure market quality and shelf life extension.

Post-harvest handling entails a series of processes, such as cleaning, sorting, grading, and packaging. These processes are essential to have the produce reach consumers in the best possible condition. Effective sorting and grading contribute to standardizing quality, whereas proper packaging using breathable packaging materials plays an important role in preserving freshness and minimizing microbial spoilage during transportation and distribution. Ineffective handling at these stages can lead to bruising, microbial contamination, or loss of moisture, eventually resulting in considerable post-harvest losses. Postharvest storage is a subject of much focus, given the perishability of beetroot. Storage in favorable conditions, viz., low temperature and high relative humidity, is essential in reducing the respiration rate and rate of dehydration. Cold storage, storage in a controlled atmosphere, and the employment of MAP are some of the approaches described in the chapter to maximize efficiency and extend the shelf life of stored products. Reduction in loss alongside maintaining nutrients as well as bioactive components forms one of its aims. New technologies and treatments are slowly rewriting beetroot post-harvest handling. Uses of edible coatings, ozone technology, nanotechnology-based packaging, and intelligent sensors are being investigated to prolong freshness, track quality, and enhance safety. In addition, mechanization breakthroughs during harvest and mechanical grading systems are adding up to greater efficiency and less dependence on labor. These new methods, though still new in most of the world, hold great potential for heightening the quality of beetroot that reaches end consumers.

In spite of all these advances, there are a number of challenges still encountered in beetroot post-harvest management, particularly in developing countries. Poor infrastructure, low coverage of cold chain systems, poor training of farmers, and high vulnerability to spoilage are significant barriers. All these can only be overcome by employing a multi-disciplinary intervention involving farmer training, policy support, investment in infrastructure, and adoption of sustainable technology. In summary, the commercialization of beetroot can only be achieved if there is the enhancement of harvesting technologies, effective handling of post-harvest produce, and the introduction of innovative post-harvest storage and packaging. Postharvest losses can only be minimized with a view to maintaining food safety and security, in addition to creating economic benefits alongside the provision of health-promoting functional foods. Sustained R&D of post-harvest technology, coupled with coordination among stakeholders, will go a long way toward making beetroot a hardy and high-value crop in world supply chains.

References

Abarna, S. (2010). *Production Guidelines for Beetroot*. Directorate Plant Production, Vegetable Crops Division; Directorate Agricultural Information Services, Department of Agriculture, Forestry & Fisheries, Pretoria, South Africa.

Ahmad, M. S., & Siddiqui, M. W. (2015). *Postharvest quality assurance of fruits: Practical approaches for developing countries* (pp. 1–265). Springer. https://doi.org/10.1007/978-3-319-21197-8

Akan, S., Horzum, Ö., & Akal, H. C. (2022). The prevention of physicochemical and microbial quality losses in fresh-cut red beets using different packaging under cold storage conditions. *LWT*, *155*. https://doi.org/10.1016/j.lwt.2021.112877

Akan, S., Tuna Gunes, N., & Erkan, M. (2021). Red beetroot: health benefits, production techniques, and quality maintaining for food industry. *Journal of Food Processing and Preservation*, 45(10). https://doi.org/10.1111/jfpp.15781

Arazi, H., & Eghbali, E. (2021). Possible effects of beetroot supplementation on physical performance through metabolic, neuroendocrine, and antioxidant mechanisms: a narrative review of the literature. *Frontiers in Nutrition*, 8. https://doi.org/10.3389/fnut.2021.660150

Batoqui, P., Biondo, F., Boeing, J. S., Oliveira Barizão, É., Evelazio De Souza, N., Matsushita, M., Celestino De Oliveira, C., Boroski, M., & Vergílio Visentainer, J. (n.d.). Evaluation of beetroot (Beta vulgaris L.) leaves during its developmental stages: a chemical composition study. *Food Science and Technology*, Campinas, 34(1), 94–101.

Bernasconi, A., Gamper Cardinali, C., Koller, M., Flury, P., & Schärer, H.-J. (2023). Reduce the post-harvest losses in organic beetroot production. In: The International Society for Plant Pathology, The French Phytopathological Society (Eds.), *12th International Congress of Plant Pathology. Book of Abstract ICPP 2023, 20–25 August 2023, Lyon, France*, p. 856

Biji, K. B., Ravishankar, C. N., Mohan, C. O., & Srinivasa Gopal, T. K. (2015). Smart packaging systems for food applications: a review. *Journal of Food Science and Technology*, 52(10), 6125–6135. https://doi.org/10.1007/s13197-015-1766-7

Carlos De Melo Gonçalves, A., Iarley Da Silva, T., Sebastião De Melo Filho, J., Gomes De Moura, J., Vieira De Sousa, L., Leno, M., Veras, M., Pereira, W. E., Carlos, Á., Neto, G., Dias, T. J., De, J., & Alves De Albuquerque, A. (2019). Postharvest quality of beetroots grown under different irrigation depths and ascorbic acid doses. *Journal of Agricultural Science*, 11(16), 180. https://doi.org/10.5539/JAS.V11N16P180

Chen, L., Zhu, Y., Hu, Z., Wu, S., & Jin, C. (2021). Beetroot as a functional food with huge health benefits: antioxidant, antitumor, physical function, and chronic metabolomics activity. *Food Science and Nutrition*, 9(11), 6406–6420. https://doi.org/10.1002/fsn3.2577

Du, L., Huang, X., Li, Z., Qin, Z., Zhang, N., Zhai, X., Shi, J., Zhang, J., Shen, T., Zhang, R., & Wang, Y. (2025). Application of smart packaging in fruit and vegetable preservation: a review. *Foods*, 14(3). https://doi.org/10.3390/foods14030447

Hoffmann, C. M., Engelhardt, M., Gallmeier, M., Gruber, M., & Märländer, B. (2018). Importance of harvesting system and variety for storage losses of sugar beet Bedeutung von Erntetechnik und Sorte für die Lagerungsverluste von Zuckerrüben. *Sugar Industry*, 143(8), 474–484.

Jammer, A., Albacete, A., Schulz, B., Koch, W., Weltmeier, F., van der Graaff, E., Pfeifhofer, H. W., & Roitsch, T. G. (2020). Early-stage sugar beet taproot development is characterized by three distinct physiological phases. *Plant Direct*, 4(7). https://doi.org/10.1002/pld3.221

Kader, A. (2002). *Postharvest technology of horticultural crops*. https://books.google.co.in/books?hl=en&lr=&id=O1zhx2OWftQC&oi=fnd&pg=PA1&dq=Kader,+A.+A.+(2002).+Postharvest+Technology+of+Horticultural+Crops.+University+of+California+Agriculture+and+Natural+Resources.&ots=4jw_60ElEP&sig=9wG0v-I4IaZV_DAbIXj8IWyYCqw

Kitinoja, L., & Kader, A. A. (2002). *Small-Scale Postharvest Handling Practices: A Manual for Horticultural Crops (4 th Edition)*. https://www.fao.org/4/ae075e/ae075e00.htm

Kujala, T. S., Loponen, J. M., Klika, K. D., & Pihlaja, K. (2000). Phenolics and betacyanins in red beetroot (Beta vulgaris) root: distribution and effect of cold storage on the content of total phenolics and three individual compounds. *Journal of Agricultural and Food Chemistry*, 48(11), 5338–5342. https://doi.org/10.1021/jf000523q

Kumari, V. S., & Divakar, S. (2017a). Quality analysis of raw jackfruit based noodles. *Asian Journal of Dairy and Food Research*, 36(01). https://doi.org/10.18805/ajdfr.v36i01.7458

Lammers, P. S., & Schmittmann, O. (2024). Evolution of sugar beet harvesters'quality over the last decades, with special consideration of the methodical aspects of the test procedure. *Agricultural Engineering International: CIGR Journal*, 26(2). https://cigrjournal.org/index.php/Ejounral/article/view/9061

Mafe, A. N., Edo, G. I., Makia, R. S., Joshua, O. A., Akpoghelie, P. O., Gaaz, T. S., Jikah, A. N., Yousif, E., Isoje, E. F., Igbuku, U. A., Ahmed, D. S., Essaghah, A. E. A., & Umar, H. (2024). A review on food spoilage mechanisms, food borne diseases and commercial aspects of food preservation and processing. *Food Chemistry Advances*, 5. https://doi.org/10.1016/j.focha.2024.100852

Mampa, S. S., Maboko, M. M., Soundy, P., & Sivakumar, D. (2017). Nitrogen application and leaf harvesting improves yield and nutritional quality of beetroot. *HortTechnology*, 27(3), 337–343. https://doi.org/10.21273/HORTTECH03599-16

Mkhari, T., Kaseke, T., & Fawole, O. A. (2023a). Encapsulation of betalain-rich extract from beetroot postharvest waste using a binary blend of gum Arabic and maltodextrin to promote a food circular bioeconomy. *Frontiers in Nutrition*, 10, 1235372. https://doi.org/10.3389/FNUT.2023.1235372

Nesakumar, N., Baskar, C., Kesavan, S., Rayappan, J. B. B., & Alwarappan, S. (2018). Analysis of moisture content in beetroot using Fourier transform infrared spectroscopy and by principal component analysis. *Scientific Reports*, *8*(1). https://doi.org/10.1038/s41598-018-26243-5

Olsson, H., Al-Saadi, J., Oehler, D., Pergolizzi, J., & Magnusson, P. (2019). Physiological effects of beetroot in athletes and patients. *Cureus*. https://doi.org/10.7759/cureus.6355

Paulauskienė, A., Šileikienė, D., Karklelienė, R., Tarasevičienė, Ž., & Česonienė, L. (2023). Quality research of the beetroots (Beta vulgaris L., ssp. vulgaris var. conditiva Alef.) grown in different farming systems applying chemical and holistic research methods. *Sustainability*, *15*(9). https://doi.org/10.3390/su15097102

Piasna-Słupecka, E., Leszczyńska, T., Drozdowska, M., Dziadek, K., Domagała, B., Domagała, D., & Koronowicz, A. (2023). Young shoots of red beet and the root at full maturity inhibit proliferation and induce apoptosis in breast cancer cell lines. *International Journal of Molecular Sciences*, 24(8). https://doi.org/10.3390/ijms24086889

Rud, V. P., Terokhina, L. A., Vitoptova, V. A., Shablia, O. S., & Sidora, V. V. (2024). Table beet. Opportunities and prospects of production. *Vegetable and Melon Growing*, 74, 103–113. https://doi.org/10.32717/0131-0062-2023-74-103-113

Rudra, S. G., Basu, S., & Chanda, A. (Eds.). (2022). *Food Storage, Spoilage and Shelf Life: Recent Developments and Insights*. Frontiers Media SA. https://doi.org/10.3389/978-2-88976-850-9

Schouten, S.P., & van Schaik, A.C.R. (1981). Storage of red beets. *Acta Horticulturae*, *116*, 3. https://doi.org/10.17660/ActaHortic.1981.116.3.

Seadh, S., Abdel-Moneam, M., Ibrahim, M., & Mohamed, I. (2021). Study the losses in sugar beet roots after harvesting and reducing it by using different storage methods, covering and spraying treatments during storage periods. *Journal of Plant Production*, 12(5), 559–566. https://doi.org/10.21608/jpp.2021.178931

Singh, A., Kaur, A., Srivastava, P. K., & Kumar, D. (2024). Integrated pest management in beetroot cultivation: a systematic revie. *World Journal of Advanced Research and Reviews*, 21(1), 1687–1698. https://doi.org/10.30574/wjarr.2024.21.1.0184

Singh, B., & Hathan, B. S. (2017). Effect of different packaging materials on the storage study of beetroot powder. *Asian Journal of Dairy and Food Research*, 36(01). https://doi.org/10.18805/AJDFR.V36I01.7460

Stoica, F., Râpeanu, G., Rațu, R. N., Stănciuc, N., Croitoru, C., Țopa, D., & Jităreanu, G. (2025). Red beetroot and its by-products: a comprehensive review of phytochemicals, extraction methods, health benefits, and applications. *Agriculture*, 15(3). https://doi.org/10.3390/agriculture15030270

Šuput, D., Rakita, S., Spasevski, N., Tomičić, R., Dragojlović, D., Popović, S., & Hromiš, N. (2024). Dried beetroots: optimization of the osmotic dehydration process and storage stability. *Foods*, *13*(10). https://doi.org/10.3390/foods13101494

Tayyab, M., Wakeel, A., Mubarak, M. U., Artyszak, A., Ali, S., Hakki, E. E., Mahmood, K., Song, B., & Ishfaq, M. (2023). Sugar beet cultivation in the tropics and subtropics: challenges and opportunities. *Agronomy*, 13(5). https://doi.org/10.3390/agronomy13051213

Ukuku, D., Rudra, S. G., Basu, S., & Chanda, A. (n.d.). *Open Access Edited and Reviewed by Editorial: Food Storage, Spoilage and Shelf Life: Recent Developments and Insights*. https://www.foodsafetynews.com/2022/05/staph-blamed-for-foodborne-

Viskelis, J., Nevidomskis, S., Bobinas, C., Urbonaviciene, D., Bobinaite, R., Karkleliene, R., & Viskelis, P. (2019). *Evaluation of beetroot quality during various storage conditions*, pp. 170–175. https://doi.org/10.22616/foodbalt.2019.048

Wills, R. B. H., & Golding, J. B. (2016). Quality evaluation and management. *Postharvest: An Introduction to the Physiology and Handling of Fruit and Vegetables*, pp. 189–215. https://doi.org/10.1079/9781786391483.0189

Yahia, E., & Carrillo-Lopez, A. (2018). *Postharvest physiology and biochemistry of fruits and vegetables*. https://books.google.co.in/books?hl=en&lr=&id=lMlaDwAAQBAJ&oi=fnd&pg=PP1&dq=Benkeblia,+N.+(2018).+Postharvest+Physiology+and+Biochemistry+of+Fruits+and+Vegetables.+Elsevier.&ots=8ytSnr439k&sig=kRXWMkVFXBZ_01x765FVbMaGcog

Yasaminshirazi, K., Hartung, J., Fleck, M., & Graeff-Hönninger, S. (2021). Impact of cold storage on bioactive compounds and their stability of 36 organically grown beetroot genotypes. *Foods*, *10*(6). https://doi.org/10.3390/foods10061281

4

Beetroot Processing and Technology

Varsha Kanojia, Rishi Richa, and Shabnum Showkat

4.1 Introduction

Beetroot (*Beta vulgaris* L.), a member of the Chenopodiaceae family, is celebrated for its striking crimson color (Kovarovič et al., 2017). It is also known by various names such as spinach beet, beet, chard, white beet, sea beet, garden beet, and Chukander. This root vegetable is highly valued for its medicinal properties and positive impact on human health. It can be eaten raw, boiled, steamed, or roasted, making it a versatile addition to the diet. Red beetroot is particularly abundant in essential minerals like manganese, magnesium, iron, sodium, potassium, and copper (Mathangi, 2019). It is commonly used in salads, pickles, soups, and other culinary dishes (Kumar, 2015). One of the key components of beetroot is betalains, which are water-soluble nitrogenous pigments (hydrophilic and stable) categorized into two groups of betacyanin hue red-violet and betaxanthin hue yellow-orange Chhikara (Kushwaha, Sharma, et al., 2019). Betanin, the primary betacyanin, makes up about 90% of the red pigment in beetroot. These pigments can be utilized in the meat, jam, jellies, sweets, breakfast cereals, soups, desserts, and tomato pastes food industry as a colorant labeled "red beetroot" because of their non-toxic, non-carcinogenic, and non-poisonous nature. In Europe and the USA, the beetroot pigment is approved as a food additive (E162) (Chandran et al., 2014; Kavalcová et al., 2015; Dhiman et al 2021). Beetroot extracts are rich in betalains and phenolic compounds, which provide antimicrobial properties. These compounds have shown effectiveness against gram-positive bacteria as well as gram-negative bacteria. This makes beetroot extracts a promising ingredient for antibacterial films, sprays, and lotions (Kumar & Brooks, 2018). Beetroot juice is an excellent source of polyphenols and exhibits significant antioxidant activity and helps to lower blood pressure levels in men (Wootton-Beard & Ryan, 2011; Coles & Clifton, 2012; Clifford et al., 2015). It has the ability to enhance athletic performance. Studies suggest that beetroot juice can improve exercise endurance and muscle efficiency, especially during moderate-intensity workouts. These benefits are attributed to the presence of nitrates, which reduce the oxygen cost of exercise and extend the time to exhaustion (Ormsbee et al., 2013). Consuming 0.25 L of beetroot juice daily can enhance maximal running efficiency over nine weeks (Satyanand et al., 2014; Dhiman et al 2021). However, it is advisable to dilute beetroot juice before consumption, as undiluted juice may cause side effects such as burning sensations, diarrhea, or dizziness (SAPJE, 2023).

Fresh and unpasteurized beetroot juice has a limited storage life of approximately 24–72 hours when stored under refrigeration (Sokołowska et al., 2015). To enhance the shelf stability of beetroot and its juice, various thermal processing methods such as blanching,

DOI: 10.1201/9781003527084-4

autoclaving, and spray drying are employed (Prieto-Santiago et al., 2020). Microbial stability can be maintained by spray drying, under which the juice pass through elevated temperatures, reducing its water activity and moisture content to below 5%, (Tontul & Topuz, 2017; Dhiman et al., 2021). However, conventional thermal techniques often result in the degradation of betalains, the key bioactive compounds in beetroot (Prieto-Santiago et al., 2020). To overcome these limitations, several advanced and non-thermal technologies have been introduced, including pulsed electric field processing high pressure processing (HPP), high pressure carbon dioxide treatment irradiation (Junqueira-Goncalves et al., 2011), microwave processing, and ultrasound treatment (Sokołowska et al., 2017; Liu et al., 2010; Kaur & Singh, 2014; Loginova et al., 2011; Bozkir et al., 2019). These techniques are particularly effective for microbial and enzymatic inactivation, especially of peroxidase (POD) and polyphenol oxidase (PPO), which contribute to pigment degradation. Moreover, they facilitate the extraction of bioactive compounds and the production of beetroot powder without compromising the integrity of betalains. Traditional methods for pigment extraction, such as grinding, pressing, boiling, and maceration, often suffer from limitations such as low efficiency, poor pigment stability, and extended processing times. In contrast, modern technologies have demonstrated higher efficiency, faster processing speeds, and lower energy and solvent consumption while enhancing cell permeability and mass transfer during extraction (Nirmal et al., 2021). Processing beetroot into a variety of value-added products—including yogurt, candy, snacks, ice cream, jelly, and cream cheese spread—offers consumers diverse ways to enjoy its nutritional and therapeutic benefits (Chhikara, Kushwaha, Sharma, et al., 2019). Additionally, different processing techniques can be employed to develop beetroot-based products: juice is obtained using a centrifuge blender, powder is produced via spray drying, chips are prepared through freeze drying, and cooked beetroot is made by boiling (Vasconcellos et al., 2016). Beyond extending shelf life, value addition and processing contribute to increased market potential and economic viability (Pant & Chinwan, 2014). Given the vast potential of beetroot for processing and product diversification, this study aims to comprehensively examine the various techniques available for preservation, extraction, drying, and formulation of innovative beetroot-based products. The insights provided will be valuable for food manufacturers and processors seeking effective methods to enhance the utility of beetroot in the food industry.

4.2 Bioactive Compounds

Beetroot is a highly nutritious vegetable packed with a variety of bioactive compounds. It is an excellent source of essential vitamins, including A, C, E, K, and B-complex, as well as minerals such as potassium, zinc, sodium, phosphorus, calcium, and magnesium. Betalains include betacyanins (e.g., betanin and isobetanin) and betaxanthins (e.g., vulgaxanthin I and II). Additionally, beetroot is rich in amino acids like valine threonine, methionine, cysteine, and isoleucine, along with natural sugars (Chhikara et al., 2019; Domínguez et al., 2017; Thiruvengadam et al., 2024). Other phenolic acid derivatives found in beetroot include chlorogenic acid, epicatechin, and 4-hydroxybenzoic acid, while flavonoids like betagarin, betavulgarin, cochliophilin A, and dihydroisorhamnetin contribute to its health benefits (Georgiev et al., 2010). Beetroot also contains volatile compounds such as isopentanol, isovaleraldehyde, furfural, ethanol, and dimethyl sulfide (Tan & Hamid, 2021; Thiruvengadam et al., 2024). Betalains, the nitrogen-containing pigments in beetroot, are responsible for its bioactive properties. These compounds have a cyclic amine-reactive group that contributes to their reducing potential (Tan & Hamid, 2021). Beetroot juice has also demonstrated acetylcholinesterase inhibitory activity, suggesting potential benefits for managing Alzheimer's disease. Betacyanin content in beetroot juice is 2.2 times higher than betaxanthins, with phenolic acids like chlorogenic, caffeic, p-coumaric, and sinapic acids identified through high-performance liquid chromatography analysis (Deseva et al., 2020). Extraction methods for beetroot involve maceration with ethanol-citric acid solvent, followed by filtration and evaporation, revealing phenolics, tannins, flavonoids, and saponins with strong antioxidant properties (Lembong, Utama, & Saputra, 2019). Powdered beetroot undergoes extraction, centrifugation, and filtration to isolate bioactive compounds such as saponins, quinones, flavonoids, phenols, coumarins, steroids, anthocyanins, and betacyanins, with betanin being the primary pigment (Zia, Sunita, & Sneha, 2021). These extracts exhibit antibacterial properties against a wide range of pathogens. Ethanol extracts of beetroot have a total polyphenol content ranging from 820.10 to 1,280.56 mg/kg, with antioxidant activity between 19.63% and 29.82% (Kavalcova et al., 2015). Beetroot extracts also contain alkaloids, tannins, phlobatannins, anthraquinones, flavonoids, carbohydrates, saponins, and cardiac glycosides. While ethanol extracts show antibacterial properties, their anti-helminthic activity is limited (Nwankwo & Bamidele, 2021). The primary betalain in red beet is betanin, a potent antioxidant and betanidin-5-O-beta-glucoside with phenolic and cyclic amine groups (Babarykin et al., 2019). Betaine (trimethylglycine), another significant compound, acts as an osmolyte, protecting cells, proteins, and enzymes from environmental stressors like dehydration and extreme temperatures. Insufficient dietary betaine intake may lead to conditions such as steatosis and plasma dyslipidemia (Babarykin et al., 2019). Methanolic extracts of beet leaves contain quercetin, kaempferol, and rutin, with rutin being the predominant polyphenol (61%). Beet leaves also contain chlorophylls, carotenoids, and betalains, enhancing their nutritional and functional value (Babarykin et al., 2019).

4.3 Beetroot Processing Techniques

4.3.1 Encapsulation

Encapsulation is a method used to enclose bioactive compounds within a protective material, safeguarding them from environmental factors such as moisture, light, heat, oxidation, and mechanical stress. This technique is essential in maintaining the stability and prolonging the shelf life of bioactive compounds. Various encapsulation techniques include spray drying, freeze drying, melt extrusion, coacervation, and electrospraying (Dhiman, Suhag, Singh, & Prabhakar, 2021). Spray drying, a commonly used encapsulation method, converts a liquid feed into dry solid particles by atomizing it in a hot drying medium. Due to its cost-effectiveness and flexibility, it is widely employed in the food and beverage industries to encapsulate active compounds within a protective matrix, ensuring their stability (Barbosa-Cánovas et al., 2005). Key parameters such as inlet air temperature, feed flow rate, and carrier type significantly impact the final product's quality (Ochoa-Martinez et al., 2015). For the preparation of beetroot microcapsules, beetroot juice was mixed with a 30% gum Arabic solution in a 3:1 ratio and spray-dried at an atomization pressure of 0.5 MPa. The degradation of betalains is primarily attributed to hydrolytic reactions, which are accelerated by increased moisture content in spray-dried products. This may facilitate oxygen diffusion into the encapsulated material, leading to oxidation (Serris & Biliaderis, 2001; Cai & Corke, 2000). In another study, beetroot juice concentrate was encapsulated using whey protein concentrate, which formed smooth, non-sticky films, allowing for higher powder recovery during the spray drying process. The efficiency of encapsulation was significantly influenced by inlet air temperature and whey protein concentration, with increased drying temperatures leading to reductions in antioxidant activity, betalain content, total phenolic content (TPC), and bulk density (Bazaria & Kumar, 2016a). Using maltodextrin combined with dextrose and gum Arabic for beetroot juice encapsulation through spray drying resulted in an encapsulation efficiency (EE) of up to 73.27%. The resulting powder had a very low moisture content (1.26%–2.88%), ensuring microbiological stability (Bazaria & Kumar, 2016b). Additionally, feed flow rate influenced hygroscopicity (Bazaria & Kumar, 2018). The choice of carrier material also plays a crucial role in determining the stability of beetroot pigments. When encapsulated using gum Arabic, the resulting powder retained a higher content of violet pigment while showing lower yellow pigment levels. This formulation exhibited enhanced stability over time due to reduced hygroscopicity. However, maltodextrin proved more effective in preserving natural pigments, making it a preferable option for food pigments (Janiszewska, 2014). Another study encapsulated beetroot extract in maltodextrin and gum Arabic matrices using spray drying and reported an EE of up to 96%, with a moisture content of 1.6%. The EE was influenced by inlet air temperature and wall material concentration (Poornima & Sinthiya, 2020). A study optimizing the spray drying process for beetroot juice showed that packed bulk density decreased with higher drying temperatures and maltodextrin concentrations. The moisture content of the resulting powder was influenced by a feed flow rate, with higher flow rates reducing evaporation due to shorter contact time. Additionally, when the drying temperature was increased from 140°C to 160°C, betalain content decreased from 24.23 to 15.76 mg/100 g of dry matter. Optimal conditions for producing high-quality beetroot powder with minimal moisture content and better betalain retention were found to be an inlet temperature of 149°C, maltodextrin concentration of 20%, and a feed flow rate of 10 mL/min (Singh & Hathan, 2017).

4.3.2 Freeze Drying or Lyophilization

Freeze drying or lyophilization is a preferred technique for preserving heat-sensitive materials. This process involves two primary stages: first, the material is frozen, and then the ice is sublimated under low pressure, resulting in dehydration. Since the entire drying process occurs under low temperature and pressure conditions, it effectively retains volatile compounds (Barbosa-Cánovas et al., 2005) A study demonstrated the successful encapsulation of beetroot pomace extract within a soy protein matrix using freeze drying. The samples were initially frozen at –18°C and subsequently dried at 40°C for 48 hours, achieving an EE of up to 86.14%. The encapsulation process significantly enhanced the release of polyphenolic compounds in intestinal fluids compared to gastric fluids (Šaponjac et al., 2016). Similarly, in another study, Šaponjac et al. (2020) encapsulated beetroot juice using soy protein and subjected it to freeze drying at 40°C and 1 Pa pressure, resulting in an EE of 92.48%. This method ensured high retention of betalains, flavonoids, and total phenolics. Further research by Chranioti et al. (2015) involved lyophilizing beetroot extracts by first freezing the samples at –30°C and then drying them at 1.7 Pa and 57°C for 48 hours. Five different wall materials were tested, and the combination of maltodextrin and gum Arabic was found to be the most effective in microencapsulation, providing thermal stability. Additionally, incorporating encapsulated beetroot extracts into formulations revealed that wall materials composed of gum Arabic and modified starch contributed to greater color stability than other tested materials (Chranioti et al., 2015). Pre-treatments such as blanching, ultrasonication, and different freezing techniques (fast freezing, slow freezing, and combined freezing) prior to freeze drying can enhance product quality in terms of water activity, porosity, structural integrity, and shrinkage. Among these, ultrasonication proved most effective in lowering water activity, reducing it to 0.08, whereas control samples subjected to different freezing methods exhibited water activity levels between 0.21 and 0.24. In contrast, blanching resulted in an increase in water activity (Šaponjac et al., 2016).

Blanching has been found to enhance porosity while minimizing shrinkage when compared to ultrasonication (Ciurzynska et al., 2021). In another study, beetroot extract was successfully encapsulated using pumpkin protein isolate via both spray drying and freeze drying, achieving EEs of 75% and 92%, respectively. Due to its non-toxic nature, pumpkin protein isolate presents a promising alternative to synthetic drugs, which may have undesirable side effects (Čakarević et al., 2019). From the reviewed studies, it is evident that the yield and EE of beetroot during freeze drying are influenced by several factors, including processing conditions, wall-to-core ratio, and the choice of wall material. The stability of encapsulated beetroot is significantly affected by storage factors such as temperature and relative humidity (Serris & Biliaderis, 2001). While freeze drying is known for its ability to preserve bioactive compounds and enhance product stability, it is also an energy-intensive and time-consuming process. To further improve EE, electrospraying presents a viable alternative, with the potential to achieve efficiencies of up to 99%. This technique allows for the non-thermal, rapid encapsulation of bioactive compounds in a single step, producing particle sizes smaller than 1 μm while requiring lower energy consumption (Dhiman, Suhag, Singh, & Prabhakar, 2021).

4.3.3 Freezing

Freezing serves as an effective preservation method by slowing down various degradation processes in food. At sub-zero temperatures, microbial activity is significantly reduced, limiting spoilage and extending shelf life. Additionally, enzymatic and oxidative reactions are inhibited, further preserving the quality of the food. The formation of ice crystals within the product also impacts the availability of water for chemical reactions, as water transitions into a solid state, decreasing its participation in deteriorative processes (Singh & Heldman, 2008). A study conducted by Paulikienė and Raila (2014) examined the freezing of red beets within a controlled climatic chamber at temperatures ranging from –2°C to –22°C. The results demonstrated that air velocity played a crucial role in moisture loss, with higher airflow accelerating the freezing rate. Additionally, a decrease in temperature from –2°C to –22°C resulted in greater weight loss in the beetroot samples (Paulikienė & Raila, 2014). Using freezing as a pretreatment before drying, particularly with a blast freezer at –20°C, was found to disrupt the microstructure of beetroot. This alteration enhanced mass transfer efficiency and shortened the drying time. Furthermore, freezing contributed to increased levels of ascorbic acid (AA), betalains, and total polyphenols. These improvements were attributed to structural changes in the beetroot tissue, which may have facilitated interactions among various components, leading to the release of bioactive compounds in their free forms (Vallespir, Cárcel, et al., 2018).

Freezing as a pretreatment before drying has been found to significantly reduce the drying time. Research indicates that freezing beetroot before drying can lead to a drying time reduction of up to 17%, although this process may also cause some color changes in the product (Vallespir, Rodríguez, et al., 2018). A similar study by Vallespir et al. (2019) found that the drying time was reduced by 11%–17% in frozen beetroot samples compared to fresh ones. This reduction was attributed to the microstructural damage caused by ice crystal formation, which enhanced the mass transfer rate and facilitated moisture removal. Dalmau et al. (2019) studied the impact of various freezing techniques, including conventional freezing at –20°C in a cold chamber, ultra-freezing at -80°C, and immersion in liquid nitrogen at -196°C, on the bioaccessibility of beetroot bioactives during in vitro gastric digestion. While freezing led to a decline in TPC and AA levels before digestion, it also improved the release of these compounds, increasing their bioaccessibility by approximately 20% compared to non-frozen samples. This increase in bioavailability was likely due to structural damage induced by freezing, which exposed the bioactive compounds more effectively. However, among the freezing methods studied, immersion in liquid nitrogen caused the highest loss of TPC. Based on these findings, freezing not only extends the shelf life of beetroot but also improves the bioavailability of its beneficial compounds. Furthermore, its role as a pretreatment for drying enhances process efficiency by reducing drying time, making it a valuable technique for beetroot preservation and processing.

4.3.4 High Temperature Processing

In order to guarantee food safety and consumer health after consumption, commonly consumed beetroot products undergo heat processing. Microbes and enzymes in beetroot and its derivatives are effectively rendered inactive by thermal methods. However, the nutritional value, sensory qualities, and quality of food are all negatively impacted by thermal processing. In various studies, the effects of thermal processing methods such as pasteurization (Kathiravan et al., 2014), blanching (Chaudhary & Kumar, 2020), autoclaving (Prieto-Santiago et al., 2020), and sterilization (Gumienna et al., 2014) on the shelf life extension of beetroot and beetroot products, as well as their negative effects, have been investigated. Beetroot juice, puree, and whole peeled beetroots were autoclaved for 10–60 minutes at 120°C in a recent study by Prieto-Santiago et al. (2020). All three products' betalain content dropped as autoclave time increased, reaching 82.3%, 68.4%, and 73.8%, respectively. For beetroot juice, whole beetroot, and beetroot puree, the highest ratio of betacyanin/betaxanthin was recorded after 10, 20, and 30 minutes, respectively. The ready to drink beetroot juice batch thermal pasteurization method was standardized in a study by Kathiravan et al. (2014). In contrast to processing for 900 seconds, it was discovered that pasteurizing juice at 96°C for 720 seconds totally inactivated the local microflora while retaining the betalain content and antioxidant. Betacyanin and betaxanthin degraded up to 39.9% and 12.28% after thermal treatment, and their AA decreased as well. Beetroot juice underwent a color shift as a result of thermal treatment, which was explained by the breakdown of betacyanin by heat (Kathiravan et al., 2014). When fermented with Saccharomyces cerevisiae, sterilized sugar beet pulp (121°C for 15 minutes) produced a higher yield of ethanol than pasteurized sugar beet pulp. All microbiological impurities were removed by sterilization, which also raised the production of ethanol. Pasteurized sugar beet pulp had lower sucrose losses than sterilized pulp (Gumienna et al., 2014).

When beetroot slices were conventionally thermally blanched at 90°C for 7 minutes, the activity of POD and PPO was reduced by roughly 63% and 75%, respectively, in comparison to the unblanched sample. Following blanching, a notable rise in TPC and betanin was noted. Heat-induced vegetable cell wall breakdown, which releases chemicals from the cells, was the cause of this increase (Paciulli et al., 2016). It recommends using a gentle heat treatment for a shorter period of time in order to produce beetroot products with a higher nutritional content and safety. Additionally, after blanching, AA levels dropped from 6.6 to 6.3 mg/100 g FW, according to Paciulli et al. (2016).

One useful technique for improving the safety of beetroot and its derivatives is thermal processing. Nevertheless, this method may cause a number of chemical and physical changes that might change the organoleptic properties and perhaps reduce the amounts or bioavailability of specific nutrients. In particular, heat processing may cause beetroot's vital nutrients, color, AA, and bioactive compounds to degrade. Specifically, the loss of water-soluble vitamins and other healthy substances is linked to water blanching. As a result, novel processing methods might be useful substitutes, which will be discussed in later parts.

4.3.5 Non-Thermal Processing

4.3.5.1 Pulsed Electric Field (PEF)

PEF has gained attention recently due to its ability to induce non-thermal permeabilization of cell membranes. When an external electric field ranging from 1 to 10 Kv/cm is applied for microseconds to milliseconds, a potential difference of about 0.3–1 V is generated across the cytoplasmic membrane, leading to the formation of pores in the plasma membrane (Gómez Galindo & Henriksson, 2012; Porras-Parral et al., 2012). This treatment increases the porosity of plant cells, thereby improving juice extraction (Fincan et al., 2004; Kannan, 2011; Nowacka et al., 2019). At sufficiently high electric field intensities, the cell membrane may either temporarily destabilize or rupture irreversibly through electroporation. This pore formation is theorized to occur due to ion polarization across the membrane, which compresses the non-conductive membrane to the point of rupture (Fincan et al., 2004). The growing interest in PEF is attributed to several advantages, including its ability to reduce microbial load, lower enzymatic activity, and preserve bioactive compounds in plant tissues due to the lower temperature (Bevilacqua et al., 2017; El-Belghiti et al., 2005; Toepfl et al., 2005). In the processing and preservation of beetroot and its derivatives, PEF has become increasingly popular (see Table 4.1). According to Shynkaryk et al. (2008), the use of PEF pretreatment promotes moisture diffusivity, resulting in a decrease of drying temperatures by 20°C –25°C due to tissue damage produced by electroporation. It has been shown, meanwhile, that samples exposed to PEF pretreatment experience further tissue structure degradation, and post-drying rehydration does not return the texture to its initial condition. Because of their coloring qualities and health advantages, natural pigments are becoming more and more popular for use in culinary applications.

According to Loginova et al. (2011), PEF treatment considerably enhanced betalains' extraction while cutting down on extraction time. In particular, around 90% of the total betanin was extracted in 300 minutes after five pulses of 2 μs were administered at 7 kV/cm. This is five times faster than samples that did not get PEF treatment. In a different study, Eshtiaghi and Knorr (2002) showed that sugar beet cell membranes may be efficiently disrupted in less than 20 seconds by strong PEF treatment (20 pulses, ~2.5 kV/cm, 5 μF, 1–6 Hz). Unlike conventional thermal disintegration techniques, which call for temperatures between 75°C and 80°C, this approach enables the extraction of sugar beet at room temperature after PEF pretreatment.

Several similar studies have also shown that PEF enhances extraction, whether for compounds like betanines (Lopéz et al., 2009), juice (Kannan, 2011), or sugar (El-Belghiti et al., 2005). For instance, when PEF was applied at 4.38 kV/cm, using 20 pulses and 4.10 kJ/kg of

TABLE 4.1

Non-Thermal Technologies for Processing of Beetroot

S.no.	Technique	Processing and Operating Conditions	Findings	References
1.	Pulsed electric field	Extraction at 4,380 V/cm 20, pulses	Extraction of betanin and vulgaxanthin	Nowacka et al. (2019), Dhiman et al (2021)
2.	HPP	Juice processing at 600 MPa for 3 minutes Blanching of beetroot slices at 650 MPa for 3, 7, 15, and 30 minutes	The shelf life of the juice has been extended to over 21 days at 4°C storage, with the nutritional, physicochemical, and sensory attributes remaining well-preserved There was a partial inactivation of PPO and POD increase in betanine content by up to six times compared to the raw product	Ubeira-Iglesias et al. (2019) Paciulli et al. (2016)
3.	γ-Irradiation	Extraction at 4–5 kGy	Microbial preservation was effective for almost two weeks at 5°C	Junqueira-goncalves et al. (2011)
	Microwave	Pigment extraction from beetroot peel at 229 and 384 MW	The extraction of betalains reached a successful the levels of betacyanin and betaxanthin exceeding those obtained through conventional methods	Singh et al. (2017)
5.	Ultrasounds	Betalain extraction at 165 W, 25 kHz, 37°C, and 52°C for 90 minutes Pectin extraction at 25 kHz and 1,000 rpm Pretreatment before oven drying at 24 kHz	Higher phenolic compounds and antioxidant activity The extraction yield improved while the processing time was reduced by approximately 1.9 and 1.7 times, respectively Drying time was decreased by 22.2%, and neither color nor water activity was adversely affected by the duration of treatment	Righi Pessoa da Silva et al. (2018) Velyamov et al. (2019) Soquetta et al. (2017)
6.	Cold atmospheric pressure plasmas	Beetroot juice treatment	Plasma-treated beetroot juice demonstrated the ability to inhibit the proliferation of cancer cell lines	Dzimitrowicz et al. (2021)

energy, there was a 329% increase in betanin yield and a 244% increase in vulgaxanthin yield, compared to the control (Nowacka et al., 2019). PEF offers a very promising method for processing beets because of its capacity to inactivate enzymes and bacteria while reducing heat harm to product quality. As was previously mentioned, the higher extraction rate acts as a useful pretreatment to produce high-quality beetroot juice more quickly. Furthermore, the negative effect on microbes helps to lower the microbial burden in the finished product.

4.3.5.2 High Pressure Processing

High hydrostatic pressure, often known as HPP, is a recent development in the food processing sector. It is used all over the world to prolong the shelf life of fruit and vegetable juices, maintain their natural flavor and nutritional content, and preserve their nutritional value (Sokołowska et al., 2017). HPP has shown significant promise as a substitute method for heat treatments in guaranteeing the quality and safety of minimally processed food items (Hogan et al., 2005). Without overheating, it renders harmful and spoiling microbes inactive. Vegetative cells are rendered inactive by pressures between 100 and 800 MPa; however, spores and enzymes are not affected. Accordingly, a variety of food products can be sterilized by combining heat treatment with HPP (Martinez-Monteagudo & Balasubramaniam, 2008). Because continuous operation at such high pressures is challenging, HPP applications are often hydrostatic batch systems. Le Chatelier's principle, which asserts that if a system in equilibrium is disturbed, it attempts to neutralize this disturbance, is the first of the two general principles of HPP described by Hogan et al. (2005). The isostatic principle, which asserts that pressure is quickly and uniformly distributed across a sample, is the second. Gram-negative bacteria are only moderately sensitive to high pressures, whereas molds and yeasts are the most sensitive microorganisms. Conversely, gram-positive bacteria and their spores are extremely resilient and need extremely high pressures to be rendered inactive (Bevilacqua et al., 2017).

Table 4.1 summarizes the utilization of HPP for beetroot juice. Sokołowska et al. (2017) found that there was a decrease in the overall number of microorganisms by 3.8–4.5 log cfu/mL. in beetroot juice after being subjected to HPP treatment for around ten minutes at pressures of 300, 400, and 500 MPa, with 10% pigment loss and more stability in betaxanthins than betacyanins. Paciulli et al. (2016) evaluated the effects of HPP at 650 MPa on beetroot slices during different processing times from 3–30 minutes in a different research. Peroxidase (25%) and polyphenol oxidase (10%) were somewhat inactivated by HPP, according to their findings. Additionally, compared to raw samples, the betanine content of HPP-treated samples rose six times, and HPP samples had improved chewiness and hardness in comparison to those that underwent heat treatment.

All things considered, HPP improved the items' nutritional value. Ubeira-Iglesias et al. (2019) found that beetroot juice processed using HPP (600 MPa, 3 minutes) had no appreciable changes in sensory quality and comparable physicochemical and nutritional qualities to freshly squeezed juice, including bioactive compounds and antioxidant qualities. Only minor nutritional losses were seen during storage, and the HPP-processed juice's microbiological shelf-life was found to be 21 days when refrigerated (Ubeira-Iglesias et al., 2019). In the treatment of beetroot and its derivatives, HPP is a major breakthrough since it efficiently preserves vital nutrients and bioactive substances while lowering microbial contamination and increasing the product's shelf life.

4.3.5.3 Irradiation

Food products pass through ionizing radiation, such as gamma rays, electron beams, and X-rays, which is known as food irradiation (Chemat & Vorobiev, 2019). It does not entail heating the product and increases the permeability of cell walls; therefore, it is sometimes referred to as cold pasteurization. It can be used for sterilization, pathogen reduction, ripening control, disinfestation, sprouting suppression, and shelf life extension. There are no nutritional, microbiological, or toxicological issues with radiation doses under 10 kGy, which are regarded as safe (Fellows, 2000). In contrast to the control group, which underwent extraction without any previous treatment, gamma radiation (as described in Table 4.1) has been used as a pretreatment for the solid–liquid extraction of betanin, leading to enhanced cell wall permeability and extraction efficiency.

The stability of the extracted betanin decreased in comparison to the control, even though increasing the dosage rate improved the extraction degree of betanin and increased the degradation constant (Nayak et al., 2006). Beetroot extract's aerobic psychotropic bacteria, yeasts, and molds were successfully suppressed by a 5 kGy gamma radiation exposure, maintaining the pigment structure for at least 12 days while kept at 5°C. Later, ham-flavored cream cheese was colored with this irradiation beetroot extract (Junqueira-Goncalves et al., 2011). Exposure to gamma radiation affects the texture of beets. Reduced cohesion, hardness, springiness, chewiness, and gumminess were the result of increasing the dosage rate to 12 kGy. However, the textural damage shown in irradiated beetroot slices was reduced by pretreatment with a 1% calcium chloride solution (Nayak et al., 2007). The potential of irradiation and its optimization for beetroot chemical extraction and preservation require more investigation.

4.3.5.4 Ultrasound Processing

Ultrasound refers to sound waves with frequencies exceeding 20 x 103 Hz. When ultrasonic waves are applied to liquids, cavitation is created, which raises the temperature and pressure (Siddique et al., 2025) Ultrasounds have successfully aided in the extraction of betalains in beetroot processing (Righi Pessoa da Silva et al., 2018). Solid–liquid extraction, which uses heat and organic solvents, is a common technique for removing bioactive chemicals from plant materials. These traditional methods, however, have a number of shortcomings, such as lengthy processing periods, the use of dangerous chemicals, and the possibility of bioactive substances degrading during extraction.

As a popular substitute technique, ultrasound-assisted extraction (UAE) breaks down the cell wall by cavitating, which increases mass transfer, energy transfer, and solvent penetration and raises the extraction rate, reducing thermal stress (Chemat et al., 2016; Soquetta et al., 2017). Sivakumar et al. (2009) and Righi Pessoa da Silva et al. (2018) claim that ultrasound greatly increased the colorant extraction efficiency from beets. The use of a 1:1 mixture of water and ethanol with an ultrasound power of 80 W for three hours increases the yield of colorants by 8% as compared to magnetic stirring and conventional methods (orbital shaker and metabolic shaker).

Pectin could also be extracted from red beetroot pomace using UAE and agitated enzymatic extraction (Velyamov et al., 2019). When ultrasonic waves were used, the extraction time was shorter and pectin yield was higher, 1.7 and 1.9 times, respectively, than when enzymatic extraction was used. As a result, ultrasonography has been useful in maximizing the extraction of several beetroot components. Moreover, beetroot drying has also been done using ultrasonic waves. Low air velocities (below 4 m/s) and low temperatures (usually below 50°C) make it difficult for conventional convective drying systems to dry items efficiently (Bozkir et al., 2019). By increasing the drying rate, ultrasonic incorporation into the drying process can help to alleviate these limits. Similar to the action of a sponge under constant pressure and relaxation (the "sponge effect"; Fuente-Blanco et al., 2006), the mechanical energy generated by ultrasound causes periodic compressions and expansions. These expansions and compressions may exert forces greater than the surface tension holding water molecules in fluid capillaries, creating microscopic channels that permit matter to move. The microstructure of beetroot was damaged by ultrasound treatment, which also decreased the drying time by 36%–43%. Additionally, the effective diffusion coefficient (60%–73%) and external mass transfer coefficient (28%–49%) increased significantly (Vallespir, Cafcel, et al., 2018). According to recent research, beets may be dried with ultrasonic assistance in around 55% less time and at a pace that is almost three times faster than with conventional hot air drying techniques. The average temperature of the drying process, which was observed at 55°C, remained similar to that of hot air drying even though the temperature of the beetroot samples rose during ultrasound-assisted drying. Ultrasound-assisted drying successfully maintained the beetroot's natural color in terms of product quality, with betanin retention being around 15% greater than that attained by hot air drying.

Furthermore, water activity assessments showed that samples dried with hot air had values of 0.382, and those helped by ultrasound had values of 0.255 (Szadzińska et al., 2019). As a result, when combined with traditional techniques for beetroot chemical extraction and drying procedures, ultrasound becomes a powerful processing tool. The yield and quality of bioactive chemicals extracted from beets are greatly increased when ultrasound is combined with other methods, increasing their potential for use in a variety of dietary applications.

4.3.6 Microwave Processing

Microwaves are electromagnetic radiation with frequencies ranging from 300 MHz to 300 GHz. The usage of microwaves in the food business has increased recently, and microwave processing technologies are now employed in baking, pasteurization, sterilizing, thawing, blanching, frying, baking, drying, and extracting bioactive components (Orsat et al., 2017). The water in food is the main factor that promotes dipolar rotation. Meda et al. (2017) have been used in a number of food processing procedures, including the extraction of bioactive components from beetroot, boiling, blanching, and drying (Table 4.1). In both laboratory and industrial settings, microwave drying is acknowledged as one of the quickest and most efficient food preservation processing procedures, whether it is used alone or in combination with other drying processes. Compared to conventional food processing methods, this approach has several benefits, such as shorter processing times, consistent volumetric heating, an automated and self-regulating system, increased productivity, less energy use, and compatibility with other processing techniques.

Plant-based foods can be dried by microwave-assisted drying (Singh et al., 2013). Osmotically pre-treated beetroot in sugar-salt solution was first dried using a convective tray drier and then microwave-dried at different air temperatures from 55°C to 75°C and microwave powers ranging from 540 to 1,080 W, according to Kaur and Singh (2014).) Higher temperature (75 °C) and higher microwave power (1,080 W) drying resulted in a 44.28% reduction in drying time due to a higher drying rate, which could be explained by the higher energy input, the quick heat penetration of microwaves, and the forced gas removal. This condition led to higher total soluble solids, increased water activity, and enhanced hardness compared to fresh samples at 55°C air temperature and 1,080 W power and helped enhance the rehydration properties of beetroot, resulting in less case hardening and shrinkage (Kaur & Singh, 2014; Dhiman et al., 2021).

Microwave vacuum drying is another drying method of beetroot (Figiel, 2010; Lech et al., 2015). Since the ideal drying conditions are reached at lower temperatures and low pressures, one of the key advantages of vacuum microwave drying is its capacity to preserve bioactive chemicals. The overall drying time, shrinkage, thermal degradation of the compound, and oxidation were greatly reduced in vacuum microwave (VM) drying as compared to traditional drying (Figiel, 2010). Higher microwave power and the sample's larger specific surface area work together to boost the drying rate and reduce the drying time. Regardless of moisture content, puffing was seen in beetroot samples after vacuum microwave drying. After the samples were taken out of the dryer, this puffing behavior decreased under normal pressure.

Beetroot has also been blanched in a microwave to deactivate POD and PPO (Latorre et al., 2011). At higher microwave power levels, beetroot shrank and lost more weight after being blanched in the microwave. In particular, after microwave treatment at 100 and 200 W for 5 minutes, respectively, weight reductions of 9.4% and 24.7% as well as shrinkage rates of 6% and 17% were noted. Enzyme activity was reduced by 90% after 5 minutes at the 200 W setting. PPO demonstrated higher resistance than POD across all microwave power treatments.

Betanines and betaxanthins, among other beetroot chemicals, have also been extracted by microwave treatment. At 400 W with a 100% duty cycle for 100–150 seconds, the maximum production of betanines was achieved.

There are notable differences in the extraction periods required to obtain the highest yields of betaxanthins and betanines. However, pigment damage resulted from the microwave extraction's fast heating. L-AA was added to the extraction solvent in a two-step microwave extraction process to improve betalains' stability, reduce pigment deterioration, and increase the yield. The results showed that when beetroot was exposed to 400 W for 130 seconds at a concentration of 0.040 mol/L AA, the production of betanine rose to 187.67 mg/100 g.

Using betaxanthins, the maximum yield (125.43 mg/100 g beetroot) could be achieved at 400 W/100% for 130 seconds without the addition of AA (Cardoso-ugarte et al., 2014). Singh et al. (2017) found that microwaves were used for the extraction of pigments from beetroot peels. Using different solvents such as ethanol and citric acid at pH from 4.74 to 5.2, betanin concentration can be achieved. The findings imply that the best solvent for removing betalains from beetroot peels is ethanol.

When it came to preserving the polyphenol and betalain content of beetroot, microwave treatment at 960 W for 7 minutes was shown to be more beneficial than the conventional boiling (100°C, 30 minutes) and roasting (180°C, 90 minutes) procedures for making mayonnaise. By adding microwaved beetroot, mayonnaise's texture qualities and oxidative stability were both greatly improved throughout a four-week storage period at 4°C. Microwave processing is a recommended method for upcoming applications due to its safe and easy use, as well as its promising ability to improve beetroot properties. The protective impact of microwaved beetroot's antioxidant capacity was on par with that of commercially available synthetic antioxidants (Raikos et al., 2016). The creation of dried beetroot goods, such beetroot chips, is probably going to start with this technique.

4.3.7 Cold Plasma Processing (CPP)

A new nonthermal technique for food processing, cold plasma technology, has major benefits for inactivating germs found in food items. Cost-effectiveness and little change to the product's natural properties are the hallmarks of cold plasma food sterilizing. Cold plasmas have been successfully used to reduce food allergies, inactivate anti-nutritional factors, and enhance the barrier qualities of packing materials (Ekezie et al., 2017). CPP helps to increase the bioavailability of vital metals like iron (Fe), manganese (Mn), copper (Cu), calcium (Ca), and zinc (Zn). It also increases the ferric ion-reducing power, which is mainly caused by the evaporation of water.

Throughout the procedure, the amounts of carbs and betalains were mostly constant. Significantly, the treated juice's TPC rose, most likely as a result of cell membrane deterioration and the breaking of covalent connections between phenolic components. The processed juice stops the cancer cell lines from growing, but it does not affect healthy cells (Dzimitrowicz et al., 2021). As a result, cold plasma technology offers a practical way to raise the beetroot quality of juice. Future studies should concentrate on increasing shelf life and deactivating enzymes and microbes, even if there has not been much study done on the use of cold plasma in beetroot processing.

4.4 Utilization of Beetroot in Value-Added Food Products as Nutrient Enrichment

4.4.1 Juices, Sports Drinks, and Smoothies

It is used for natural sweetness and vibrant color. Both unpasteurized, freshly squeezed beetroot juice and pasteurized varieties are often consumed. Freshly squeezed juice is the best choice for maintaining flavor, fragrance, and nutritional value because pasteurization's heat treatment can lower the amount of betalain. Nonetheless, pasteurization is necessary to guarantee the juice's stability and security. Fresh, clean beetroots are ground to get the juice, and any pulp or solid particles are then removed via clarifying. Black pepper and ginger can be added to ready to drink beetroot and then thermally processed for 30 minutes at 90°C–100°C and then placed in sterile glass bottles, which helped to maintain their physicochemical characteristics for two weeks at room temperature (Dhiman et al., 2020).

Hemmatinafar et al. (2023) observed that beetroot juice contains nitrate, which boosts the stamina and performance of athletes. He also investigated the effect of beetroot sports drinks on female volleyball players and found that it is suitable as a sports drink.

4.4.2 Fermented Beverages

Beet kvass, a probiotic drink, combines beetroot with fermentation for gut health benefits. In India, this fermented juice is known as Kanji. Jakubczyk et al. (2024) created a fermented beetroot beverage featuring six distinct variations of fermented beetroot juice (Beta vulgaris L.). These variations include the traditional version, as well as those enhanced with garlic, horseradish, acerola, a salt-free option, and an iodized variant. The flavoring agents were incorporated prior to the fermentation of the beetroots.

4.4.3 Dairy Products

Yogurts, ice creams, and cheese spreads are a few dairy products. As a natural colorant and flavoring agent, they combine well with other ingredients for a unique, nutrient-dense spread. They are discussed below.

4.4.4 Flavored Yogurt

Yogurt is a good source of probiotics. In order to make beetroot-flavored yogurt, Kumar and Ramasamy (2016) mixed 150 g of skim milk with 850 mL of distilled water. The pasteurization procedure is carried out at temperatures between 80°C and 85°C for 15 minutes, and then the temperature is decreased to 42°C–45°C. To stop fermentation at this lowered temperature, a starting culture is added, fully mixed, and then chilled for 12 hours at 4°C.

The pH of the resultant yogurt ranges from 6.5 to 7.8. Yogurt's general nutritional profile is improved by adding beetroot juice, which also raises its mineral level. Similarly, Yadav et al. (2016) prepared the beetroot-based yogurt using dried beetroot powder with good sensory scores.

4.4.5 Flavored Ice Cream

Beetroot juice can be added to flavored ice cream. This delicacy is made by combining butter (80% fat), skim milk powder (95% solids not fat), and milk (5% fat and 9.5% solids not fat). The skim milk powder, butter, and stabilizers are added after the milk has been heated to between 55°C and 60°C. The mixture was homogenized at pressures, then pasteurized, cooled overnight, reheated, and mixed with beetroot juice (Manoharan et al., 2012). In another study, Evstigneeva et al. (2020) used beetroot puree to create beetroot-based ice cream. In addition to adding nice color, beetroot enhanced its content with healthy elements.

4.4.6 Snacks and Confectionery

4.4.6.1 Chips and Energy Bars

Beetroot can be dehydrated or baked into chips as a healthy snack alternative (Singh & Shengupta, 2020). Compared to conventionally made beetroot chips, multigrain beetroot crisps have higher nutritional advantages, making them an innovative healthy snacking option. These crisps are distinguished by their high fiber and low calorie content, which help to prolong fullness and decrease the frequency of snacking. Iron and calcium supplements are especially helpful in treating osteoporosis-related problems linked to obesity. Energy bars can be used as a natural binder and flavor enhancer in protein and granola bars. Tangariya et al. (2023) developed a legume and oil seeds-based snack bar, which was formulated with beetroot powder incorporated in snack bars.

4.4.6.2 Candies

Its natural sweetness and color make it a substitute for synthetic dyes. Sakhare et al., (2019) developed beetroot candy. Blanched beetroot pieces were placed in a bowl, and an equal amount of sugar was added, maintaining a 1:1 ratio. The beetroot pieces were arranged in alternating layers with sugar for a duration of 24 hours at room temperature, specifically between 25°C and 30°C. Following this steeping period, the syrups were drained and concentrated by heating. Once heated, the syrup was allowed to cool before being reintroduced to the beetroot pieces. Over the course of several days, the syrup concentration was gradually increased by 10°Brix daily, reaching a final concentration of 75°Brix for both candy types. Additionally, 2% citric acid was added to each syrup. The beetroot slices were immersed in the 75°Brix sugar solution for one week, allowing the slices and syrup to reach equilibrium. Afterward, the slices were drained of any excess syrup and dried in a tray dryer at 60°C for 10 hours to achieve the desired moisture content. The finished candies were then packaged in LDPE pouches (500–600 μm) and stored at room temperature until they were ready for sensory evaluation and chemical analysis. Shreya Bhattarai and Kusma (2021) developed sugar and honey-based beetroot candies.

4.5 Nutritional Supplements

4.5.1 Powder and Functional Foods

Powder can be used in shakes, capsules, and food fortification for its rich antioxidant and nitrate content. (Brzezińska-Rojek et al., 2023). Techniques, including spray drying and freeze drying, have been used to create stable microcapsules and dehydrated powder from beetroot juice, with encapsulation efficiencies of up to 92%. Functional food is incorporated in products aimed at improving cardiovascular health or athletic performance (Chen et al., 2021).

4.5.2 Jam Beetroot Jam and Jelly

Beetroot-based jam and jelly are innovative, nutrient-enriched value-added food products that combine the vibrant color, natural sweetness, and health benefits of beetroot. These products not only serve as delicious spreads but also cater to the growing consumer demand for nutrient-enriched and natural food options. Perumpuli et al. (2018) developed beetroot jam. Initially, it was cleaned, peeled, and cut into slices. Subsequently, they were processed in a food processor to create a beetroot pulp. The preparation of beetroot jam involved cooking the beetroot pulp at a temperature range of 80°C–100°C for 15 minutes, during which pectin (2% w/w), sugar (35% w/w), artificial flavor, and citric acid (0.5% w/w) were incorporated until 46°Brix was reached.

4.6 Pigments and Dyes from Beetroot

Natural pigments, particularly betalains, are widely used as natural dyes in food, cosmetics, textiles, and the pharmaceutical industry. These pigments are eco-friendly, non-toxic, and biodegradable, making them an excellent alternative to synthetic dyes.

Betalains, betacyanins, and betaxanthins are key pigments in beetroot. Betalains are water-soluble; betacyanins provide the red to violet color (e.g., betanin). While betaxanthins contribute yellow to orange hues. It is stable in acidic to neutral pH (3–7) and sensitive to high heat, light, and oxygen. Naturally, pigments are present in beetroot at concentrations of 0.8–1.3%.

Fresh or dried beetroot is blended in solvents such as water, ethanol, or ethanol–water mixtures. Heat gently (below 50°C) to enhance pigment release. Filtration was done. Filter the mixture through a muslin cloth or fine sieve to obtain the pigment extract. Evaporate excess solvent using vacuum evaporation or freeze-drying for concentrated dye. Adding antioxidants like AA to protect pigments from oxidation. Nirmal et al. (2021) informed about emerging technologies for beetroot pigment extraction, and Zia et al. (2021) extracted natural color from pigment.

4.7 Conclusion

Beetroot is a vegetable recognized for its significant nutritional and therapeutic properties, offering various health benefits such as anti-carcinogenic, antihypertensive, antioxidative, anti-diabetic, and anti-obesity effects because of different bioactive compounds.

Carotenoids, saponins, betalains, phenolics, flavonoids, and AAs are key bioactive compounds. There are different technologies under thermal and non-thermal, which can be used for processing beetroot. Among them, HPP, PEF, ultrasonification, and microwave processing are examples of advanced procedures that have proven effective in inactivating microbes and enzymes while maintaining the quality of beetroot. There can be different value-added products such as jam, jelly, candy, yogurt, functional foods, chips, powder, colorant, and ice cream can be formed and extend shelf life. Therefore, beetroot processing can reduce post-harvest losses and promote sustainable utilization.

References

Babarykin, D., Smirnova, G., Pundinsh, I., Vasiljeva, S., Krumina, G., & Agejchenko, V. (2019). Red beet (Beta vulgaris) impact on human health. *Journal of Biosciences and Medicines*, 7(3), 61–79. http://doi.org/10.4236/jbm.2019.73007

Barbosa-Canovas, G. V., Ortega-Rivas, E., Juliano, P., & Yan, H. (2005). Encapsulation processes. In *Food Powders: Physical Properties, Processing, and Functionality* (pp. 199–218). Springer. https://doi.org/10.1007/0-387-27613-0

Bazaria, B., & Kumar, P. (2016a). Effect of whey protein concentrate as drying aid and drying parameters on physicochemical and functional properties of spray dried beetroot juice concentrate. *Food Bioscience*, 14, 21–27. https://doi.org/10.1016/j. fbio.2015.11.002

Bazaria, B., & Kumar, P. (2016b). Effect of dextrose equivalency of maltodextrin together with Arabic gum on properties of encapsulated beetroot juice. *Journal of Food Measurement and Characterization*, 11, 156–163. https://doi.org/10.1007/s11694-016-9382-4

Bazaria, B. & Kumar, P. (2018). Optimization of spray drying parameters for beetroot juice powder using response surface methodology (RSM). *Journal of the Saudi Society of Agricultural Sciences*, 17(4): 408–415.

Bevilacqua, A., Petruzzi, L., Perricone, M., Speranza, B., Campaniello, D., Sinigaglia, M., & Corbo, M. R. (2017). Nonthermal technologies for fruit and vegetable juices and beverages: Overview and advances. *Comprehensive Reviews in Food Science and Food Safety*, 1–61. https://doi.org/10.1111/1541-4337.12299

Bhandarkar, N. S., Brown, L., & Panchal, S. K. (2019). Chlorogenic acid attenuates high-carbohydrate, high-fat diet–induced cardiovascular, liver, and metabolic changes in rats. *Nutrition Research*, 62, 78–88 https://doi.org/10.1016/j.nutres.2018.11.002

Bozkir, H., Rayman Ergün, A., Serdar, E., Metin, G., & Baysal, T. (2019). Influence of ultrasound and osmotic dehydration pretreatments on drying and quality properties of persimmon fruit. *Ultrasonics Sonochemistry*, 54, 135–141. https://doi.org/10.1016/j.ultsonch.2019.02.006

Brzezińska-Rojek, J., Sagatovych, S., Malinowska, P., Gadaj, K., Prokopowicz, M., & Grembecka, M. (2023). Antioxidant capacity, nitrite and nitrate content in beetroot-based dietary supplements. *Foods*, 12(5), 1017. https://doi.org/10.3390/foods12051017

Cai, Y. Z., & Corke, H. (2000). Production and properties of spray dried Amaranthus betacyanin pigments. *Journal of Food Science*, 65(7), 1248–1252. https://doi.org/10.1111/j.1365-2621.2000.tb10273.x

Čakarević, J., Šeregelj, V., Tumbas Šaponjac, V., Ćetković, G., Čanadanović Brunet, J., Popović, S., Kostić, M. H., & Popović, L. (2019). Encapsulation of beetroot juice: A study on the application of pumpkin oil cake protein as new carrier agent. *Journal of Microencapsulation*, 37(2):121-133 https://doi.org/10.1080/02652048.2019.1705408

Cardoso-ugarte, G. A., Sosa-morales, M. E., Ballard, T., Liceaga, A., & Martín- Gonzalez, M. F. S. (2014). Microwave-assisted extraction of betalains from red beet (Beta vulgaris). *Lebensmittel-Wissenschaft und -Technologie- Food Science and Technology*, 59, 276–282. https://doi.org/10.1016/j.lwt.2014.05.025

Chandran, J., Nisha, P., Singhal, R. S., & Pandit, A. B. (2014). Degradation of colour in beetroot (Beta vulgaris L.): A kinetics study. *Journal of Food Science & Technology*, 51, 2678–2684. https://doi.org/10.1007/s13197-012-0741-9

Chaudhari, S. N., & Nikam, M. P. (2015). Development and sensory analysis of beetroot jelly. *International Journal of Science and Research*, 4(10), 827–830. Retrieved from https://www.ijsr.net/archive/v4i10/SUB158883.pdf.

Chaudhary, V., & Kumar, V. (2020). Study on drying and rehydration characteristics of tray dried beetroot (Beta vulgaris L.) and functional properties of its powder. *Chemical Science Review and Letters*, 9(33), 98–108. https://doi.org/10.37273/chesci.CS082050061

Chemat, F., & Vorobiev, E. (Eds.). (2019). *Green food Processing Techniques: Preservation, Transformation and Extraction*. Academic Press.

Chemat, F., Rombaut, N., Sicaire, A. G., Meullemiestre, A., Fabiano-Tixier, A. S., & Abert-Vian, M. (2016). Ultrasound assisted extraction of food and natural products. Mechanisms, techniques, combinations, protocols and applications. A review. *Ultrasonics Sonochemistry*, 34, 540–560. https://doi.org/10.1016/j.ultsonch.2016.06.035

Chen, L., Zhu, Y., Hu, Z., Wu, S., & Jin, C. (2021). Beetroot as a functional food with huge health benefits: Antioxidant, antitumor, physical function, and chronic metabolomics activity. *Food Science & Nutrition*, 9(11), 6406–6420.

Chhikara, N., Kushwaha, K., Jaglan, S., Sharma, P., & Panghal, A. (2019). Nutritional, physicochemical, and functional quality of beetroot (Beta vulgaris L.) incorporated Asian noodles. *Cereal Chemistry*, 96(1), 154–161. https://doi.org/10.1002/cche.10126

Chhikara, N., Kushwaha, K., Sharma, P., Gat, Y., & Panghal, A. (2019). Bioactive compounds of beetroot and utilization in food processing industry: A critical review. *Food Chemistry*, 272, 192–200. https://doi.org/10.1016/j.foodchem.2018.08.022

Cho, Y. R., Chang, J. Y., & Chang, H. C. (2007). Production of gamma-aminobutyric acid (GABA) by Lactobacillus buchneri isolated from kimchi and its neuroprotective effect on neuronal cells. *Journal of Microbiology and Biotechnology*, 17(1), 104–109. https://europepmc.org/abstract/MED/18051360

Chranioti, C., Nikoloudaki, A., & Tzia, C. (2015). Saffron and beetroot extracts encapsulated in maltodextrin, gum Arabic, modified starch and chitosan: Incorporation in a chewing gum system. *Carbohydrate Polymers*, 127, 252–263. https://doi.org/10.1016/j.carbpol.2015.03.049

Ciurzynska, A., Falacinska, J., Kowalska, H., Kowalska, J., Galus, S., Marzec, A., & Domian, E. (2021). The effect of pretreatment (blanching, ultrasound and freezing) on quality of freeze-dried red beets. *Foods*, 10(1), 132. https://doi.org/10.3390/foods10010132

Clifford, T., Howatson, G., West, D. J., & Stevenson, E. J. (2015). The potential benefits of red beetroot supplementation in health and disease. *Nutrients*, 7(4), 2801–2822. https://doi.org/10.3390/nu7042801

Coles, L. T., & Clifton, P. M. (2012). Effect of beetroot juice on lowering blood pressure in free-living, disease-free adults: A randomized, placebo-controlled trial. *Nutrition Journal*, 11(1), 1–6. https://doi.org/10.1186/1475-2891-11-106

Dalmau, M. E., Llabrés, P. J., Eim, V. S., Rosselló, C., & Simal, S. (2019). Influence of freezing on the bioaccessibility of beetroot (Beta vulgaris) bioactive compounds during in vitro gastric digestion. *Journal of the Science of Food and Agriculture*, 99(3), 1055–1065. https://doi.org/10.1002/jsfa.927

Deseva, I., Stoyanova, M., Petkova, N., & Mihaylova, D. (2020). Red beetroot juice phytochemicals bioaccessibility: An in vitro approach. *Polish Journal of Food and Nutrition Sciences*, 70(1), 45–53. https://doi.org/10.31883/pjfns/116590

Dhiman, A., Chawla, D., Thakur, D., Chauhan, D., Bharara, C., & Singh, A. (2020). Development and optimization of ready to serve (RTS) beetroot drink. *Asian Journal of Dairy and Food Research*, 39(1), 49–57. https://doi.org/10.18805/Ajdfr.DR-1504

Dhiman, A., Suhag, R., Singh, A., & Prabhakar, P. K. (2021). Mechanistic understanding and potential application of electrospraying in food processing: A review. *Critical Reviews in Food Science and Nutrition*, 1–30. https://doi.org/10.1080/10408398.2021.1926907

Domínguez, R., Cuenca, E., Maté-Muñoz, J. L., García-Fernández, P., Serra-Paya, N., Estevan, M. C. L., Herreros, P.V., & Garnacho-Castaño, M. V. (2017). Effects of beetroot juice supplementation on cardiorespiratory endurance in athletes. A systematic review. *Nutrients*, 9(1), 1–18. https://doi.org/10.3390/nu9010043

Ekezie, F. G. C., Sun, D. W., & Cheng, J. H. (2017). A review on recent advances in cold plasma technology for the food industry: Current applications and future trends. *Trends in Food Science & Technology*, 69, 46–58. https://doi.org/10.1016/j.tifs.2017.08.007

El-Belghiti, K., Rabhi, Z., & Vorobiev, E. (2005). Kinetic model of sugar diffusion from sugar beet tissue treated by pulsed electric field. *Journal of the Science of Food and Agriculture*, 85, 213–218. https://doi.org/10.1002/jsfa.1944

Eshtiaghi, M. N., & Knorr, D. (2002). High electric field pulse pretreatment: Potential for sugar beet processing. *Journal of Food Engineering*, 52, 265–272. https://doi.org/10.1016/S0260-8774(01)00114-5

Evstigneeva, T., Iakovchenko, N., Kuzmicheva, N., & Skvortsova, N. (2020). Applying beetroot as food ingredient in ice-cream production. *Agronomy Research*, 18(S3), 1662–1672. https://doi.org/10.15159/AR.20.115

Fellows, P. J. (2000). Irradiation. In P. J. Fellows (Ed.), *Food Processing Technology*. CRC Press.

Ferrentino, G., & Spilimbergo, S. (2011). High pressure carbon dioxide pasteurization of solid foods: Current knowledge and future outlooks. *Trends in Food Science & Technology*, 22(8), 427–441. https://doi.org/10.1016/j.tifs.2011.04.009

Figiel, A. (2010). Drying kinetics and quality of beetroots dehydrated by combination of convective and vacuum-microwave methods. *Journal of Food Engineering*, 98, 461–470. https://doi.org/10.1016/j.jfoodeng.2010.01.029

Fincan, M., DeVito, F., & Dejmek, P. (2004). Pulsed electric field treatment for solid-liquid extraction of red beetroot pigment. *Journal of Food Engineering*, 64, 381–388. https://doi.org/10.1016/j.jfoodeng.2003.11.006

Fuente-Blanco, S., Sarabia, E. R. F., Acosta-Aparicio, V. M., Blanco-Blanco, A., & Gallego- Juárez, J. A. (2006). Food drying process by power ultrasound. *Ultrasonics*, 44(SUPPL), 523–527. https://doi.org/10.1016/j.ultras.2006.05.181

Gómez Galindo, F., & Henriksson, P. (2012). Design of equipment for pulsed electric field processing. In J. Ahmed, & M. S. Rahman (Eds.), *Handbook of Food Process Design* (pp. 1078–1106). First edit. https://doi.org/10.1002/9781444398274.ch

Georgiev, V. G., J.Weber, E. M.Kneschke, P. N.Denev, T.Bley, & A. I.Pavlov. 2010. Antioxidant activity and phenolic content of betalain extracts from intact plants and hairy root cultures of the red beetroot Beta vulgaris cv. Detroit dark red. *Plant Foods for Human*, 65(2), 105–111. https://doi.org/10.1007/s11.30-010-0156-6.

Guadarrama-Lezama, A. Y., Cruz-Olivares, J., Martınez-Vargas, S. L., Carrillo-Navas, H., Roman-Guerrero, A., & Perez-Alonso, C. (2014). Determination of the minimum integral entropy, water sorption and glass transition temperature to establishing critical storage conditions of beetroot juice microcapsules by spray drying. *Revista Mexicana de Ingeniería Química*, 13(2), 405–416.

Gumienna, M., Szambelan, K., Jeleń, H., & Czarnecki, Z. (2014). Evaluation of ethanol fermentation parameters for bioethanol production from sugar beet pulp and juice. *Journal of the Institute of Brewing*, 120(4), 543–549. https://doi.org/10.1002/jib.101

Hemmatinafar, M., Zaremoayedi, L., Koushkie Jahromi, M., Alvarez-Alvarado, S., Wong, A., Niknam, A., Suzuki, K., Imanian, B. & Bagheri, R. (2023). Effect of beetroot juice supplementation on muscle soreness and performance recovery after exercise-induced muscle damage in female volleyball players. *Nutrients*, 15(17), 3763. https://doi.org/10.3390/nu15173763

Hogan, E., Kelly, A. L., & Sun, D. W. (2005). High pressure processing of foods: An overview. In D. W. Sun (Ed.), *Emerging Technologies for Food Processing, Emerging Technologies for Food Processing* (pp. 1–27). Elsevier.

Janiszewska, E. (2014). Microencapsulated beetroot juice as a potential source of betalain. *Powder Technology*, 264, 190–196. https://doi.org/10.1016/j.powtec.2014.05.032

Jakubczyk, K., Melkis, K., Janda-Milczarek, K. & Skonieczna-Żydecka, K. (2024). Phenolic compounds and antioxidant properties of fermented beetroot juices enriched with different additives. *Foods*, 13(1), 102. https://doi.org/10.3390/foods13010102.

Junqueira-goncalves, M. P., Cardoso, L. P., Pinto, M. S., Pereira, R. M., Soares, N. F., & Miltz, J. (2011). Irradiated beetroot extract as a colorant for cream cheese. *Radiation Physics and Chemistry*, 80, 114–118.

Kannan, V. (2011). *Extraction of Bioactive Compounds from Whole Red Cabbage and Beetroot Using Pulsed Electric Fields and Evaluation of their Functionality*. University of Nebraska-Lincoln.

Kathiravan, T., Nadanasabapathi, S., & Kumar, R. (2014). Standardization of process condition in batch thermal pasteurization and its effect on antioxidant, pigment and microbial inactivation of Ready to Drink (RTD) beetroot (Beta vulgaris L.) juice. *International Food Research Journal*, 21(4), 1305–1312.

Kaur, K., & Singh, A. K. (2014). Drying kinetics and quality characteristics of beetroot slices under hot air followed by microwave finish drying. *African Journal of Agricultural Research*, 9(16), 1238–1243. https://doi.org/10.5897/AJAR2013

Kavalcova, P., Bystrická, J., Tomáš, J., Karovičová, J., Kovarovič, J., & Lenková, M. (2015). The content of total polyphenols and antioxidant activity in red beetroot. *Potravinarstvo*, 9(1), 77–83. https://doi.org/10.5219/441

Kumar, D., & Ramasamy, D. (2016). Studies on development of yogurt flavoured with beetroot juice (Beta vulgaris L.). *International Research Journal of Biological Sciences*, 3(2), 108–117.

Kumar, S., & Brooks, M. S. L. (2018). Use of red beet (Beta vulgaris L.) for antimicrobial applications—A critical review. *Food and Bioprocess Technology*, 11(1), 17–42. https://doi.org/10.1007/s11947-017-1942-z

Kumar, Y. (2015). Beetroot: A super food. *Internal Journal of Engineering Studies and Technical Approach*, 1(3), 1–7.

Latorre, M. E., Bonelli, P. R., Rojas, A. M., & Gerschenson, L. N. (2011). Microwave inactivation of red beet (Beta vulgaris L. var. conditiva) peroxidase and polyphenoloxidase and the effect of radiation on vegetable tissue quality. *Journal of Food Engineering*, 109(4), 676–684. https://doi.org/10.1016/j.jfoodeng.2011.11.026

Lech, K., Figiel, A., Wojdyło, A., Korzeniowska, M., Serowik, M., & Szarycz, M. (2015). Drying kinetics and bioactivity of beetroot slices pretreated in concentrated chokeberry juice and dried with vacuum microwaves. *Drying Technology: International Journal*, 33(13), 1644–1653. https://doi.org/10.1080/07373937.2015.1075209

Lembong, E., Utama, G. L., & Saputra, R. A. (2019). Phytochemical Test, Vitamin C Content and Antioxidant Activities Beet Root (Beta vulgaris Linn.) Extracts as Food Coloring Agent from Some Areas in Java Island. *IOP Conference Series: Earth and Environmental Science*, 306, 012010. https://doi.org/10.1088/1755-1315/306/1/012010

Li, S. S., Cao, H., Shen, D. Z., Chen, C., Xing, S. L., Dou, F. F., & Jia, Q. L. (2020). Effect of Quercetin on Atherosclerosis Based on Expressions of ABCA1, LXR-α and PCSK9 in ApoE-/-Mice. *Chinese journal of integrative medicine*, 26(2), 114–121.

Liu, R., Choi, H. S., Zhen, X., Kim, S. L., Kim, J. H., Ko, Y. C., … & Lee, D. S. (2020). Betavulgarin isolated from sugar beet (Beta vulgaris) suppresses breast cancer stem cells through Stat3 signaling. *Molecules*, 25(13), 2999.

Liu, X., Gao, Y., Xu, H., Hao, Q., Liu, G., & Wang, Q. (2010). Inactivation of peroxidase and polyphenol oxidase in red beet (Beta vulgaris L.) extract with continuous high pressure carbon dioxide. *Food Chemistry*, 119(1), 108–113. https://doi.org/10.1016/j.foodchem.2009.06.002

Loginova, K. V., Lebovka, N. I., & Vorobiev, E. (2011). Pulsed electric field assisted aqueous extraction of colorants from red beet. *Journal of Food Engineering*, 106, 127–133. https://doi.org/10.1016/j.jfoodeng.2011.04.019

López, N., Puértolas, E., Condón, S., Raso, J., & Alvarez, I. (2009). Enhancement of the extraction of betanine from red beetroot by pulsed electric fields. *Journal of Food Engineering*, 90, 60–66. https://doi.org/10.1016/j.jfoodeng.2008.06.002

Madhu, B., Srinivas, M. S., Srinivas, G., & Jain, S. K. (2019). Ultrasonic technology and its applications in quality control, processing and preservation of food: A review. *Current Journal of Applied Science and Technology*, 32(5), 1–11. https://doi.org/10.9734/cjast/2019/46909

Manoharan, A., Ramasamy, D., Dhanalashmi, B., Gnanalashmi, K. S., & Thyagarajan, D. (2012). Organoleptic evaluation of beetroot juice as natural color for strawberry flavor ice cream. *Indian Journal of Medicine and Healthcare*, 1(1), 34–37.

Martinez-Monteagudo, S., & Balasubramaniam, V. M. (2008). Fundamentals and applications of high- pressure processing technology. In G. V. Barbosa-Canovas (Ed.), *High Pressure Processing of Foods: Principles, Technology and Applications, high Pressure Processing of Foods.*

Mathangi, S., & Balasaraswathi, M. (2019). Formulation of horsegram cake enriched with beetroot powder: *International Journal of Applied Home Science*, 6(1):61–65.

Meda, V., Orsat, V., & Raghavan, V. (2017). Microwave heating and the dielectric properties of foods. In M. Regier, K. Knoerzer, & H. Schubert (Eds.), *The Microwave Processing of Foods*. Second Edi. https://doi.org/10.1016/B978-0-08-100528-6.00002-4.

Nayak, C. A., Chethana, S., Rastogi, N. K.˜ A., & Raghavarao, K. S. M. S. (2006). Enhanced mass transfer during solid–liquid extraction of gamma-irradiated red beetroot. *Radiation Physics and Chemistry*, 75, 173–178. https://doi.org/10.1016/j.radphyschem.2005.03.015

Nayak, C. A., Suguna, K., Narasimhamurthy, K., & Rastogi, N. K. (2007). Effect of gamma irradiation on histological and textural properties of carrot, potato and beetroot. *Journal of Food Engineering*, 79, 765–770.

Nirmal, N. P., Mereddy, R., & Maqsood, S. (2021). Recent developments in beetroot pigment extraction techniques and their food applications. *Food Chemistry*, 356 (September 2020), 129611. https://doi.org/10.1016/j.foodchem.2021.129611

Nowacka, M., Tappi, S., Wiktor, A., Rybak, K., Miszczykowska, A., Czyzewski, J., Drozdzal, K., Witrowa-Rajchert, D., & Tylewicz, U. (2019). The impact of pulsed electric field on the extraction of bioactive compounds from beetroot. *Foods*, 8(244), 1–12. https://doi.org/10.3390/foods8070244

Nwankwo, C. C., & Bamidele, D. T. 2021. Antimicrobial and anti- helminthic activities of beetroot plant. *GSC Biological and Pharmaceutical Sciences*, 15(3), 93–101. https://doi.org/10.30574/gscbps.2021.15.3.0123.

Ochoa-Martinez, L. A., Garza-Juarez, S. E., Rocha-Guzman, N. E., Morales-Castro, J., & Gonzalez-Herrera, S. M. (2015). Functional properties, color and betalain content in beetroot-orange juice powder obtained by spray drying. *Research Review: Journal of Food and Dairy Technology*, 3(2), 30–36.

Ormsbee, M., Lox, J., & Arciero, P. (2013). Beetroot juice and exercise performance. *Nutrition and Dietary Supplements*, (November), 27. https://doi.org/10.2147/nds.s52664

Orsat, V., Raghavan, G. S. V., & Krishnaswamy, K. (2017). Microwave technology for food processing: An overview of current and future applications. In M. Regier, K. Knoerzer, & H. Schubert (Eds.), *The Microwave Processing of Foods* (pp. 100–116). https://doi.org/10.1016/C2014-0-04182-2. Second Edi.

Paciulli, M., Medina-Meza, I. G., Chiavaro, E., & Barbosa-Ćanovas, G. V. (2016). Impact of thermal & high pressure processing on quality parameters of beetroot (Beta vulgaris L.). *Lebensmittel Wissenschaft und -Technologie- Food Science and Technology*, 68, 98–104. https://doi.org/10.1016/j.lwt.2015.12.029

Pant, S., & Chinwan, D. (2014). Value addition of food products. *International Journal of Engineering & Science Research*, 4(12), 946–949.

Paulikiene, S., & Raila, A. (2014). Study on the efficiency of the vegetables freezing process. *Environmental Research, Engineering and Management*, 3(69), 68–74.

Perumpuli, P. A. B. N., Fernando, G., Kaumal, M., Arandara, M., & Silva, S. (2018). Development of low sugar vegetable jam from beetroot (Beta vulgaris L.): Studies on physicochemical sensory and nutritional properties. *International Journal of Theoretical & Applied Sciences*, 10(2), 22–27.

Pitalua, E., Jimenez, M., Vernon-Carter, E. J., & Beristain, C. I. (2010). Antioxidative activity of microcapsules with beetroot juice using gum Arabic as wall material. *Food and Bioproducts Processing*, 88, 253–258. https://doi.org/10.1016/j.fbp.2010.01.002

Poornima, K., & Sinthiya, R. (2020). Encapsulation of beetroot extract using spray drying. *International Journal for Research in Applied Science and Engineering Technology*, 5, 346–352. https://doi.org/10.22214/ijraset.2017.4064

Porras-Parral, G., Miri, T., Bakalis, S., & Fryer, P. J. (2012). The effect of electrical processing on mass transfer in beetroot and model gels. *Journal of Food Engineering*, 112, 208–217. https://doi.org/10.1016/j.jfoodeng.2012.04.005

Prieto-Santiago, V., Cavia, M. M., Alonso-Torre, S. R., & Carrillo, C. (2020). Relationship between color and betalain content in different thermally treated beetroot products. *Journal of Food Science & Technology*, 57, 3305–3313. https://doi.org/10.1007/s13197-020-04363-z

Raikos, V., McDonagh, A., Ranawana, V., & Duthie, G. (2016). Processed beetroot (Beta vulgaris L.) as a natural antioxidant in mayonnaise: Effects on physical stability, texture and sensory attributes. *Food Science and Human Wellness*, 5(4), 191–198. https://doi.org/10.1016/j.fshw.2016.10.002

Righi Pessoa da Silva, H., da Silva, C., & Bolanho, B. C. (2018). Ultrasonic-assisted extraction of betalains from red beet (Beta vulgaris L.). *Journal of Food Process Engineering*, 41(6), 1–6. https://doi.org/10.1111/jfpe.12833

Sakhare, K. S., Sawate, A. R., Kshirsagar, R. B. & Taur, A. T. (2019). Studies on technology development, organoleptic evaluation and proximate composition of beetroot candy by using different sweeteners. *Journal of Pharmacognosy and Phytochemistry*, 8(2), 766–769.

Salem, A. M., Ragheb, A. S., Hegazy, M. G., Matboli, M., & Eissa, S. (2019). Caffeic acid modulates miR-636 expression in diabetic nephropathy rats. *Indian Journal of Clinical Biochemistry*, 34, 296–303.

Salimi, A., Bahiraei, T., Ahdeno, S., Vatanpour, S., & Pourahmad, J. (2021). Evaluation of cytotoxic activity of betanin against U87MG human glioma cells and normal human lymphocytes and its anticancer potential through mitochondrial pathway. *Nutrition and Cancer*, 73(3), 450–459.

Saponjac, V. T. T., Čanadanović-Brunet, J. M., Ćetković, G. S., Jakišić, M. V., Vulić, J. J., Stajčić, S. S., & Šeregelj, V. N. (2020). Optimisation of beetroot juice encapsulation by freeze-drying. *Polish Journal of Food and Nutrition Sciences*, 70(1), 25–34. https://doi.org/10.31883/pjfns/115153

Satyanand, V., Mahaboobvali, S., Phanikrishna, B., Narayanasamy, D., Lilly, N., Mujeer, S., & Salma, S. (2014). A study of beet root derived dietary nitrate efficacy on performance of runners. *Indian Journal of Basic and Applied Medical Research*, 3(2), 690–695.

Sawicki, T., Wiczkowski, W., Hrynkiewicz, M., Bączek, N., Hornowski, A., Honke, J., & Topolska, J. (2021). Characterization of the phenolic acid profile and in vitro bioactive properties of white beetroot products. *International Journal of Food Science and Technology*, 56(2), 629–638. https://doi.org/10.1111/ijfs.14710

Serris, G. S., & Biliaderis, C. G. (2001). Degradation kinetics of beetroot pigment encapsulated in polymeric matrices. *Journal of the Science of Food and Agriculture*, 81(8): 691–700. https://doi.org/10.1002/jsfa.864

Shynkaryk, M. V., Lebovka, N. I., & Vorobiev, E. (2008). Pulsed electric fields and temperature effects on drying and rehydration of red beetroots. *Drying Technology: International Journal*, 26(6), 695–704. https://doi.org/10.1080/07373930802046260

Siddique, Q., Shahi, N. C., Richa, R., Showkat, S., Gupta, A. K., Naik, B., & Kumar, V. (2025). Insights into Extraction methods of pigments from Celosia cristata L. flower: A review. *Journal of Agriculture and Food Research*, 101599. https://doi.org/10.1016/j.jafr.2024.101599

Sinegre, T., Milenkovic, D., Bourgne, C., Teissandier, D., Nasri, Y., Dannus, L. T., … & Lebreton, A. (2020). Impact of epicatechin on the procoagulant activities of microparticles. *Nutrients*, 12(10), 2935.

Singh, A., Ganesapillai, M., & Gnanasundaram, N. (2017). Optimizaton of extraction of betalain pigments from Beta vulgaris peels by microwave pretreatment. In *IOP Conference Series: Materials Science and Engineering PAPER*. https://doi.org/10.1088/1757-899X/263/3/032004

Singh, B., & Hathan, B. S. (2017). Process optimization of spray drying of beetroot Juice. *Journal of Food Science & Technology*. https://doi.org/10.1007/s13197-017-2659-8

Singh, N., & Sengupta, R. (2020). Multigrain beetroot crisps: A healthy snacking option. *Asian Journal of Medicine and Health*, 18(3), 19–25.

Singh, R. P., & Heldman, D. R. (2008). Food freezing. In R. P. Singh, & D. R. Heldman (Eds.), *Introduction to Food Engineering* (pp. 501–541). Academic Press.

Singh, S. K., & Singh, A. (2017). Microwave extraction of betalains from beetroot peel. *Food Science & Nutrition*, 5(4), 755–764. https://doi.org/10.1002/fsn3.424

Singh, S., Gaikwad, K., Omre, P. K., & Kumbhar, B. K. (2013). Microwave convection drying characteristics of beet root (Beta vulgaris L.) using modeling equations for drying. *Journal of Food Processing & Technology*, 4(9). https://doi.org/10.4172/2157-7110.1000263

Sivakumar, V., Anna, J. L., Vijayeeswarri, J., & Swaminathan, G. (2009). Ultrasound assisted enhancement in natural dye extraction from beetroot for industrial applications and natural dyeing of leather. *Ultrasonics Sonochemistry*, 16(6), 782–789. https://doi.org/10.1016/j.ultsonch.2009.03.009

Sokołowska, B., Koss-Mikołajczyk, I., & Mikołajczyk, T. (2017). The effect of high pressure processing on the quality of beetroot juice. *LWT - Food Science and Technology*, 86, 474–479. https://doi.org/10.1016/j.lwt.2017.08.022

Sokołowska, B., Porębska, I., Nasiłowska, J., Niezgoda, J., & Dekowska, A. (2015). Inactivation of the native microflora in beetroot juice by high pressure carbon dioxide combined with temperature. *Polish Journal of Natural Sciences*, 30(1), 81–89.

Sokołowska, B., Skapska, S., Fonberg-Broczek, M., Niezgoda, J., Rutkowska, M., Chotkiewicz, M., Dekowska, A., & Rzoska, S. J. (2013). The effect of high hydrostatic pressure on the survival of Saccharomyces cerevisiae in model suspensions and beetroot juice. *High Pressure Research*, 33(1), 165–171. https://doi.org/10.1080/08957959.2013.769047

Sokołowska, B., Skąpska, S., Niezgoda, J., Rutkowska, M., Dekowska, A., & Rzoska, S. J. (2014). Inactivation and sublethal injury of Escherichia coli and Listeria innocua by high hydrostatic pressure in model suspensions and beetroot juice. *High Pressure Research*, 34(1), 147–155. https://doi.org/10.1080/08957959.2013.877458

Sokołowska, B., Wózniak, Ł., Skąpska, S., Porębska, I., Nasiłowska, J., & Rzoska, S. J. (2017). Evaluation of quality changes of beetroot juice after high hydrostatic pressure processing. *High Pressure Research*, 37(2), 214–222. https://doi.org/10.1080/08957959.2017.1302443

Soquetta, M. B., Schmaltz, S., Righes, F. W., Salvalaggio, R., & Terra, L. M. (2017). Effects of pretreatment ultrasound bath and ultrasonic probe, in osmotic dehydration, in the kinetics of oven drying and the physicochemical properties of beet snacks. *Journal of Food Processing and Preservation*, 42(1), 1–9. https://doi.org/10.1111/jfpp.13393

Szadzińska, J., Mierzwa, D., Pawłowski, A., Musielak, G., Pashminehazar, R., & Kharaghani, A. (2019). Ultrasound- and microwave-assisted intermittent drying of red beetroot. *Drying Technology*, 38(1–2), 93–107. https://doi.org/10.1080/07373937.2019.1624565

SAPJE. (2023). *The 10 benefits of beetroot juice for your health*. www.sap.je/en/the-10-benefits-of-beetroot-juice-for-your-health-sapje/ (Accessed: 14 November 2025).

Tan, M. L., & Hamid, S. B. S. (2021). Beetroot as a potential functional food for cancer chemoprevention, a narrative review. *Journal of Cancer Prevention*, 26(1), 1–17. https://doi.org/10.15430/JCP.2021.26.1.1

Tangariya, P., Awasthi, P., & Sahoo, A. (2023). Effect of beetroot powder incorporation on the textural, sensory and nutritional quality of legume and oil seeds based snack bar. *Asian Journal of Dairy and Food Research*, 42(1), 117–122.

Thiruvengadam, M., Chung, I. M., Samynathan, R., Chandar, S. H., Venkidasamy, B., Sarkar, T., & Simal-Gandara, J. (2024). A comprehensive review of beetroot (Beta vulgaris L.) bioactive components in the food and pharmaceutical industries. *Critical Reviews in Food Science and Nutrition*, 64(3), 708–739. https://doi.org/10.1080/10408398.2022.2108367

Toepfl, S., Heinz, V., & Knorr, D. (2005). Overview of pulsed electric field processing for food. In D. W. Sun (Ed.), *Emerging Technologies for Food Processing, Emerging Technologies for Food Processing* (pp. 69–91). Elsevier.

Tontul, I., & Topuz, A. (2017). Spray drying of beetroot juice and its effects on microbial stability. *Drying Technology*, 35(2), 179–187. https://doi.org/10.1080/07373937.2016.1193322

Tontul, I., & Topuz, A. (2017). Spray-drying of fruit and vegetable juices: Effect of drying conditions on the product yield and physical properties. *Trends in Food Science & Technology*, 63, 91–102. https://doi.org/10.1016/j.tifs.2017.03.009

Ubeira-Iglesias, M., Wilches-Pérez, D., Cavia, M. M., Alonso-Torre, S., & Carrillo, C. (2019). High hydrostatic pressure processing of beetroot juice: Effects on nutritional, sensory and microbiological quality. *High Pressure Research*, 1–15. https://doi.org/10.1080/08957959.2019.1666842

Vallespir, F., Ćarcel, J. A., Marra, F., Eim, V. S., & Simal, S. (2018). Improvement of mass transfer by freezing pretreatment and ultrasound application on the convective drying of beetroot (Beta vulgaris L.). *Food and Bioprocess Technology*, 11, 72–83. https://doi.org/10.1007/s11947-017-1999-8

Vallespir, F., Rodríguez, Ó., Eim, V. S., Rosselló, C., & Simal, S. (2018). Freezing pretreatments on the intensification of the drying process of vegetables with different structures. *Journal of Food Engineering*, 239(May), 83–91. https://doi.org/10.1016/j.jfoodeng.2018.07.008

Vallespir, F., Rodríguez, Ó., Eim, V. S., Rosselló, C., & Simal, S. (2019). Effects of freezing treatments before convective drying on quality parameters: Vegetables with different microstructures. *Journal of Food Engineering*, 249, 15–24.

Vasconcellos, J., Conte-Junior, C., Silva, D., Pierucci, A. P., Paschoalin, V., & Alvares, T. S. (2016). Comparison of total antioxidant potential, and total phenolic, nitrate, sugar, and organic acid contents in beetroot juice, chips, powder, and cooked beetroot. *Food Science and Biotechnology*, 25(1), 79–84. https://doi.org/10.1007/s10068-016-0011-0

Velyamov, S. M., Jingilbaev, S. S., & Akterian, S. G. (2019). Ultrasound-assisted and agitated enzymatic extraction of pectin from red beet (*Beta vulgaris* L. var. *conditiva*) roots. *Bulgarian Journal of Agricultural Science*, 25(1), 196–202.

Winter, A. N., Brenner, M. C., Punessen, N., Snodgrass, M., Byars, C., Arora, Y., & Linseman, D. A. (2017). Comparison of the neuroprotective and anti-inflammatory effects of the anthocyanin metabolites, protocatechuic acid and 4-hydroxybenzoic acid. *Oxidative medicine and cellular longevity*, 2017(1), 6297080.

Wootton-Beard, P. C., & Ryan, L. (2011). A beetroot juice shot is a significant and convenient source of bioaccessible antioxidants. *Journal of Functional Foods*, 3, 329–334. https://doi.org/10.1016/j.jff.2011.05.007

Yadav, M., Masih, D., & Sonkar, C. (2016). Development and quality evaluation of beetroot powder incorporated yogurt. *International Journal of Science, Engineering and Technology*, 4(4), 582–586.

Yuceli, S., Yazici, G. N., Mammadov, R., Suleyman, H., Kaya, M., & Ozdogan, S. (2020). The effect of rutin on experimental traumatic brain injury and edema in rats. *In vivo*, 34(5), 2453–2460.

Zhang, Z. H., Wang, L. H., Zeng, X. A., Han, Z., & Brennan, C. S. (2019). Non-thermal technologies and its current and future application in the food industry: A review. *International Journal of Food Science and Technology*, 54, 1–13. https://doi.org/10.1111/ijfs.13903

Zhao, L., Sun, J., Shi, S., Qin, X., Zhang, K., & Xu, J. (2020). Kaempferol protects retinal ganglion ceils from high-glucose-induced injury by regulating vasohibin-1. *Neuroscience Letters*, 716, 134633.

Zia, P., Sunita, M., & Sneha, S. (2021). Extraction of natural colour from beet root (Beta vulgaris) its phytochemical analysis and anti- bacterial activity. *EAS Journal of Nutrition and Food Sciences*, 3(4), 80–85. https://doi.org/10.36349/easjnfs.2021.v03i04.002

5

Bioactive Compounds in Beetroot

Hafsa Tahir, Sadia Rafiq, Zara Arshad, and Momina Javed

5.1 Introduction

Based on the recent studies, beetroot is considered a "super food" for its nutritional properties. Though categories like functional and non-functional foods are already being studied in food science, the term "superfood" is now making its way in the nutrition world. This brings us to the concept: what exactly is a superfood?

According to NIH (National Institute of Health), the "superfood" class was first introduced in the 1990s, representing a distinct category of food having superior nutritional values. Certainly, there are several definitions of "superfood"; however, to put it simply, it characterizes food based on its richness in macro- and micronutrients that have a direct beneficial impact on human health and are bestowed with healing properties. Research shows that, initially, it was believed that "superfood" is one produced naturally, without any addition of complementary elements, and particularly home-grown using traditional agricultural methods. Nowadays, due to its immense demand, it is either being produced via intensive agricultural practices or displaced by other crops (because of issues like monoculture methods).

Beetroot (*Beta vulgaris*), being consumed for more than 2,000 years, has its ancient origin in Europe in the middle of the sixteenth century. The currently produced genotypes of beetroot are derived from eastern Mediterranean regions. Earlier, the root crops were trimmed and bunched, and then they were loosely packed for distribution. Considering its incremental usage over time, beet is being marketed as both fresh and ready-to-cook. In supermarkets, processed beet covers the main proportion where it's sold as a pickle (both in slices and whole) in glass jars.

Although beetroot is famous for its bright-red color due to the presence of anthocyanin, there are yellow and white varieties as well. The particular red-beets have gained popularity both in the domestic and in the industrial domains because of their functional properties and bright colors, which make them perfect for organic colorant or additive (Akan et al., 2021). Moreover, as the concept of "rainbow food" is introduced, which focuses on having all or multiple colors of food in a meal, beetroots (especially the red-beet) hold a significant value in everyday meal planning. Definitely, color isn't everything that makes beets so special; the presence of fundamental nutritional compounds makes a difference. Solely analyzing the functional characteristics of red beet shows that it is rich in antioxidants, vitamins, carotenoids, flavonoids, minerals, ascorbic acid, and polyphenols, along with highly effective bioactive compounds like betalains. Further details of each bioactive component are described in the topics yet to be

DOI: 10.1201/9781003527084-5

discussed (Hadley & Fordham, 2003). The therapeutic effects of these components make beetroot one of the top functional foods.

5.2 Nutritional Profile of Beetroot

The nutrient profile of beetroot indicates the abundance of macro- and micronutrients, making it effective for daily consumption, as an energy booster in pre-workouts, and in fighting against cardiovascular problems. According to the USDA, approximately 100 g of beet (raw) contains 9.56 g of carbohydrates, 1.61 g of protein, and 0.17 g of lipids (total fat) (Table 5.1) (Food Data Central, 2019). The nutritional profile of beetroot varies based on processing types, the details of which are mentioned in the table below (Mirmiran et al., n.d.). For a clear understanding, we can find the exact percentage of macronutrients in 100 g of beet (raw) by using the following formula.

$$\text{Percentage} = (\text{Nutrient amount}/\text{Total amount of macronutrients}) \times 100$$

TABLE 5.1

Nutritional Profile of Beetroot (Raw, Cooked, Canned, and Fresh Juice)

Parameters	Raw	Cooked	Canned	Fresh Juice
Water, g	87.58	87.06	90.96	–
Energy, kcal	43	44	31	30
Protein, g	1.61	1.68	0.91	1.02
Total fats, g	0.17	0.18	0.14	0
Carbohydrate, g	9.56	9.96	7.21	6.6
Fiber, g	2.8	2	1.8	0
Sugars, g	6.76	7.96	5.51	6.6
Calcium, mg	16	16	15	0
Iron, mg	0.8	0.79	1.82	0
Magnesium, mg	23	23	17	–
Phosphorus, mg	40	38	17	–
Potassium, mg	325	305	148	–
Sodium, mg	78	77	194	93
Zinc, mg	0.35	0.35	0.21	–
Vitamin C, mg	4.9	3.6	4.1	0
Thiamin, mg	0.031	0.027	0.01	–
Riboflavin, mg	0.04	0.04	0.04	–
Niacin, mg	0.334	0.331	0.157	–
Folate, µg	109	80	30	–
Total phenolic content (a)	255	238	192	225
Total flavonoid content (b)	260	261	173	126

Source: Food Data Central (2019).

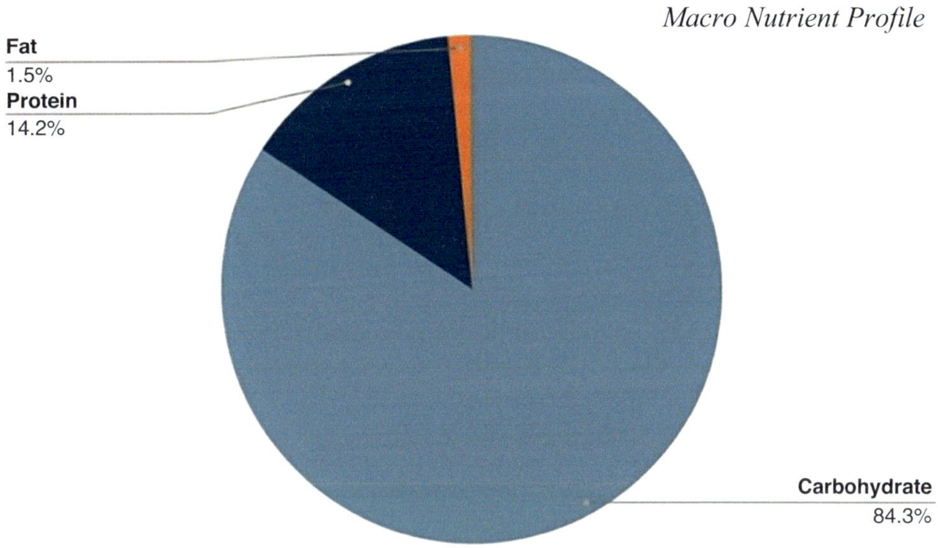

FIGURE 5.1
Percentage nutrient profile.

Whereas, the total amount of macronutrients for the above scenario (100 g beet) can be calculated as:

$$\text{Total} = 9.56(\text{Carbohydrate}) + 1.61(\text{Protein}) + 0.17(\text{Fat}) = 11.34\ \text{g}$$

After calculating the individual percentages, we get that 100 g of beet (raw) gives 84.3% CHO, 14.2% protein, and 1.5% fat (Figure 5.1).

The micronutrient composition of beetroot indicates the presence of essential minerals like potassium, copper, sodium, iron, calcium, magnesium, phosphorus, zinc, and manganese with the highest bioavailability factor. Furthermore, it has been found that the sugar, vitamin C, and flavonoids contained in different commercial organic and conventional beetroot juices varied from 1.73–7.85, 10.75–20.36, and 2.02–2.36 mg (per 100 g), respectively (Kazimierczak et al., 2016). Approximately 70%–100% of the phenolic content of beetroot is comprised of betalains, which are only present in 0.8–1.3 g/L of fresh beetroot juice (60% betacyanins and 40% betaxanthins) (Wruss et al., 2015). Among inorganic nutrients, beetroot (juice) contains the highest amount of dietary nitrate (NO_3), which has reportedly increased its importance in pre-exercise nutrition and in the management of cardiovascular diseases. Although NO_3 is believed to be inert in nature, in the human body (mainly in the oral cavity), it has the ability to convert into NO_2 through enzymatic activity. Nitric oxide (in reduced form) then enters the blood and causes vasodilation, which results in increased blood flow toward muscles (active musculature). The subsequent rise in kinetic activity helps athletes and runners (mid-level) perform better, improve running speed, and minimize the rate of perceived exertion (Murphy et al., 2012). According to Baiao et al., in beetroot, the concentration of NO_3 increases in the following order: seed, fruit, bulb, tuber, root, stem, leaf, and petiole. As the consumption of CHO-rich food before high-endurance activities significantly improves performance, the higher CHO content and NO_3 in beetroot (juice) come as a perfect choice for athletes and trainers (Ormsbee et al., 2013).

The comparative analysis of beetroot with other root vegetables (e.g., carrots, turnips, and sweet potatoes) has shown that while all of them are solely rich in starch (carbohydrates), beetroot provides a great balance of macro and micronutrients. Especially, the protein content in beetroot is comparatively higher than others, which showcases its non-empty-caloric nature. Moreover, the high amount of dietary fibers in beetroot makes it a gut-friendly functional food due to its role in improving bowel movements and healthy gut microbiota (Wang et al., 2023). Another reason behind the superiority of beetroot is its long history in traditional medicine. It was used for liver detoxification, blood purification, and as a remedy for fever decades ago. Surprisingly, the current studies have manifested the presence of antioxidants, bioactive components like phenols, betacyanin and betalains, and micronutrients like iron, Vit C, and B9, indicating the connection of its usage in traditional medicine (Hamedi et al., 2019).

As the name of this chapter suggests, all major bioactive compounds in beetroot, for example, betalains, phenolic components, carotenoids, minerals, and vitamins, and other bioactive components like nitrates, saponins, and polyphenols are discussed in detail using science-backed research studies. Their potential structures, modes of action, and usages in health science are elaborated, clearly describing micro- and macronutrient content. This chapter will not only give a clear understanding of the bioactive components available in beetroot but also enlighten on its applications in dealing with multiple diseases.

5.3 Betalains

5.3.1 Betalain Biosynthesis

Betalains are water-soluble nitrogen-containing pigments found in the vacuoles of beetroot cells and are among the most significant organic components of the plant. These pigments are classified into two major groups: red-violet betacyanins and yellow-orange betaxanthins, both of which share betalamic acid as a core structural component. The type of betalain is determined by the compound that combines with betalamic acid (Clement et al., 1992).

The biosynthesis of betalains begins with the amino acid tyrosine, which undergoes hydroxylation to form L-3,4-dihydroxyphenylalanine (L-DOPA). L-DOPA is subsequently converted into dopaquinone and betalamic acid. Betalamic acid then condenses with various amino acids to form betaxanthins or with cyclo-DOPA to produce betacyanins.

Goldman and Austin (2000) examined the genetic regulation of these pigments and concluded that betalain pigmentation is controlled by two loci, R and Y. Red pigmented beets (betacyanin-rich) exhibit dominant alleles at both loci (R-Y-), while yellow rooted beets (rich in betaxanthins) possess the rrY- genotype. White beets, lacking pigmentation, contain recessive alleles at the Y locus, highlighting its essential role in betalain synthesis.

Further insights were provided by Wang et al. (2017), who analyzed betalain accumulation relative to tyrosine levels across various beet cultivars. Their findings showed that red beets contain five to eight times more betalains than yellow varieties. Interestingly, there was an inverse correlation between tyrosine and betalain concentrations, suggesting inefficient tyrosine utilization in yellow beets. Sugar beets exhibited both minimal betalain and tyrosine levels. The study concluded that while a higher tyrosine supply is essential for betacyanin synthesis in red beets, efficient utilization of tyrosine is necessary for betaxanthin production in yellow beets.

The image associated with this section illustrates the genes **BvCYP76AD1** and **BvCYP76AD5**, which are involved in the biosynthesis of betacyanins and betaxanthins, respectively. Tyrosine biosynthesis is linked to two arogenate dehydrogenases, **ADHα** and **ADHβ**, while dashed arrows represent spontaneous reactions (Figure 5.2).

5.3.2 Factors Affecting Betalain Stability and Bioavailability

The stability and bioavailability of betalains in beetroot are influenced by multiple **pre- and post-harvest factors**, including cultivation practices, storage conditions, and processing techniques.

Cultivation Conditions: Optimal betalain content is achieved under favorable water availability, nutrient-rich soil, suitable climate, and proper maturity stage of the beetroot.

Storage Conditions: According to the *Plant Archives Journal*, high storage temperatures (~40°C) significantly reduce betalain content, while cooler conditions (~10°C) help preserve their stability and nutritional value.

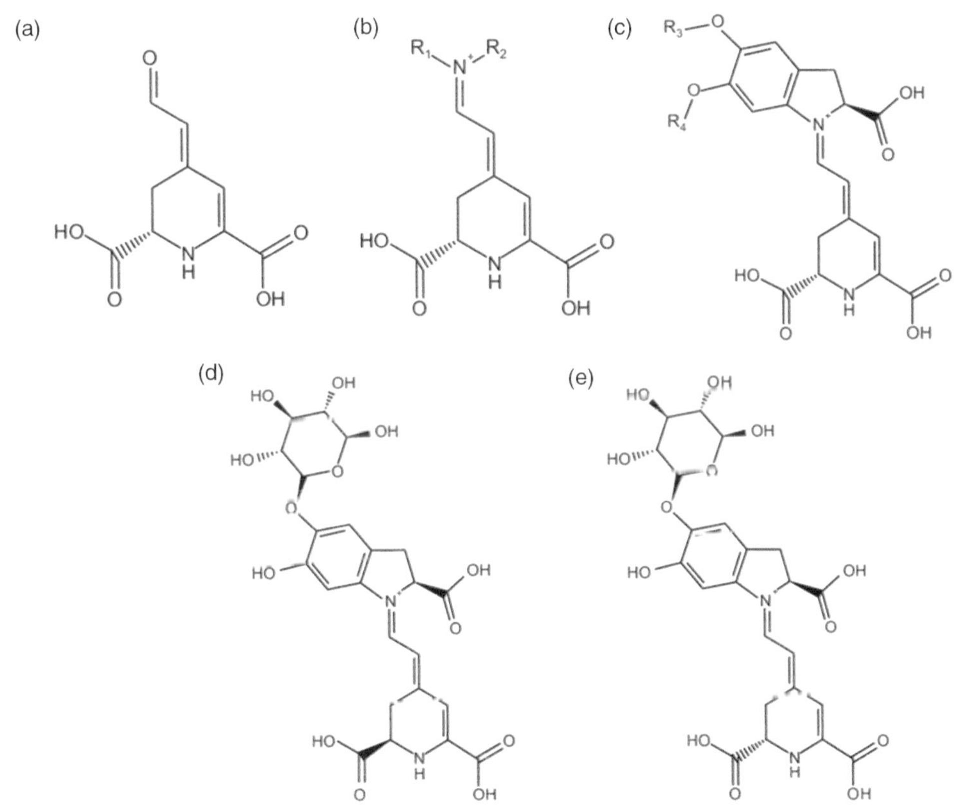

FIGURE 5.2
Chemical structures of (a) betalamic acid; (b) betaxanthin, where R1 is an amine or amino acid and R2 is normally hydrogen; (c) betacyanins (when R3 and R4 are hydrogen, the structure corresponds to betanidin); (d) betanin; and (e) isobetanin. (Tan and Hamid, 2021.)

Processing Techniques: Betalain stability is particularly sensitive to high-temperature processing. Vieira Teixeira da Silva et al. (2019) observed that **betanin**, a major betalain, remains stable for up to 20 days at 4°C and neutral pH (pH 7). Under frozen conditions, its shelf life can be extended up to 275 days. However, temperatures above 50°C impair betalain bioactivity. Boiling beets, for instance, can reduce betalain content by approximately **51%**, due to the degradation of glycosides and the formation of dehydrogenated and decarboxylated derivatives (Fu et al., 2020). The resultant pigment changes from red-violet to yellowish-brown, limiting betalains' use as natural colorants.

Antioxidant Activity: Heat-induced degradation of betalains enhances the formation of phenolic derivatives, increasing antioxidant capacity despite pigment loss.

pH Sensitivity: Betalains are most stable in a pH range of **5–6** (Castellar et al., 2003), making them suitable for low-acid foods. Below pH 3, betalain stability and color intensity decrease significantly (Stintzing et al., 2004).

Light Exposure: Light, particularly UV and visible light, destabilizes betalains by exciting pigment chromophores, leading to increased molecular reactivity. Jackman et al. (1996) and Cai et al. (1998) reported that light exposure in the range of 2.200–4.400 lux results in considerable betalain degradation.

Bioavailability: Studies using rat models have shown that betanin is metabolized primarily in the **stomach lining (74%)**, with further breakdown in the **colon (60%)** and **small intestine (35%)**. Only about **2.7%** of betanin is excreted unchanged in urine and feces (Krantz et al., 1980). Bioavailability may be enhanced through complexation with antioxidant metals such as **selenium**, which can stabilize the pigment during digestion (Figure 5.3).

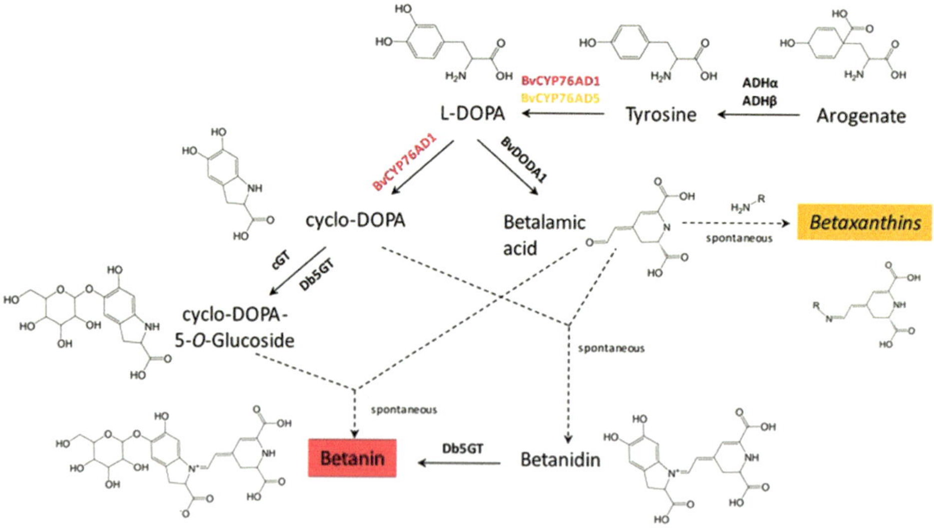

FIGURE 5.3
The figure highlights the involved genes (BvCYP76AD1 and BvCYP76AD5 for betacyanins and betaxanthins, respectively) in the biosynthesis of betalains.

5.3.3 Real-World Applications of Betalains

Betalains offer organoleptic benefits (color and sensory attributes) that are critical for food acceptability, alongside health-promoting properties that make them valuable functional ingredients.

Due to growing health concerns over synthetic dyes, betalains from beetroot provide a natural alternative. Their water solubility makes them suitable for supplements, and their antioxidant capacity helps in prolonging shelf life.

Therapeutically, betalains have demonstrated antimicrobial properties—with Hayek and Ibrahim (2012) reporting that betalain extracts inhibit the growth of E. coli. Cardiovascular benefits are also well-documented; Jajja et al. (2014) showed that oral consumption of beet juice lowers both systolic and diastolic blood pressure.

Despite promising anticancer, anti-inflammatory, cardioprotective, and hepatoprotective properties, betalains remain underutilized in the nutraceutical industry. Stability concerns—related to pH, temperature, oxygen, light, water activity, and storage—continue to limit their industrial application. Nevertheless, advances in preservation techniques and stabilization strategies are expanding the scope of betalains in health and food sectors.

5.3.4 Future Perspectives on Betalain Research

Ongoing research has identified alternative betalain-rich sources beyond beetroot, including *Amaranth* species, cactus fruits (*Opuntia*, *Eulychnia*, and *Hylocereus*), and others. However, some promising plants like *Phytolacca americana* Linn, although rich in betalains, are excluded from food applications due to their toxic components.

Recent advancements in genetic engineering are helping regulate betalain biosynthesis. Nishihara et al. (2024) discuss the use of biotechnology to enhance pigment expression in non-native hosts. Furthermore, Hernández et al. (2021) emphasize that improved beet cultivation practices and genetic selection can lead to increased betalain content, opening new opportunities for personalized nutrition and wellness products.

5.4 Phenolic Compounds

Phenolic compounds, commonly referred to as polyphenols, are a diverse class of secondary metabolites ubiquitously found in plants and plant-derived foods and beverages. These compounds exhibit extensive structural variability, ranging from simple molecules such as gallic acid and vanillin to complex polymeric structures, including flavonoids, stilbenes, and tannins. Notably, over 8,000 flavonoids have been identified, with ongoing discoveries enriching this repertoire. Although the terms "polyphenols" and "phenolic compounds" are often used interchangeably, polyphenols specifically refer to molecules containing at least two phenolic rings.

According to Bate-Smith and Swain, polyphenols are water-soluble compounds with molecular weights between 500 and 3,000 Da, characterized by their ability to precipitate proteins, gelatin, and alkaloids. Haslam further refined this definition, suggesting that polyphenols typically contain 12–16 hydroxyl groups across 5–7 aromatic rings per 1,000 Da. These compounds are generally derived from shikimate-phenylpropanoid or polyketide pathways and are distinct from nitrogen-containing secondary metabolites

(Cheynier, 2012). Although the upper limit for molecular weight was originally capped at 4,000 Da, newer studies suggest that soluble proanthocyanidins in apples and grapes may reach 10,000–20,000 Da in hydroalcoholic solutions.

Phenolic compound distribution varies based on plant species, organs, and developmental stages. Their structure is tightly linked to functional roles in plants. Anthocyanins, for example, impart red and blue hues to flowers and fruits, aiding in pollination and seed dispersal. They are also found in young leaves, possibly deterring herbivory. Flavonols protect plant tissues from UV radiation, while tannins such as proanthocyanidins, gallotannins, and ellagitannins contribute to plant defense mechanisms against herbivores, fungi, and viruses due to their protein-precipitating abilities (Silva et al., 2023).

In food processing, phenolic compounds undergo transformations that affect the sensory and functional quality of plant-based foods. Anthocyanins are vital pigments in red fruit juices and wines. Enzymes like polyphenol oxidase, laccase, and peroxidase catalyze oxidation reactions, leading to browning during cutting, bruising, drying, or fermentation processes like black tea production. These reactions also influence taste and flavor, contributing significantly to consumer acceptance. Understanding phenolic structure-function relationships and the effects of processing is crucial for improving food quality (Chhikara et al., 2019).

Modern analytical tools, particularly HPLC coupled with UV–Vis spectrophotometry and mass spectrometry, facilitate the identification of low-molecular-weight phenolics. However, due to the polydisperse nature of tannins—especially proanthocyanidins—specialized methods are needed. Ellagitannins, primarily studied by Japanese researchers, are a highly diverse group with over 1,000 characterized structures (Chen et al., 2021). However, since recent reviews have comprehensively addressed ellagitannins, they will not be discussed further here (Chhikara et al., 2019).

During black tea fermentation, oxidation of flavan-3-ols results in the formation of complex polyphenols such as ethyl-acetate soluble theaflavins and water-soluble thearubigins. Benzotropolone theaflavins and their derivatives—including methylated theaflavins, theaflavates, theaflavic acids, and bistheaflavins—have been extensively studied. Thearubigins, however, remain less well-characterized, although advances in analytical methods are beginning to reveal their structural diversity (Georgiev et al., 2010).

In winemaking, phenolic compounds undergo various extraction and transformation processes that influence the wine's color, flavor, and health-promoting attributes. Understanding these biochemical reactions is fundamental to optimizing production techniques and product quality (Thiruvengadam et al., 2024).

5.4.1 Classification of Phenolic Compounds

Phenolic compounds are organic molecules that contain at least one hydroxyl group attached to an aromatic ring. These compounds are typically glycosides stored in the vacuoles of plant cells and are synthesized as secondary metabolites under both normal and stress conditions. Polyphenols are characterized by multiple hydroxyl groups on one or more benzene rings, and they may range from monomeric to highly polymerized structures. Their classification is based on the basic carbon skeleton:

C6 Class: A single benzene ring

C6–Cn Class: A benzene ring linked to a carbon chain (one to seven atoms)

C6–Cn–C6 Class: Two benzene rings joined by a carbon chain

Polyphenols are generally divided into two main categories: flavonoids (C6–C3–C6) and non-flavonoids. Subclassification is based on the number of phenol units, substitution patterns, and inter-unit linkages (Sauceda et al., 2017).

Phenolic Acids: These non-flavonoids include benzoic acid derivatives (C1–C6) and hydroxycinnamic acid derivatives (C3–C6), often bound to polyphenols or plant structural components (Velderrain-Rodríguez et al., 2014).

Stilbenes and Lignans: Stilbenes are defined by two phenyl groups connected by a two-carbon chain and occur in free, glycosylated, or cis/trans forms. Lignans are products of the phenylpropanoid pathway, consisting of C6–C3 units condensed at specific positions. Lignans are abundant in seeds, legumes, and vegetable oils and possess antioxidant, anticancer, and estrogenic properties (Bohn, 2014).

Flavonoids: The major group of phenolics in fruits and vegetables, flavonoids have a 15-carbon structure with two aromatic rings (A and B) linked via a heterocyclic ring (C). The substitution pattern of these rings results in subclasses like flavones, flavonols, flavanones, and anthocyanins. Glycosylation and other modifications give rise to a wide range of colors in plant tissues (Nayik & Gull, 2020).

5.5 Carotenoids in Beetroot

Carotenoids are a group of natural fat-soluble pigments synthesized by plants, including beetroot (*Beta vulgaris*). Although beetroot is more commonly known for its high betalain content, it also contains carotenoids such as β-carotene and lutein, which contribute to its nutritional value and antioxidant potential (Rebecca et al., 2014). Carotenoids are isoprenoid molecules made up of eight isoprene units and contain 40 carbon atoms. These compounds in beetroot are primarily associated with antioxidant and protective roles, especially β-carotene and lutein.

5.5.1 Bioavailability

The bioavailability of carotenoids in beetroot is influenced by factors such as the plant matrix, food processing, dietary fat intake, and digestion. Goñi et al. (2006) reported that lutein exhibited high bioavailability (up to 79%) in the small intestine, while lycopene and β-carotene were more bioavailable in the large intestine. Though beetroot is not a rich source of lycopene, its β-carotene and lutein content may become more bioavailable when consumed with fat or subjected to heat processing. Methods such as homogenization and cooking can enhance the release of carotenoids from beetroot tissues. Oil-based formulations of beetroot extracts may also improve carotenoid absorption compared to whole beetroot (van het Hof et al., 2000). Stable-isotope techniques are being used to improve the assessment of carotenoid absorption (Parker et al., 1999).

5.5.2 Protective Mechanism

Carotenoids in beetroot, especially β-carotene, function as antioxidants by quenching singlet oxygen and scavenging free radicals. β-Carotene helps prevent lipid peroxidation by neutralizing peroxyl radicals, offering protection to cell membranes and lipids, particularly under oxidative stress. In vitro and in vivo studies have demonstrated that β-carotene mitigates lipid damage, although its effect may vary based on oxygen levels—excess oxygen can lead to pro-oxidant activity (Bendich & Olson, 1989). The effectiveness of carotenoids in beetroot is further modulated by their tissue concentrations and synergistic action with other antioxidants such as α-tocopherol.

5.6 Vitamins and Minerals

5.6.1 Overview of Vitamins and Minerals in Beetroot

Vitamins and minerals are organic compounds and play a role in the maintenance and regulation of different biological functions that take place in the body. These micronutrients take part in several cellular functions together, like tissue development, DNA synthesis, energy metabolism, the function of the brain and spinal cord, and oxygen transportation, and support normal growth and development. They boost the immune system and thus help the body's cells to fight against different infections. (Ali et al., 2024). If they become deficient in the body, it can cause harmful effects on one's health; therefore, these essential micronutrients should be included in the diet on a daily basis to protect against various kinds of diseases.

Vegetables and fruits are an important source of vitamins and minerals and are significant constituents of our diets, providing us with a marked amount of nutrients, especially minerals, vitamins, and fiber. Beetroots are used as a vegetable and are a very good source of essential vitamins and minerals, have zero level of cholesterol, and are low in calories. Pharmacists use beetroot's extract or juice to make conventional supplements or medicine, as it contains a good amount of vitamins and minerals and proves beneficial in the cure of several medical conditions caused by nutrient deficiency because of its anti-inflammatory as well as antioxidant effects. (Liliana & Nistor, 2020).

Vitamins and minerals act together and synergize their effect by harmonizing with each other. For example, Vitamin B6 and magnesium show a synergistic effect for the relief of anxiety-related premenstrual symptoms (De Souza et al, 2004). This study evaluated Vitamin B6 and magnesium, which function synergistically in the treatment of premenstrual symptoms. Similarly, magnesium with phosphorus performs a synergistic function as well because there are many enzymes that are activated by magnesium. Consumption of beetroot can prove to be beneficial for relieving anxiety-related premenstrual symptoms because beetroot contains a beneficial amount of magnesium as well as vitamin B6. This study assists us in taking vitamins and minerals together rather than taking them separately for the treatment of different health-related problems. Beetroot is capable of performing this function due to its great content of vitamins and minerals, which synergize their effect by coordinating with each other.

5.6.2 Vitamin Concentration of Beetroot

Vitamins are organic in nature and can't be produced by the human body naturally or can be produced only in a very small quantity. We have to consume vitamins through a diet rich in essential vitamins to meet the requirements of our body to perform biologically important functions. Those factors that direct the need for vitamins for the body are based on the age of the person, gender, physical attributes, physiological condition of that person, and the extrinsic factors.

Beetroots are a great source of biologically important vitamins, including fat-soluble and water-soluble vitamins. Beetroots are rich in fat-soluble vitamins A, C, E, and K. In addition to this, beetroots also contain water-soluble B-vitamins, including B1-thiamine, B2-riboflavin, and B3-niacin, which are involved in the energy metabolism converting food into energy and support the body during exercise and physical activity. Beetroot also contains vitamin B6-pyridoxine and B12-cyanocobalamin, contributing to the development of brain function and supporting cognition. Beetroot is an excellent source of vitamin B9-folate, which is significant for tissue development, DNA synthesis, fetal development, and neural tube formation. Eating raw beetroot daily can fulfil up to 27% of the daily folate requirement of the body. Additionally, beetroot contains vitamin B7-biotin and vitamin B5-pantothenic acid, which aid in the regulation of different hormones, maintain skin health, and support hair growth. Beetroots are rich in vitamin C, which is also a renowned antioxidant that reduces oxidative stress in the body; it is found in the top part of the beetroot. Eating 100 g of beetroot greens can fulfil 50% of the daily RDA requirement of vitamin C, which ultimately prevents cancer formation and strengthens the immune function. In beetroot, vitamins K and A are found in great amounts, which are essential for protein production and for the maintenance of bone health. Adding beetroot to the diet significantly promotes well-being and overall health and supports an energetic lifestyle. (Usman & Davidson, 2015). Odoh and Okoro found the vitamin concentration in 100 g of beetroot. The composition of 100 g of beetroot, according to their research, is given in Table 5.2.

The USDA National Nutrient Database for Standard Reference, which is one of the great places of origin for the composition of food data, also shows similar values of vitamins in 100 g of beetroot as mentioned in Table 5.2. The distribution of micronutrients is assorted in different parts of beetroot. Many of the vitamins found in tubers and other plants are also present in the leaves; for example, folate is found in the roots, and vitamins A and B3 are present in leaves (Sentkowska & Krystyna, 2023). Table 5.2 shows the vitamin composition in the beet's root and its leaves. Rossetto et al. (2009) conducted a study for the detection of ascorbic acid (Vitamin C) in the different parts of beetroot. This study shows that the vitamin C or ascorbic acid concentration is greater in the beet's pulp and lesser in the leaves of beetroot. Ascorbic acid concentration is also influenced by the technique of

TABLE 5.2

Composition of 100 g of Beetroot

Vitamins	Average Value
Vitamin A	2.6 mcg
Vitamin K	3.2 mcg
Vitamin C	4.36 mg
Vitamin E	0.18 mg
Vitamin B2	0.034 mg
Vitamin B3	0.03 mg
Vitamin B5	0.151 mg
Vitamin B6	90 mg

Source: Odoh and Okoro (2013).

TABLE 5.3
Composition of Vitamins in Beet Leaves and Roots

Vitamins	Source	Unit	Amount
Vitamin K	Leaves	mg	280
Vitamin C	Root	mg	10.01
Vitamin A	Root	μg	2
	Leaves	mg	3.93
Vitamin B1	Root	mg	0.031
Vitamin B2	Root	mg	0.027
Vitamin B3	Root	mg	0.331
Vitamin B5	Root	mg	0.145
Vitamin B6	Root	mg	0.067
Vitamin B9	Root	μg	80

Source: Razzak et al. (2024).

cultivation because it varies by the part of the plant; for example, for the beets cultivated using organic manure, the order of beetroot's parts is as follows: pulp > skin > leaf > stalk; and for the beetroots that are grown traditionally, it is pulp > stalk > skin > leaf. This indicates that the beetroots that are grown with organic manure have a greater concentration of vitamin C in every part of the beetroot as compared to those grown with traditional fertilizers. It is also seen that the concentration of B vitamins is higher in organic juice as compared to those that come from beets that are cultivated by conventional methods, as in the case of pyridoxal (vitamin B6) and nicotinamide (vitamin B3); their concentration is threefold higher in the organic juice compared to other juices. These vitamins are seen lower than the detection limit (0.03 mg/L) in other juices. We should have also noted that the possible disparity between the biologically active compounds and the vitamin content is due to their different diversity, cultivation method, and growing conditions (Table 5.3).

5.7 Minerals Composition of Beetroot

Like vitamins, minerals are also not produced by the human body naturally; we must have to consume them through a balanced diet. In the beetroot, minerals not only contribute to the body's normal physiological functions but also show bioactive anti-inflammatory and anti-oxidative effects; thus, minerals in the beetroot give both therapeutic and nutritional benefits. Beetroot is a nutrient-rich vegetable and serves as a great source of different bioactive minerals such as iron, copper, zinc, calcium, sodium, potassium, and manganese, which strengthen its properties through several metabolic processes and maintain the overall health of the human body. Minerals are major constituents in maintaining cardiovascular health. Minerals like iron, copper, zinc, and selenium have great antioxidant functions, which protect the cells from oxidative damage and support the cardiovascular system by maintaining blood pressure (Mohammadifard et al., 2019; Noushin et al., 2017). Selenium is considered one of the significant trace elements for both animals and humans because it has a crucial role in important metabolism pathways, like in the metabolism of thyroid hormone and in the immune system (Kieliszek, 2019). The deficiency of these elements can cause adverse effects on cardiovascular health and cause cancers (Sentkowska and Krystyna, 2020). Iron, copper,

TABLE 5.4

Minerals in 100 g of Beetroot

Minerals	Average Value (mg/100 mg)
Calcium	12.20±1.20
Potassium	30.12±0.29
Iron	0.75±1.20
Zinc	0.21±1.01
Sodium	72.58±1.12
Copper	0.09±0.47
Manganese	0.79±1.98

and zinc also support the immune system. Additionally, beetroot is also a great source of magnesium, potassium, and phosphorus, which are involved in cognitive functions and neurotransmitter synthesis. According to the amount required by the human body, minerals are divided into two categories: macrominerals, which are required in larger quantities, like calcium, potassium, and phosphorus, and microminerals (iron, copper, and zinc), which are required in smaller quantities. The mineral composition of beetroot (mg/100 g), according to Odoh and Okoro, is shown in Table 5.4. The study reveals that beetroot contains only 45 kcal in a serving of 100 g (Usman & Devidson, 2015).

Abdo et al. (2020) analyzed the content of minerals in the beetroot juice, and the research reveals that the juice of beetroot has the highest levels of all the observed minerals, excluding magnesium (217.6 mg/100 g) and calcium (412.52 mg/100 g). It is found that all the previous studies on the beetroot juice concentration indicate the potassium concentration ranged from 635 mg/100 g to 3,053.7 mg/100 g, which is considered enough to maintain the cardiovascular system and blood pressure in the long run (Siervo et al., 2013). Also, iron shows the greatest concentration (911.65 mg/100 g) in beetroot juice, which plays a role in the prevention or cure of anemia (Neha et al., 2018). Keep in mind that mineral compositions in beetroot juice may vary depending on the variety of beetroot.

In the upper Austria, a study was conducted for the analysis of different beetroot varieties by Wruss et al. (2015), which included seven popular varieties of beetroot, and the results obtained were similar to those reported by previous studies (Petek et al., 2012). It is important to note that the mineral composition of beetroot varies depending on the variety of beetroot and which part of the beetroot is used because mineral composition changes according to the part of the plant and its condition of growing. In the leaves, the copper ion concentration is more than two times higher as compared to that of tubers (0.075 and 0.191 mg/100 g of beetroots) (Ceclu & Nistor, 2020). Likewise, as for iron, there is a great difference, as there is 2.57 g of iron present in the leaves of beetroot and 0.80 g in the tubers (per 100 g of beetroot), and other minerals like sodium, potassium, calcium, and magnesium are found in higher concentrations in the leaves. Table 5.5 shows the concentration of minerals in leaves and roots.

5.7.1 Minerals and Vitamins Bioavailability in Beetroot

There are several important factors identified by research that influence the vitamin and mineral bioavailability in beetroot and are important for the absorption of nutrients. Our body may not get enough benefits from beetroot, impacted by some hidden factors that affect the absorption of minerals and vitamins, including interactions between different nutrients, different individual characteristics, and environmental conditions. There are

TABLE 5.5

Composition of Minerals in Beet Leaves and Roots

Minerals	Source	Unit	Amount
Calcium (Ca)	Leaves	mg	2,220
	Root	mg	16
Potassium(K)	Root	µg	325
	Leaves	mg	1,440
Magnesium(Mg)	Root	mg	23
	Leaves	mg	350
Phosphorus (P)	Root	mg	40
	Leaves	mg	330
Iron (Fe)	Root	mg	0.80
	Leaves	mg	16.90
Sodium (Na)	Root	mg	78
Zinc (Zn)	Root	mg	0.35

Source: Maira et al. (2024).

certain factors that boost the absorption of micronutrients and some hinder them, like inhibitors and promoters. The interaction between different nutrients enhances the absorption, like vitamin–vitamin interaction, as vitamin A absorption is enhanced by vitamin E, thus increasing the bioavailability of vitamin A (Sklan & Donoghue, 1982). Likewise, in mineral–vitamin interaction, as iron absorption is intensified by Vitamin C, it increases iron bioaccessibility (Lynch & James, 1980). The absorption of mineral content in beetroot is inhibited or hindered by some anti-nutritional elements and causes complications, which decrease mineral bioavailability in the body. These anti-nutritional compounds include oxalic acid, which causes inhibition of magnesium, calcium, tannins, and phytates, which inhibit zinc, iron, and saponins (Igual et al., 2023). Soluble oxalates, by chelating mineral content, form the insoluble salt of calcium oxalates. According to (Bielik & Kolisek, 2021), gut microbiome influences the absorption of nutrients, as healthy microbiota increases nutrients' bioaccessibility. Bioavailability of minerals and vitamins is also affected by environmental conditions, such as climate change, the kind of fertilizers that were used, and the handling process (Moraghan & Mascagni, 1991). Processing and cooking also destroy the micronutrient content in beetroot because most of the nutrients leach out, resulting in the loss of several vitamins and minerals (Luisa, 2012). By keeping all circumstances in mind, we must select beetroot that is cultivated by organic fertilization, adopt a healthy lifestyle, and incorporate beetroot in our diet as a rich source of micronutrients.

5.8 Other Compounds in Beetroot

5.8.1 Overview of Other Bioactive Compounds in Beetroot

Beetroot is an excellent source of bioactive micronutrients; besides this, beetroot also contains non-nutrient compounds like nitrates, which are chemical substances, and these compounds are not essential nutrients but are of great importance in providing health benefits; among these, the chief compound is nitrate, which converts into nitric oxide and performs anti-inflammatory, anti-oxidative, as well as vasodilatory functions and aids in management of cardiovascular diseases (Baião et al., 2020; Brzezińska-Rojek et al., 2023). Beetroots are rich in nitrates and have a 100% oral bioavailability (van velzen et al., 2008).

5.8.2 Nitrates

The main source of nitrates is vegetables, and beetroot, in particular, is a great source of nitrates. Estimated they are provide about 80%–85% of the consumed nitrates availability, and in other food like meat, processed foods are much more less as compared to beetroot (van Velzen et al., 2008). One serving of the beetroot contains more nitrate concentration as compared to nitrates consumed endogenously (Lundberg et al., 2009). In the circulation Dietary nitrates and nitrites is not useful to do any physiological functions until these compounds are not being derived from nitric oxide which is considered as an important nitrovasodilator (Lundberg & Weitzberg, 2008). In the pathways of nitrite, nitric oxide is enhanced by the dietary consumption of nitrates (Webb et al., 2008). Nitric oxide is considered a good precursor for the cardiovascular system by lowering blood pressure and relaxing the blood vessels. Nitric oxide enhances the vascular functions and supports the healthy flow of blood (Bian et al., 2008). If high amounts of nitrates are consumed, they can cause toxicity as well, and nitrate intoxication is mainly caused by the groundwater uptake, with the highest concentration of nitrates in it. In humans, the most lethal dose of nitrate ions recorded is 67–833 mg/kg of body weight (Boink et al., 1999). Thus, consumption of nutritional and non-nutritional compounds should be in adequate amounts for proper physiological functions.

5.9 Conclusion

Beetroot (*Beta vulgaris*) stands out as a potent functional food, rich in diverse bioactive compounds, including betalains, phenolic compounds, carotenoids, vitamins, minerals, and dietary nitrates. These compounds not only contribute to its vibrant color and flavor but also provide significant health benefits, such as antioxidant, anti-inflammatory, antimicrobial, cardioprotective, and hepatoprotective effects. While betalains and nitrates are key components with promising therapeutic potential, the overall nutritional profile of beetroot supports its role in managing chronic diseases, enhancing athletic performance, and addressing micronutrient deficiencies. Despite challenges related to bioavailability and stability—especially during processing and storage—advancements in genetic engineering, optimized cultivation, and formulation technologies offer new pathways for expanding its use in the food, pharmaceutical, and nutraceutical industries. Collectively, the evidence underscores beetroot's status as a versatile superfood with wide-ranging health applications.

References

Abdo, E., El-Sohaimy, S., Shaltout, O., Abdalla, A., & Zeitoun, A. (2020). *Nutritional evaluation of beetroots (Beta vulgaris L.) and its potential application in a functional beverage. Plants*, 9(12), 1752. https://doi.org/10.3390/plants9121752

Akan, S., Gunes, N., & Erkan, M. (2021). Red beetroot: health benefits, production techniques and quality maintaining for food industry. *Journal of Food Processing and Preservation*, 14. https://doi.org/10.1111/jfpp.15781

Ali, A., Mac Dionys Rodrigues da Costa, E., Magalhães, P. & Martins, A.M.C. (2024). Biological importance of vitamins and minerals. In H. Ullah, A. Rauf, & M. Daglia (Eds.), *Nutraceuticals: a holistic approach to disease prevention* (p. 63). Berlin/Boston: Walter de Gruyter GmbH & Co KG.

Baião, D., Silva, D., Mere Del Aguila, E., & Paschoalin, V. (2017). Nutritional, bioactive and physicochemical characteristics of different beetroot formulations. *Food Additives*, 6(6), 21–43.

Baião, D. S. S., da Silva, D. V. T., & Paschoalin, V. M. F. (2020). *Beetroot, a remarkable vegetable: its nitrate and phytochemical contents can be adjusted in novel formulations to benefit health and support cardiovascular disease therapies.* **Antioxidants**, 9(10), 960. https://doi.org/10.3390/antiox9100960

Bendich, A., & Olson, J. A. (1989). Biological actions of carotenoids 1. *The FASEB Journal*, 3(8), 1927–1932.

Bian, K., Doursout, M.-F., & Murad, F. (2008). Vascular system: role of nitric oxide in cardiovascular diseases. *The Journal of Clinical Hypertension*, 10(4), 304–310.

Bielik, V., & Martin, K. 2021. Bioaccessibility and bioavailability of minerals in relation to a healthy gut microbiome. *International Journal of Molecular Sciences*, 22(13), 6803. https://doi.org/10.3390/ijms22136803

Bohn, T. (2014). Dietary factors affecting polyphenol bioavailability. *Nutrition Reviews*, 72(7), 429–452.

Boink, A. B. T. J., Dormans, J. A. M. A., Speijers, G. J. A., & Vleeming, W. (1999). *Effects of nitrates and nitrites in experimental animals.* In W. S. Wilson & A. S. Ball (Eds.), *Managing risks of nitrates to humans and the environment* (pp. 317–326). Cambridge, UK: Royal Society of Chemistry.

Brzezińska-Rojek, J., Sagatovych, S., Malinowska, P., Gadaj, K., Prokopowicz, M., & Grembecka, M. (2023). Antioxidant capacity, nitrite and nitrate content in beetroot-based dietary supplements. *Foods*, 12(5), 1017. https://doi.org/10.3390/foods12051017

Cai, Y., Sun, M., & Corke, H. (1998).Colourant properties and stability of Amaranthus betacyanin pigments. *Journal of Agricultural and Food Chemistry*, 46, 4491–4495. https://doi.org/10.1021/jf980457g.

Castellar, M. R., Obón, J. M., Alacid, M., & Fernández-López, J. A. (2003). Color properties and stability of betacyanins from Opuntia fruits. *Journal of Agricultural and Food Chemistry*, 51, 2772–2776. https://doi.org/10.1021/jf021045h

Ceclu, L. & Nistor, O.-V. (2020). Red beetroot: composition and health effects – a review. *Journal of Nutritional Medicine and Diet Care*, 6, 043. https://doi.org/10.23937/2572-3278.1510043.

Chen, L., Zhu, Y., Hu, Z., Wu, S., & Jin, C. (2021). Beetroot as a functional food with huge health benefits: antioxidant, antitumor, physical function, and chronic metabolomics activity. *Food Science & Nutrition*, 9(11), 6406–6420.

Cheynier, V. (2012). Phenolic compounds: from plants to foods. *Phytochemistry Reviews*, 11(2), 153–177. https://doi.org/10.1007/s11101-012-9242-8

Chhikara, N., Kushwaha, K., Sharma, P., Gat, Y., & Panghal, A. (2019). Bioactive compounds of beetroot and utilization in food processing industry: a critical review. Food Chemistry, 272, 192–200.

Clement, L. S., Mabry, T. J., Wyler, H., & Dreiding, A. S. (1992) Chemical review and evolutionary significance of the betalains. In H.-D. Behnke & T. J. Mabry (Eds.), *Evolution and Systematics of the Caryophyllales* (pp. 247–261). Berlin, NY: Springer Verlag.

De Souza, M. C., Walker, A. F., Robinson, P. A., & Bolland, K. (2000). A synergistic effect of a daily supplement for 1 month of 200 mg magnesium plus 50 mg vitamin B6 for the relief of anxiety – related premenstrual symptoms: a randomized, double-blind, crossover study. *Journal of Women's Health & Gender-Based Medicine*, 9(2), 131–139.

Fadlilah, S. H., Suhartomo, D. M., Yusan, R. T. & Yudhawan, I. (2023). Beetroot (Beta vulgaris L.) and its potential as an anemia treatment in pregnancy. *Medical and Health Journal*, 3(1), 114–127.

Food Data Central. (2019). *USDA FoodData Central: Raw Soybean Oil.* https://fdc.nal.usda.gov/fdc-app.html#/food-details/169145/nutrients

Fu, Y., Shi, J., Xie, S. Y., Zhang, T. Y., Soladoye, O. P., & Aluko, R. E. (2020). Red beetroot betalains: perspectives on extraction, processing, and potential health benefits. *Journal of Agricultural and Food Chemistry*, 68, 11595–11611. https://doi.org/10.1021/acs.jafc.0c04241

Georgiev, V. G., Weber, J., Kneschke, E.-M., Denev, P. N., Bley, T., & Pavlov, A. I. (2010). Antioxidant activity and phenolic content of betalain extracts from intact plants and hairy root cultures of the red beetroot Beta vulgaris cv. Detroit Dark Red. *Plant Foods for Human Nutrition*, 65(2), 105–111. https://doi.org/10.1007/s11130-010-0156-6

Goldman, I. L. (2020). The root vegetables: beet, carrot, parsnip, and turnip. https://cabidigitallibrary.org. https://www.cabidigitallibrary.org/doi/pdf/10.5555/20203203603

Goldman, I. L., & Austin, D. (2000) Linkage among the R, Y and BI loci in table beet. *Theoretical and Applied Genetics*, 100, 337–343.

Goñi, I., Serrano, J., & Saura-Calixto, F. (2006). Bioaccessibility of β-carotene, lutein, and lycopene from fruits and vegetables. *Journal of Agricultural and Food Chemistry*, 54(15), 5382–5387.

Hadley, P., & Fordham, R. (2003). Vegetables of temperate climates. In: Carrot, P. & Beetroot, (Eds), *Benjamin Caballero, Encyclopedia of Food Sciences and Nutrition* (Second Edition, pp. 5941–5945), Academic Press. https://doi.org/10.1016/B0-12-227055-X/01238-4

Hamedi, S., & Honarvar, M. (2019). Beta vulgaris – a mini review of traditional uses in Iran, phytochemistry and pharmacology. *Current Drug Discovery Technologies*, 16(1), 74–81. https://doi.org/10.2174/1570163815666180308142912

Hayek, S., & Ibrahim, S. A. (2012). Antimicrobial activity of xoconostle pears (Opuntia matudae) against Escherichia coli O157:H7 in laboratory medium. *International Journal of Microbiology*, 2012, 368472. https://doi.org/10.1155/2012/368472

Hernández-Aguirre, O. A., Muro, C., Hernández-Acosta, E., Alvarado, Y., & Díaz-Nava, Md. C. (2021). Extraction and stabilization of betalains from beetroot (Beta vulgaris) wastes using deep eutectic solvents. *Molecules*, 26(21), 6342. https://doi.org/10.3390/molecules26216342

Igual, M., Fernandes, Â., Dias, M. I., Pinela, J., García-Segovia, P., Martínez-Monzó, J., & Barros, L. (2023). The in vitro simulated gastrointestinal digestion affects the bioaccessibility and bioactivity of beta vulgaris constituents. *Foods*, **12**(2), 338.

Jackman, R. L., & Smith, J. L. (1996). Anthocyanins and betalains. In: Hendry, G. F., & Houghton, J. D. (Eds), *Natural Food Colourants* (pp. 244–309). London: Blackie Academic & Professional.

Jajja, A., Sutyarjoko, A., Lara, J., Rennie, K., Brandt, K., Qadir, O., & Siervo, M. (2014). Beetroot supplementation lowers daily systolic blood pressure in older, overweight subjects. *Nutrition Research*, 34, 868–875. https://doi.org/10.1016/j.nutres.2014.09.007

Kazimierczak, R., Siłakiewicz, A., Hallmann, E., Srednicka-Tober, D., & Rembiałkowska, E. (2016). Chemical composition of selected beetroot juices in relation to beetroot production system and processing technology. *Notulae Botanicae Horti Agrobotanici Cluj-Napoca*, 44(2), 491–498. https://doi.org/10.15835/nbha44210372

Kieliszek, M. (2019). Selenium–fascinating microelement, properties and sources in food. *Molecules*, 24(7), 1298.

Krantz, C., Monier, M., & Wahlstrom B. (1980). Absorption, excretion, metabolism and cardiovascular effects of beetroot extract in the rat. *Food and Cosmetics Toxicology*, 18, 363–366. https://doi.org/10.1016/0015-6264(80)90190-X

Liliana, C., & Nistor, O.-V. (2020). Red beetroot: composition and health effects – a review. *Journal of Nutritional Medicine and Diet Care*, 6, 043.

Luisa, B. G. (2012). *Food Processing and Nutrition*. Amsterdam: Elsevier.

Lundberg, J. O., & Weitzberg, E. (2008). The nitrate-nitrite-nitric oxide pathway in physiology and therapeutics. *Nature Reviews Drug Discovery*, 7(2), 156–167.

Lundberg, J. O., Gladwin, M. T., Ahluwalia, A., Benjamin, N., Bryan, N. S., Butler, A., Cabrales, P., Fago, A., Feelisch, M., Ford, P. C., Freeman, B. A., Frenneaux, M., Friedman, J., Kelm, M., Kevil, C. G., Kim-Shapiro, D. B., Kozlov, A. V., Lancaster, J. R. Jr., Lefer, D. J., McColl, K., McCurry, K., Patel, R. P., Petersson, J., Rassaf, T., Reutov, V. P., Richter-Addo, G. B., Schechter, A., Shiva, S., Tsuchiya, K., van Faassen, E. E., Webb, A. J., Zuckerbraun, B. S., Zweier, J. L., & Weitzberg, E. (2009). Nitrate and nitrite in biology, nutrition and therapeutics. *Nature Chemical Biology*, 5(12), 865–869. https://doi.org/10.1038/nchembio.260

Lynch, S. R., & James, D. C. (1980). Interaction of vitamin C and iron. *Annals of the New York Academy of Sciences*, 355(1), 32–44.

Mirmiran, P., Houshialsadat, Z., Gaeini, Z., Bahadoran, Z., & Azizi, F. (n.d.). Functional properties of beetroot (Beta vulgaris) in management of cardio-metabolic diseases. *Nutrition & Metabolism*, 17(1). https://doi.org/10.1186/s12986-019-0421-0

Mohammadifard, N., Humphries, K. H., Gotay, C., Mena-Sánchez, G., Salas-Salvadó, J., Esmaillzadeh, A., … Sarrafzadegan, N. (2019). Trace minerals intake: risks and benefits for cardiovascular health. *Critical Reviews in Food Science and Nutrition*, 59(8), 1334–1346. https://doi.org/10.1080/10408398.2017.1406332

Moraghan, J. T., & Mascagni, H. J. Jr. (1991). Environmental and soil factors affecting micronutrient deficiencies and toxicities. *Micronutrients in Agriculture*, 4, 371–425.

Murphy, M., Eliot, K., Heuertz, R. M., & Weiss, E. (2012). Whole beetroot consumption acutely improves running performance. *Journal of the Academy of Nutrition and Dietetics*, 112, 548–552.

Nayik, G. A., & Gull, A. (2020). *Antioxidants in Vegetables and Nuts-Properties and Health Benefits*. Singapore: Springer.

Neha, P., S. K. Jain, N. K. Jain, H. K. Jain, & Mittal, H. K. (2018). Chemical and functional properties of beetroot (Beta vulgaris L.) for product development: a review. *International Journal of Chemical Studies*, 6(3), 3190–3194.

Nishihara, M., Hirabuchi, A., Teshima, T. et al. (2024). Flower color modification in Torenia fournieri by genetic engineering of betacyanin pigments. *BMC Plant Biology*, 24, 614. https://doi.org/10.1186/s12870-024-05284-1

Odoh, U. E., & Okoro, E. C. (2013) .Quantitative phytochemical, proximate/nutritive composition analysis of Beta vulgaris Linnaeus (Chenopodiaceae). *International Journal of Current Research*, 5, 3723–3728.

Ormsbee, M. J., Bach, C. W., & Baur, D. A. (2014). Pre-exercise nutrition: the role of macronutrients, modified starches and supplements on metabolism and endurance performance. *Nutrients*, 6(5),1782–1808. https://doi.org/10.3390/nu6051782

Parker, R. S., Swanson, J. E., You, C.-S., Edwards, A. J., & Huang, T. (1999). Bioavailability of carotenoids in human subjects. *Proceedings of the Nutrition Society*, 58(1), 155–162.

Petek, M., Custic, M. H., Toth, N., Slunjski, S., Coga, L., Pavlovic, I., Karazija, T., Lazarevic, B., Cvetkovic, S. (2012). Nitrogen and crude proteins in beetroot (Beta vulgaris var. conditiva) under different fertilization treatments. *Notulae Botanicae Horti Agrobotanici*, 40, 215–219.

Razzak, M. A., Sharif, M. K., Naz, T., Rauf, M. A., Shahid, F., Shahzad, R., … Inam, A. (2024). Evaluating the bioactive compounds of beetroot and their pharmacological activities in promoting health. *European Journal of Health Sciences*, *10*(1), 13–30.

Rebecca, L. J., Sharmila, S., Das, M. P., & Seshiah, C. (2014). Extraction and purification of carotenoids from vegetables. *Journal of Chemical and Pharmaceutical Research*, 6(4), 594–598.

Rossetto, M. R. M., Vianello, F., da Rocha, S. A. & Lima, G. P. P. (2009). Antioxidant substances and pesticide in parts of beet (Beta vulgaris L.) grown under organic and conventional manure. *African Journal of Plant Science*, **3**(12), 245–253.

Sauceda, A. E. Q., Sáyago-Ayerdi, S. G., Ayala-Zavala, J. F., Wall-Medrano, A., de la Rosa, L. A., González-Aguilar, G. A., & Álvarez-Parrilla, E. (2017). Biological actions of phenolic compounds. In: E. M. Yahia (Ed.), *Fruit and Vegetable Phytochemicals: Chemistry and Human Health* (2nd Edition, pp. 125–138). Hoboken, NJ: John Wiley.

Sentkowska, A., & Krystyna, P. (2020). Determination of selenium species in beetroot juices. *Heliyon*, 6(6), e04194.

Sentkowska, A., & Pyrzyńska, K. (2023). Old-fashioned, but still a superfood—red beets as a rich source of bioactive compounds. *Applied Sciences*, 13(13), 7445.

Siervo, M., Lara, J., Ogbonmwan, I., & Mathers, J. C. (2013). Inorganic nitrate and beetroot juice supplementation reduces blood pressure in adults: a systematic review and meta-analysis. *Journal of Nutrition*, 143(6), 818–26. https://doi.org/10.3945/jn.112.170233

Silva, A., Silva, V., Igrejas, G., Aires, A., Falco, V., Valentão, P., & Poeta, P. (2023). Phenolic compounds classification and their distribution in winemaking by-products. *European Food Research and Technology*, 249(2), 207–239. https://doi.org/10.1007/s00217-022-04163-z684.

Sklan, D., & Donoghue, S. (1982). Vitamin E response to high dietary vitamin A in the chick. *The Journal of Nutrition*, 112(4), 759–765.

Stintzing, F.C., Conrad, J., Klaiber, I., Beifuss, U. & Carle, R. (2004). Structural investigations on betacyanin pigments by LC NMR and 2D NMR spectroscopy. *Phytochemistry*, 65, 415422.

Tan, M., & Hamid, S. (2021). Beetroot as a potential functional food for cancer chemoprevention, a narrative review. *Journal of Cancer Prevention*, 26, 1–17. doi:10.15430/JCP.2021.26.1.1

Thiruvengadam, M., Chung, I.-M., Samynathan, R., Chandar, S. H., Venkidasamy, B., Sarkar, T., Rebezov, M., Gorelik, O., Shariati, M. A., & Simal-Gandara, J. (2024). A comprehensive review of beetroot (Beta vulgaris L.) bioactive components in the food and pharmaceutical industries. *Critical Reviews in Food Science and Nutrition*, 64(3), 708–739.

Usman, M., & Davidson, J. (2015). *Health Benefits of Beetroot*. Mendon, MA: Mendon Cottage Books.

van het Hof, K. H., Weststrate, J. A., West, C. E., & Hautvast, J. G. (2000). Dietary factors that affect the bioavailability of carotenoids. *The Journal of Nutrition*, 130(3), 503–506.

van Velzen, A. G., Sips, A. J. A. M., Schothorst, R. C., Lambers, A. C. & Meulenbelt, J. (2008). The oral bioavailability of nitrate from nitrate-rich vegetables in humans. *Toxicology Letters*, 181(3), 177–181.

Velderrain-Rodríguez, G., Palafox-Carlos, H., Wall-Medrano, A., Ayala-Zavala, J., Chen, C. O., Robles-Sánchez, M., Astiazaran-García, H., Alvarez-Parrilla, E., & González-Aguilar, G. (2014). Phenolic compounds: their journey after intake. *Food & Function*, 5(2), 189–197.

Vieira Teixeira da Silva, D., Dos Santos Baião, D., de Oliveira Silva, F., Alves, G., Perrone, D., Mere Del Aguila, E., & Flosi Paschoalin, V.M. (2019). Betanin, a natural food additive: stability, bioavailability, antioxidant and preservative ability assessments. *Molecules*, 24, 458. https://doi.org/10.3390/molecules24030458

Wang, M., Lopez-Nieves, S., Goldman, I. L., & Maeda, H. A. (2017). Limited tyrosine utilization explains lower betalain contents in yellow than red table beet genotypes. *Journal of Agricultural and Food Chemistry*, 65, 4305–4213.

Wang, Y., Do, T., Marshall, L. J., & Boesch, C. (2023). Effect of two-week red beetroot juice consumption on modulation of gut microbiota in healthy human volunteers - A pilot study. *Food chemistry*, 406, 134989. https://doi.org/10.1016/j.foodchem.2022.134989

Webb, A. J., Patel, N., Loukogeorgakis, S., Okorie, M., Aboud, Z., Misra, S., Rashid, R., Miall, P., Deanfield, J., Benjamin, N., MacAllister, R., Hobbs, A. J., & Ahluwalia, A. (2008). Acute blood pressure lowering, vasoprotective, and antiplatelet properties of dietary nitrate via bioconversion to nitrite. *Hypertension*, 51(3), 784–790. https://doi.org/10.1161/HYPERTENSIONAHA.107.103523

Wruss, J., Waldenberger, G., Huemer, S., Uygun, P., Lanzerstorfer, P., Müller, U., et al. (2015). Compositional characteristics of commercial beetroot products and beetroot juice prepared from seven beetroot varieties grown in Upper Austria. *Journal of Food Composition and Analysis*, 42, 46–55. https://doi.org/10.1016/j.jfca.2015.03.005

6

Extraction, Characterization, and Utilization of Bioactives from Beetroot in the Food Processing Industry

Divyesh Suvedi, Sonali Khanal, Rachna Verma, and Dinesh Kumar

6.1 Introduction

Beetroot (*Beta vulgaris*) has gained significant attention over the past several years because of its high content of nutrients and the occurrence of a range of bioactive molecules such as betalains, phenolic acids, and flavonoids. Beetroot contains high levels of bioactives that are responsible for its characteristic color and healthy benefits, such as antioxidant, anti-inflammatory, and anti-cancer activity (Alwazeer et al., 2023). Increased demand for natural food ingredients from consumers has compelled the food processing industry to venture into the prospect of beetroots' bioactive compounds as functional food ingredients (Dhiman et al., 2021). An overview of beetroots' extraction of their bioactives, characterization, and application, focusing on the application in the food processing industry, will be given in the chapter. One of the significant steps that affects the quality and yield of the final product is the bioactive compounds derived from beetroot. To achieve the maximum recovery of such valuable molecules, various methods have been employed for extraction, such as solvent extraction, microwave-assisted extraction, supercritical fluid extraction, and many more, which we are going to discuss (Martins et al., 2023). Each method of extraction has both advantages and drawbacks, and the bioactivity and stability of the obtained compounds may often depend on the technique used. Characterization serves to determine chemical composition, functional properties, as well as potential health effects of beetroot bioactives.

High-performance liquid chromatography (HPLC), mass spectrometry (MS), and nuclear magnetic resonance (NMR) spectroscopy, amongst many others, are usually employed to quantify and analyze the chemicals that shall be discussed herein within this chapter (Araujo-León et al., 2024). Such a description not only allows for quality assessment of beetroot foods but also helps clarify their mechanisms of action and usage in food preparation. Together with extraction and characterization, beetroot bioactive powder manufacture and encapsulation are also increasingly being practiced in food applications. Freeze-drying and spray drying are two encapsulation technologies that maximize the stability and bioavailability of a compound such that it can be better supplemented in foods (Shofinita et al., 2023). Beetroot powder production has also been a popular means of utilizing its bioactives so that it can be incorporated easily into smoothies, baked goods,

 DOI: 10.1201/9781003527084-6

and other foods. Beetroot bioactives' use in the food industry is varied. They may find application as nature-based coloring, antioxidants, and functional additives, whose role is to enhance the nutritional content of food products while ensuring that quality is added to foods. The increasing trend toward clean-label products has also intensified the demand for natural bioactives since consumers are moving more toward substituting synthetic additives (Theba et al., 2024). Furthermore, the addition of beetroot bioactives can enhance the sensory properties of food items, that is, taste and texture, thus increasing the overall consumer acceptance. This chapter offers a comprehensive overview of the extraction techniques, characterization methods, encapsulation strategies, and varied food industry applications of beetroot bioactives, along with recent innovations and emerging trends that are revolutionizing their future in functional food development.

6.2 The Extraction Methods of Beetroot

6.2.1 Conventional Solvent Extraction

Conventional solvent extraction is a very common method employed for the isolation of bioactive compounds like betalains and phenolics from red beetroot. Aqueous ethanol is one of the most popular solvents because it effectively extracts betalains, like betacyanin and betaxanthin, with high yields of color compounds, like 952.5 mg/L in optimized conditions (Zin et al., 2020). Citric acid solutions have also been marketed as more eco-friendly substitutes for ethanol since they are biodegradable, harmless, and cheap. They yield similar betalains and total phenolics from beetroot powder with yields of 3.98 mg/g for betacyanin, 3.64 mg/g for betaxanthin, and 8.28 mg/g for total phenolics (Borjan et al., 2022).

Extract process parameters have important effects on the efficiency of extraction. Elevated temperatures might improve thermally stable phenolics extraction but may cause decomposition of sensitive products such as betalains (Borjan et al., 2022; Zin et al., 2020). The pH level of the solvent is equally important; at a lower pH for citric acid solvents, stability and efficiency of betalain extraction are enhanced. Long extraction durations may enhance the yields but should also be optimally defined so that there will be no thermal degradation of betalains (Eyshi et al., 2024).

6.2.2 Advanced Extraction Methods

Advanced extraction methods used for the extraction of beetroots are presented in Figure 6.1 and Table 6.1. Extracting bioactive compounds from beetroot, focusing on efficiency, sustainability, and industrial applications.

6.3 Characterization of Bioactives from Beetroot

Beetroot contains a high number of bioactive substances like betalains, phenolic acids, flavonoids, nitrates, and other phytochemicals. These bioactives have elevated biological activities, including antioxidant, anti-inflammatory, antimicrobial, and cardioprotective

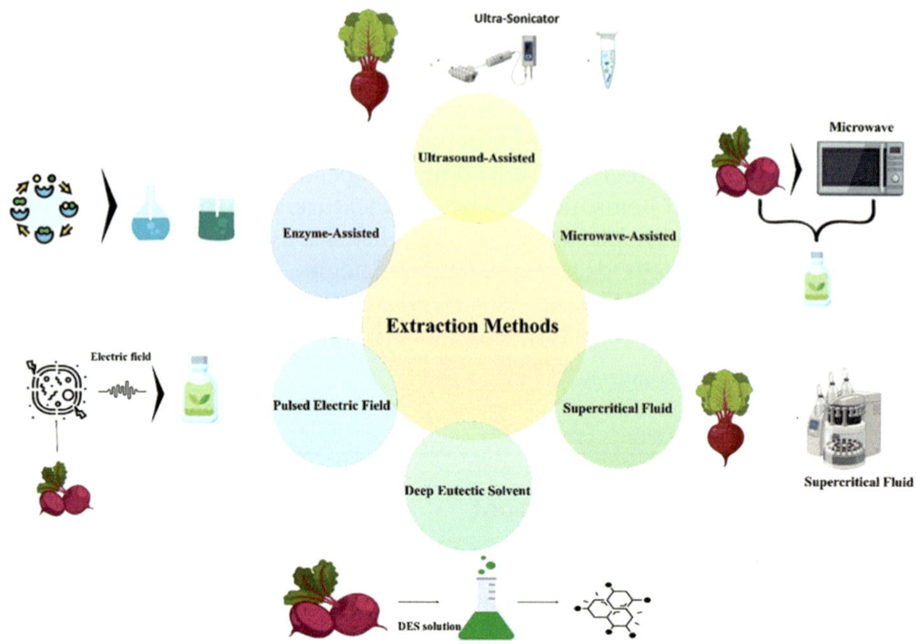

FIGURE 6.1
The different advanced extraction techniques.

TABLE 6.1

Extracting Bioactive Compounds from Beetroot, Focusing on Efficiency, Sustainability, and Industrial Applications

Extraction Techniques	Description	Advantages	References
Ultrasound-Assisted Extraction	Employ ultrasonic vibrations to break up cell walls and increase solvent access and pigment extractability	Has a high yield, lower solvent usage, and lesser extraction time	Borjan et al. (2022), Ebrahimi et al. (2022)
Microwave-Assisted Extraction (MAE)	Uses microwave energy for heating the plant matrix and the solvent to achieve better extraction and shorter processing	Has a fast process, low power consumption, and is easily scalable for industries	Da Silva et al. (2023), Stoica et al. (2025)
Supercritical Fluid Extraction	Makes use of supercritical CO_2 and co-solvents such as ethanol to isolate betalains under pressure and temperature control	Environmentally safe, selective recovery, and very high antioxidants	Borjan et al. (2022), Da Silva et al. (2023)
Deep Eutectic Solvent Extraction (DES)	Applies eutectic solvents consisting of natural substances such as urea and magnesium chloride for betalain extraction from beetroot waste	Green, stabilizes betalains for extended storage, and has a low environmental footprint	Hernández-Aguirre et al. (2021), Stoica et al. (2025)
Pulsed Electric Field	Applies brief electrical pulses to damage cell membranes, liberating intracellular pigments without generating high heat	Non-thermal technique maintains pigment stability and minimizes energy use	Chatzimitakos et al. (2023), Da Silva et al. (2023)
Enzyme-Assisted Extraction (EAE)	Applies enzymes such as cellulase or xylanase to rupture cell walls and liberate pigments efficiently	Low-temperature operation retains pigment color properties and is an efficient utilization of food wastes	Łubek-Nguyen et al. (2022), Stoica et al. (2025)

activities. Below is a description of the prominent bioactives present in beetroot as gleaned from current research (Punia Bangar et al., 2022).

6.3.1 Betalains

Betalains are nitrogen-containing pigments and are divided into two subclasses:

a. **Betacyanins**: The red pigments, most of which are represented by betanin (75%–95% of all betalains) and isobetanin.
b. **Betaxanthins**: The yellow pigments, like vulgaxanthin I and II.

Betalains are produced by the shikimate pathway and are responsible for beetroot's red color and have antioxidant, anti-inflammatory, and antimicrobial properties. Studies show significant differences in betalain content based on production systems and storage conditions. For instance, conventional beetroot varieties contained up to 1.4 times more betalains than those produced organically (Paulauskienė et al., 2023). Moreover, betalains are very light and temperature- and pH-sensitive during storage and processing (Postružnik et al., 2024).

6.3.2 Phenolic Compounds

Phenolics that occur in beetroot include gallic acid, ferulic acid, syringic acid, caffeic acid, chlorogenic acid, and vanillic acid. Phenolics are known to exhibit antioxidant activity via the scavenging of reactive oxygen species (ROS). Gallic acid is the most common phenolic in beetroot foods, with a maximum of 60.50 mg/g in cereal foods (Baião et al., 2020). Ferulic acid possesses cardiometabolic activities through oxidative stress and inflammation decrease. Phenolic content in beetroot extracts was quantified at 869.14 mg GAE/100 g using the Folin–Ciocalteu method in a study (Baião et al., 2020; Postružnik et al., 2024).

6.3.3 Flavonoids

Beetroot contains flavonoids like kaempferol, astragalin, tiliroside, and rhamnetin. They are antioxidants with the potential for the management of chronic diseases like cancer and cardiovascular disease (Punia Bangar et al., 2022; Thiruvengadam et al., 2024). Processing and production methods influence the flavonoid content.

6.3.4 Nitrates

Beetroot is a dietary source of nitrates, which are reduced to nitric oxide (NO) in the body. Beetroot is a rich source of nitrates, which are reduced to NO by the body. This compound has an important role to play in vasodilation, reducing blood pressure and enhancing endothelial function. Experiments have found that beetroot nitrate supplementation will have beneficial effects on cardiovascular function by increasing arterial compliance and inducing smooth muscle relaxation (Baião et al., 2020).

6.3.5 Other Phytochemicals

Aside from nitrates, beetroot contains other phytochemicals of varying diversity, including carotenoids, coumarins, sesquiterpenoids, and triterpenes. These are all the reasons for its antioxidant potential and nutritional importance (Chen et al., 2021; Thiruvengadam et al., 2024).

6.4 Characterization Techniques to Quantify Bioactive Compounds

Bioactive compounds of beetroot need highly complex analytical methods for effective determination and quantitation of its phytochemicals like betalains, phenolics, flavonoids, and nitrates (Punia Bangar et al., 2022c). These categories of methodologies are of the greatest significance in measuring the bioactive potential of beetroot in some of the important applications within the food, drug, and nutritional industries. What follows hereunder is an explanatory description of some of the prevalent methodologies from related evidence emanating from the pertinent literature.

6.4.1 Chromatographic Methods

6.4.1.1 High-Performance Liquid Chromatography

HPLC is usually considered to be the "gold standard" for betalains (betalacyanins and betaxanthins) and flavonoids, and phenolic acids separation and quantitation in beetroot extracts, and it utilizes a mobile phase (e.g., organic solvents and water) and stationary phase (silica-based columns) in separating compounds based on polarity and retention time (Koubaier et al., 2014; Mukherjee, 2019). Borjan et al (2022) have proved that HPLC quantitated protocatechuic acid as the major phenolic component in beetroot extracts obtained through Soxhlet extraction with 50% ethanol, with a concentration up to 18.60 mg/g. HPLC is also effective in the detection of betalain pigments like betanin and vulgaxanthin with maximum absorbance wavelengths at 535–540 nm (Harsha Sarma et al., 2023).

6.4.1.2 Gas Chromatography-Mass Spectrometry (GC-MS)

GC-MS determines volatile bioactive compounds like organic acids, terpenoids, and coumarins in beetroot extracts. A review demonstrated that GC-MS was able to detect compounds like gallic acid, ferulic acid, and syringic acid in different beetroot products, and gallic acid was found most frequently among the phenolic compounds (Bian et al., 2024).

6.4.2 Spectroscopic Techniques

6.4.2.1 UV-Visible Spectroscopy

This method measures betalains and phenolics in units of absorbance at selected wavelengths. Betanin has a maximum absorbance of 535 nm, while betaxanthins have a maximum absorbance between 460 and 480 nm (Devadiga & Ahipa, 2020; Escribano et al., 2017). Folin–Ciocalteu reagent in conjunction with UV-Vi spectroscopy can be employed to estimate the phenolic content in beetroot extracts (Lawag et al., 2023).

6.4.2.2 Fourier Transform Infrared Spectroscopy

FTIR is employed to determine functional groups like –OH, –COOH, and –NH in bioactive compounds based on their vibrational frequency analysis. FTIR has been successfully used to detect betalains and phenolics in beetroot extracts (Rotich et al., 2022).

6.4.3 Mass Spectrometry-Based Methods

6.4.3.1 *Liquid Chromatography-Mass Spectrometry (LC-MS)*

LC-MS is a union of the ability of HPLC separation and mass spectrometry to enable complete structural examination of bioactive agents, and it possesses the particular value for the identification of complex mixtures of betalains and phenolic acids found in beetroot juices (Sawicki et al., 2017). The methodology was employed in identifying trace constituents such as isobetanin and betanidin, and also the identification of major pigments such as betanin was illustrated in one study (Ekrem Parlak et al., 2024).

6.4.3.2 *Matrix-Assisted Laser Desorption/Ionization Mass Spectrometry (MALDI-MS)*

MALDI-MS is a sensitive direct profiling method for the betalains in plant tissues without sophisticated sample preparation. MALDI-MS has been used to image the pigments' distribution in beetroot slices (Montini et al., 2020).

6.4.4 Phytochemical Screening Tests

These tests are straightforward processes utilized to estimate overall classes of bioactive molecules: The Folin–Ciocalteu total phenolic content assay is conducted by treating phenolics with a phosphomolybdate reagent (Pérez et al., 2023). Aluminium Chloride Assay is used to determine flavonoids based on their ability to complex with aluminium ions (Shraim et al., 2021).DPPH Free Radical Scavenging Assay quantifies the antioxidant activity by measurement of free radical scavenging capacity of beetroot extracts (Vulić et al., 2012).

6.4.5 New Strategies

6.4.5.1 *NMR Spectroscopy*

NMR provides extensive structural information on bioactive compounds based on the examination of nuclear spin characteristics under magnetic fields. NMR was applied to validate betalains and other phytochemicals' stereochemistry in beetroot (Giampaoli et al., 2021).

6.4.5.2 *Bioautography*

This method utilizes the synergy of thin layer chromatography (TLC) and bioassays for the screening of antimicrobial or antioxidant activity against chromatograms. It has been used for screening antimicrobial betalains from beetroot extracts (Lata et al., 2019).

6.5 Beetroot Encapsulation and Powder Production

Beetroot encapsulation and powder manufacturing are state-of-the-art food processing technologies designed to ensure the protection of bioactive components (such as betanin and polyphenols) by improving stability, shelf life, and utility in food. Encapsulation saves nutrients from thermal destruction, photo destruction, and oxidation and helps them be applied in functional foods, nutraceuticals, and natural colors (Poornima et al., 2016).

6.5.1 Procedures for Beetroot Encapsulation and Powder Production

6.5.1.1 Drying Methods

6.5.1.1.1 Spray Drying

Spray drying is widely applied for the dehydration of food components because it is cost-effective, has low maintenance requirements, and has the capability to produce high-quality, stable products. Spray drying is substantially cheaper than freeze drying, with estimated costs of production being 30–50 times lower. Also, it provides a significantly shorter drying time compared to most traditional drying methods, and it can control the resulting particle size distribution precisely (Bordiga et al., 2024). The shortened processing time is especially useful in preserving bioactive compounds, for example, betalains, during drying. In the process, the moisture in the feed solution is quickly evaporated by mass and heat transfer processes induced by the high temperature of the hot air, resulting in the production of fine solid particles. Spray drying also has a good encapsulation efficiency and results in a fine, free-flowing powder that is convenient to handle and can be used in various applications (Abd Rahman et al., 2024).

6.5.1.1.2 Lyophilization or Freeze-Drying

Lyophilization or freeze drying is a dehydration method involving the removal of water from materials by the sublimation of ice in low-pressure environments. It is especially suited for the preservation of heat-sensitive food constituents. Freeze drying is done when the temperature stays below the glass transition point, and the material remains in a glassy state. This serves to avoid shrinkage of the structure and yields a very porous end product (Nowak & Jakubczyk, 2020). Betalain powders processed through freeze drying will have smaller, more uniform particle sizes than those dried through tray drying, drum drying, or continuous vacuum-belt drying. Additionally, the color characteristic (i.e., redness) of freeze-dried betalain powder has been reported to be better than sun-dried and oven-dried powders. This color maintenance improvement may be due to freeze drying having aided in the maintenance of texture as well as limiting sample shrinkage (Shofinita et al., 2023). In addition, freeze-dried betalain powders tend to have a higher chemical and mineral composition compared to those that are manufactured with standard heat-associated drying methods. With its benefits for fragile materials aside, freeze drying is an expensive process because it is high in energy requirements and takes much longer processing times compared to other methods of drying (Shofinita et al., 2023).

6.5.1.1.3 Vacuum Drying

Vacuum drying is a non-thermal drying process that avoids the degradation of delicate compounds such as betalains (Gokhale & Lele, 2014). It is a viable substitute for freeze drying, which has the benefit of drying material at low temperatures without going through a freezing step, hence saving overall drying time. Compared to freeze drying and hot air drying, vacuum drying is often less energy-intensive, and the process takes a shorter time. The operation takes place under decreased oxygen levels, which reduces pigment degradation from drying processes conducted at atmospheric pressure (Mallick & Winnett, 2014).

6.5.1.1.4 Air Drying

Although this process increases the shelf life of products, it may also cause a reduction in product quality as a result of the possible degradation of sensitive compounds (Yildirim & Umdu, 2022). There are many parameters that determine the efficiency and quality of results in air drying, including the drying temperature, the material thickness being dried, and the recirculation ratio of the air. A thicker material takes more time to dry, which causes greater pigment degradation (Yao et al., 2022). Yao et al. (2022) noted that increased drying temperature in hot air oven drying reduced the betacyanin content of dragon fruit peel (Hylocereus undatus), possibly as a result of heat breakdown. In a similar investigation, Gokhale and Lele (2014) noted that red beetroot subjected to convective air drying reduced the content of betacyanin, with betaxanthin content rising with increasing temperature between 50°C and 120°C. The betaxanthin increase was credited to improved extraction efficiency at higher temperatures. However, drying beetroot at moderate temperatures (50°C–65°C) yielded pigment retention similar to freeze drying, with betacyanin and betaxanthin contents of 5.1–4.9 and 2.4–2.3 mg/g^2 of beetroot powder, respectively.

During the constant rate period of drying, the material temperature is relatively low compared to high drying air temperatures. This is because the heat absorbed is mainly utilized for the evaporation of moisture (Fox et al., 2014). Yet, under the falling rate period, with decreasing moisture content, the sample temperature can increase more readily because more energy is available in addition to that needed for water vaporization, and temperature increase can cause increased pigment degradation at higher drying temperatures (Neikov, 2009).

At high drying temperatures, the constant drying rate stage is faster, and free surface moisture evaporates rapidly until the critical moisture content is attained. Since betalains are water-soluble pigments, evaporation of surface moisture during this period may cause surface color loss. During the following falling rate period, dissolved pigments and internal water move toward the surface, possibly increasing the yellowness and redness of the dried material and contributing to porosity (Shofinita et al., 2023).

6.5.1.1.5 Microwave Vacuum Drying

Microwave vacuum drying is an advanced technique for preserving food, where moisture is removed through the absorption of microwave energy. This generates significant vapor pressure within the core of the material, facilitating rapid drying. The primary distinction between conventional convective vacuum drying and microwave vacuum drying lies in the energy source: convective vacuum drying relies on heat transferred by circulating hot air, whereas microwave vacuum drying uses electromagnetic energy that excites water molecules through high-frequency dipolar polarization (Calín-Sánchez et al., 2020).

In microwave-assisted drying, both the power level of the magnetron and the duration of exposure play critical roles in enhancing the concentrations of betacyanin and betaxanthin (Liu et al., 2021). Dirim et al. (2019) demonstrated that increasing microwave power effectively reduces moisture content while improving color attributes such as lightness (L), redness (a*), total color difference (ΔE), chroma, and the browning index in red beet powders. Moreover, microwave processing is capable of destroying cell walls, resulting in greater extraction and retention of betalains during the final product (Nistor et al., 2017).

Integrating convective drying with vacuum-microwave drying at elevated microwave power levels not only shortens the overall drying time but also results in products with improved antioxidant properties.

6.6 Powder Production and Its Uses

Production of beetroot powder by encapsulation methods is an essential means of enhancing the stability, functionality, and applicability of beetroot-derived bioactive compounds (Stoica et al., 2025). Encapsulation is a protective technique that shields delicate phytochemicals, for example, betalains, phenolic acids, and dietary nitrates, from environmental factors like light, oxygen, moisture, and temperature that have detrimental effects on their stability and bioactivity. This technology preserves the shelf life of beetroot bioactives and renders them suitable for inclusion in different food products and nutraceutical products (Domínguez et al., 2021). The major benefits of encapsulated beetroot powder are explained below.

6.6.1 Ease of Handling and Improved Shelf Stability

Encapsulated beetroot powders show higher physicochemical stability than unencapsulated liquid or extract forms. Essential encapsulant carriers such as maltodextrin, gum arabic, and starch derivatives create film-forming matrices that protect the powder from deterioration by agents such as oxygen, water, light, and heat (Flores-Mancha et al., 2020; Yousefi et al., 2022). The result is diminished sensitivity of the powders to oxidation, deterioration, and microbial attack. The implication is a highly improved shelf life and product integrity of bioactive ingredients. In addition, encapsulated powders with lower hygroscopicity inhibit caking and agglomeration and allow for convenient handling, storage, and blending into food industrial products (Mahanti et al., 2021).

6.6.2 Uniform Distribution Within Food Matrices

Free flow and fine particle size of encapsulated beetroot powders allow for uniform dispersion in food systems. Uniformity is the cause of the uniformity of sensory properties, flavor, and intensity of color and the uniform dispersal of functional ingredients such as antioxidants and dietary nitrates (Mitrevski et al., 2023). Encapsulation techniques also hide the earthy or bitter flavor common in beetroot, thus increasing the sensory acceptability of food products that are fortified (Trindade et al., 2023).

6.6.3 Functional Food Application Flexibility

Encapsulated beetroot powder maintains functional properties and can be applied to a range of foodstuffs. It serves as a natural color additive used to contribute normal red through to purple tints, in addition to functioning as a high antioxidant capable of adding oxidative stability to food matrices. Encapsulated beetroot powders are commonly used in baked products (e.g., bread, biscuits, and cakes), dairy products (e.g., yoghurt, cheese, and smoothies), beverages (e.g., juices and energy drinks), confectionery, and dietary supplements (e.g., capsules and tablets). Aside from its technological characteristics, beetroot powder is also employed in the delivery of benefits such as antihypertensive, anti-inflammatory, and vasodilatory activities, further strengthening its use in the development of functional foods and nutraceuticals (Chen et al., 2021; Trindade et al., 2023; Tumbas Šaponjac et al., 2020).

6.6.4 Scalability and Industrial Feasibility

Encapsulation techniques used in beetroot powder production, including spray drying, freeze drying, and fluidized bed coating, are highly scalable and fully compatible with ongoing food processing operations. The techniques allow for bulk production with reproducible quality, offering economical solutions to the food processing sector (Mehta et al., 2022; Ratti, 2001). The scalability of these processes facilitates the extensive use of encapsulated beetroot powder in small-scale artisanal foods as well as large-scale commercial food production.

6.7 Utilization of Beetroot Bioactives in the Food Processing Industries

The use of bioactive compounds extracted from beetroot in food processing has become of considerable interest owing to their multifunctional activities, such as antioxidant, anti-inflammatory, and cardioprotective properties (Razzak et al., 2024). In addition, the pigments in beetroot, specifically betalains, have made it a good alternative to synthetic dyes in response to the world's clean-label and functional food trend (Sadowska-Bartosz & Bartosz, 2021).

6.7.1 Natural Colorants

Beetroot is a botanical source of natural pigments, and betalains are the main class of compounds. Betalains are nitrogenous water-soluble pigments, as discussed earlier, and they are classified into two broad groups: betacyanins (red-violet) and betaxanthins (yellow-orange). These pigments show a rich red to purple coloration and, therefore, are in great demand as natural colorants in food matrices (Sun et al., 2021).

6.7.2 Food and Beverage Applications

Betalains derived from beetroot are used intensively in the production of foodstuffs such as beverages, confectionaries, dairy products, condiments, and sauces. They are not used simply for color purposes; they also impart bio-functional properties because they exhibit antioxidative activity. Betalains are free radical scavengers, a property that stems from their planar conjugated structure and the presence of phenolic groups that confer oxidative stability to food (Trych et al., 2022).

6.7.3 Regulatory and Consumer Acceptance

Beetroot colorants are Generally Recognized as Safe by the United States Food and Drug Administration (FDA) and are authorized for use as color additives in E162 of the European Union. Consumer acceptance is skewed toward the naturally derived additives because of higher awareness regarding the harmful impacts of chemical colorants such as tartrazine and sunset yellow. Beetroot colorants thus find application to consumer demand for clean-label, sustainable, and health-promoting products ("Safety of the Proposed Extension of Use of Beetroot Red (E 162) in Foods for Special Medical Purposes in Young Children," 2016).

6.7.4 Functional Foods and Nutritional Enhancements

Apart from their use as colorants, beetroot bioactives have been promising functional food ingredients. Their addition to processed foods not only improves the organoleptic qualities but also contributes to the nutritional and therapeutic potential of such foods (Stoica et al., 2025).

6.7.5 Antioxidant Properties

Betalains and polyphenolic phytochemicals, such as flavonoids and phenolic acids present in beetroot, are strongly antioxidant in nature. These bioactives bring about the inhibition of oxidative stress by scavenging reactive oxygen species (ROS) and reactive nitrogen species (RNS), thereby preventing cellular damage and lipid peroxidation. The addition of these compounds to food products increases the functional value of products, providing potential health benefits against oxidative stress-related pathologies (Razzak et al., 2024; Stoica et al., 2025).

6.7.6 Cardiovascular Health

Beetroot is also a good source of dietary inorganic nitrates (NO_3^-), which get bioconverted into bioactive NO through the nitrate-nitrite-NO pathway. NO has a vasodilatory action, improving endothelial function and decreasing blood pressure. Several randomized controlled trials have confirmed the effectiveness of beetroot juice supplementation in decreasing systolic as well as diastolic blood pressure, thereby indicating its therapeutic application for the prevention and management of hypertension and other cardiovascular diseases (Baião et al., 2020).

6.7.7 Anti-Inflammatory Actions

Recent in vivo and in vitro studies have introduced the anti-inflammatory activity of beetroot phytochemicals. Betalains inhibit pro-inflammatory mediators such as tumor necrosis factor-alpha (TNF-α), interleukins (IL-6, IL-1β), and cyclooxygenase-2 (COX-2). These activities suggest potential applications of beetroot bioactives in food design for people suffering from chronic inflammatory diseases (Nirmal et al., 2024).

6.7.8 Gut Health and Prebiotic Action

Beetroot contains dietary fibers and oligosaccharides, which serve as prebiotics that stimulate the development and activity of health-promoting gut microbiota such as *Bifidobacteria* and *Lactobacilli*. Breakdown of these fibers in the colon results in the production of short-chain fatty acids, which have been associated with improved gut barrier function, reduced inflammation, and metabolic health. Functional foods with added beetroot fibers can thereby help ensure a healthy gut microbiome (Ko et al., 2024).

6.8 Use in Individual Food Industries

The versatility of bioactivity and functionality among beetroot compounds has enabled their use across individual food industries as well as provided technological and health advantages.

6.8.1 Bakery Products

Beetroot powder, which is a good source of dietary fiber and essential minerals, shows considerable promise as a formulation ingredient in bakery and confectionery products. The addition of higher levels of beetroot powder led to the final cake products having higher protein, fat, ash, and fiber content. Interestingly, cakes designed with the inclusion of beetroot powder in concentrations of 15% (w/w) revealed better sensory qualities, that is, greater appearance, flavor, and texture quality compared to all other inclusion percentages or control samples. Given such results, cake enriched with beetroot powder concentrations of 15% has both improved nutritional as well as sustained positive sensory quality. Daily intake of these fortified bakery foods can lead to enhanced dietary fiber consumption and general well-being (Al-Mamun et al., 2020).

6.8.2 Beverages

Functional drinks with beetroot juice or powder have become increasingly popular in sports nutrition because of their performance-improving benefits. Beetroot supplementation has been linked with enhanced oxygen efficiency utilization and reduced onset of fatigue when engaging in physical exercise. Dietary nitrates play a role in the ergogenic effects; thus, beetroot is used as a primary ingredient in pre-workout beverages and endurance drinks (Domínguez et al., 2017).

6.8.3 Snack Foods

Beetroot flakes and powders find application in the manufacture of health-focused snack foods like vegetable chips, protein bars, and extruded snacks. Color and flavor were added for enrichment, and antioxidants were added to enhance the health benefits of the product. These foods for function find consumers looking for healthier alternatives for standard high-fat, low-nutrient snack foods (Baião et al., 2020; Chen et al., 2021).

6.9 Challenges and Considerations

Although beetroot bioactives have great food processing potential, some technical and economic issues should be overcome in order to realize their maximum usage.

6.9.1 Stability Issues

Betalains are naturally unstable under some environmental conditions, for example, high pH (maximum stability between pH 4 and 5), heat, and extended exposure to light and oxygen. The degradation of betalains causes discoloration and loss of bioactivity, restricting their use in thermally processed foods. Encapsulation methods, including microencapsulation and nanoencapsulation with biopolymers (e.g., maltodextrin and gum Arabic), have been demonstrated to improve the stability and controlled release of betalains throughout processing and storage.

6.9.2 Sensory Properties

While beetroot confers desirable color properties, its strong earthy taste, contributed by geosmin and pyrazines, can be unwanted in some foods. Formulation strategies, such as flavor masking, compatibility with synergistic ingredients, and the use of natural flavor potentiators, are the most important strategies to enable consumer acceptance without any loss of nutrition. Extraction, purification, and stabilization of beetroot bioactives, particularly betalains and dietary nitrates, are complex and expensive operations. The industrial application of these methods demands optimization to maintain low costs and product quality as well as high bioactivity. Novel extraction methods, such as co-solvent-free supercritical fluid extraction, ultrasound-assisted extraction, and enzyme-assisted extraction (EAE), are being evaluated for their efficiency and eco-friendliness for industrial applications.

6.10 Innovations and Emerging Trends in the Extraction and Utilization of Beetroot Bioactives

The field of extraction and application of bioactive compounds, especially in connection with beetroot bioactives, is witnessing significant transformation. The transformation is driven by innovation in green technology, increased awareness of natural foods and functional foods among consumers, and increased regulatory demands. The ensuing sections describe current innovations and developing trends transforming the future use of beetroot bioactives in the food processing sector.

6.10.1 Green Extraction Technologies: Toward Sustainable Bioactive Recovery

With environmental sustainability, an integral part of contemporary food manufacturing, green extraction methods are in the ascendancy. These are characterized by minimal energy use, the absence or removal of hazardous solvents, and the utilization of renewable resources (Usman et al., 2023).

6.10.2 Enzyme-Assisted Extraction

EAE is the use of particular enzymes like cellulases, pectinases, and hemicellulases to break down the complex polysaccharide matrix in the cell wall of the plant. Enzymatic hydrolysis enables the liberation of intracellular bioactives like betalains, polyphenols, and dietary nitrates from beetroot tissue. EAE increases extraction yield substantially, performs under less harsh conditions (milder temperatures and neutral pH), and minimizes dependence on organic solvents, thus conforming to green chemistry principles (Fan et al., 2022).

6.10.3 Natural Deep Eutectic Solvents (NADES)

NADES are a novel group of biodegradable, non-toxic, and sustainable solvents constituted by natural primary metabolites such as sugars, amino acids, and organic acids. NADES are suggested for the selective recovery of hydrophilic and lipophilic bioactives. Their tunable polarity and viscosity properties enable the efficient recovery of betalains and phenolic compounds from beetroot matrices without compromising the functionality and integrity of these delicate molecules (Hernández-Aguirre et al., 2021).

6.10.4 PEF and HPP

Non-thermal technologies such as PEF and HPP are cutting-edge pre-treatment methods for enhancing cell membrane permeability (Vaessen et al., 2019). PEF uses pulsed high-voltage discharges in short durations, causing cell membrane electroporation, favoring the expulsion of intracellular bioactives. HPP, which is used at pressures of up to 600 MPa, affects cellular structures and facilitates mass transfer during extraction. These techniques present considerable benefits of preserving thermolabile compounds while providing high efficiency in extraction without affecting nutritional and functional quality (Jolvis Pou, 2021).

6.10.5 New Encapsulation Technologies: Enabling Beetroot Bioactives' Stability and Bioavailability

Encapsulation technologies were also shown to play a major role in safeguarding beetroot-extracted bioactives against degradation and optimizing functional performance in foods. New delivery systems are being developed for solving stability issues, controlled release, and bioavailability (Rezagholizade-shirvan et al., 2024).

6.10.6 Nanoencapsulation

Nanoencapsulation refers to the encapsulation of bioactive compounds within nanocarriers such as nanoparticles, nanoemulsions, liposomes, and solid lipid nanoparticles. Nanoencapsulation confers higher stability of betalains and polyphenols to environmental stresses, including oxidation, light, and thermal degradation. It increases solubility and gastrointestinal absorption and facilitates targeted delivery, thereby maximizing the bioefficacy of beetroot bioactives in functional foods and nutraceuticals (Asgari et al., 2020).

6.10.7 Multi-Layered Encapsulation and Layer-by-Layer (LbL) Assembly

The LbL process is the sequential deposition of oppositely charged biopolymers, which creates multilayered capsules around the core bioactives. The process is capable of fine-tuning the release profiles of encapsulated compounds against physiological stimuli such as pH change, enzyme activity, or temperature. Multilayer encapsulation can enhance the protection and stability of beetroot-derived compounds during food processing, storage, and digestion (Priyoatmojo et al., 2025).

6.10.8 Biopolymer-Based Encapsulation

The use of natural, biodegradable polymers such as chitosan, alginate, pectin, and starch as encapsulants for beetroot bioactives is gaining momentum with their safety in use and viability for clean-label applications. Biopolymers are applicable as matrices in controlled delivery formulations and also enhance the nutritional properties of foods through their inherent nutritional value, such as dietary fiber content and prebiotic activity (Martau et al., 2019).

6.11 Integration of Digital and Smart Technologies in Bioactive Processing and Application

The convergence of food processing technologies and digital technologies is ushering in a new level of precision and efficiency in the extraction and use of bioactive compounds.

6.11.1 Process Automation and Intelligent Control Systems

The use of real-time monitoring systems, artificial intelligence (AI), and machine learning algorithms makes it possible to automate and optimize extraction, purification, and encapsulation processes. Sensors that can detect temperature, pH, pressure, and bioactive concentration ensure precise control, consistency, and quality and minimize energy use in production lines (Alloun & Calvio, 2024).

6.11.2 Predictive Modeling and Computational Simulations

Predictive modeling methods are being used to model the behavior of beetroot bioactives at different processing, storage, and digestion stages. Predictive models allow for the optimization of formulation conditions and extraction procedures with less experimental burden and scaling up production processes with better predictability and reliability. (Chaari et al., 2023; Tagnamas et al., 2023)

6.11.3 Blockchain Technology for Traceability and Transparency

Blockchain systems provide end-to-end traceability from farm production to the point of final product delivery. Such transparency is essential to underpinning compliance with food safety regulations, creating consumer confidence, and verifying claims of organic origin, sustainability, and functional health benefit (Adewusi et al., 2023).

6.12 New Uses of Beetroot Bioactives in New Foodstuffs

Since there is an increased awareness of the health effects of beetroot bioactives, new uses are being discovered in various segments of the food industry.

6.12.1 Functional and Nutraceutical Beverages

Functional food beverages with beetroot extract are designed to maintain specific areas of wellness, including cardiovascular health, sports performance, and immune function. Blends with complementary bioactives (e.g., polyphenols, probiotics, and omega-3 fatty acids) are in development to generate synergistic benefits and to extend health benefits (Chen et al., 2021).

6.12.2 Plant-Based Meat Alternatives

The red natural pigments (betalains) of beetroot are being used more and more to give plant-based meat analogues a realistic, meat-like color. In addition to aesthetics, beetroot's antioxidant and anti-inflammatory activities add functional and nutritional value to these products, which appeal to health-oriented consumers (Domínguez et al., 2020; Stoica et al., 2025).

6.12.3 Nutraceuticals and Dietary Supplements

Beetroot bioactives are encapsulated and processed into capsules, tablets, and powders as dietary supplements. They are sold for their capacity to maintain cardiovascular function, reverse oxidative stress, and improve physical stamina. Clinical trials are ongoing to confirm claims and determine efficacy (Baião et al., 2020).

6.12.4 Smart Packaging and Active Films

The antioxidant and antimicrobial activity of beetroot extracts is investigated in smart packaging applications. Beetroot bioactives can be utilized in edible films and coatings that can be used to prolong the shelf life of fresh fruits and vegetables by suppressing lipid oxidation and microbial growth as a sustainable alternative to conventional packaging materials (Karnwal et al., 2025).

6.13 Future Outcomes

Sustainable Extraction Technologies: Shift toward sustainable extraction methods, that is, enzyme-assisted extraction and deep eutectic solvent-assisted extraction, will remain only on the ascending side. Besides enhancing the quality and yield of the bioactives, they arrive in harmony with rising trends toward making the food processing process more sustainable.

Advanced Encapsulation Strategies: Recent encapsulation techniques like nanoencapsulation and biopolymer-based delivery systems will continue to improve the stability and bioavailability of beetroot bioactives so that they are applicable in different food products without affecting their functional quality.

Integration of Digital Technology: The utilization of digital technology, for example, AI and blockchain, within the food processing industry will enhance the efficacy and traceability of beetroot bioactive extraction and use. Such integration will help in providing assured quality, efficient production process, and assured consumer confidence due to transparency.

Expansion of Functional Uses: The versatility of the beetroot bioactives will lead to their increased use in various food products, like plant-based meat substitutes, functional drinks, and nutraceuticals. As continued research continues to uncover additional health benefits, the market for products based on beetroot will increase significantly.

Consumer Education and Acceptance: Consumer acceptance will rise as more people become aware of the health benefits of beetroot bioactives. Consumer education programs that emphasize the nutritional and functional benefits of these natural ingredients will be instrumental in their adoption in mainstream food products.

Regulatory Support and Safety Evaluations: Sustained regulatory support for natural colorant and bioactive compound use will be necessary. Sustained safety evaluations and research on the long-term health impacts of beetroot bioactives will provide credibility and safety for food uses.

6.14 Conclusion

Extraction, characterization, and use of beetroot bioactives are a challenge to food processing innovation with good prospects. Due to their numerous health-promoting benefits and uses, beetroot bioactives have the capacity to enrich foods with nutritional and sensory attributes and address consumers' demands for natural products. As extraction and encapsulation technologies are improving day by day, with the addition of digital solutions, the future of beetroot bioactives in food processing is fantastic. Bridging stability, sensory property, and production cost issues, the sector is now able to harness the full potential of beetroot bioactives, and healthy, sustainable food items will follow.

References

Abd Rahman, N. F. N. N., Zubairi, S. I., Hashim, H., & Yaakob, H. (2024). Revolutionizing spray drying: an in-depth analysis of surface stickiness trends and the role of physicochemical innovations in boosting productivity. *Journal of Food Quality*, 2024(1), 8929464. https://doi.org/10.1155/2024/8929464

Adewusi, A. O., Chiekezie, N. R. & Eyo-Udo, N. L. (2023). Blockchain technology in agriculture: enhancing supply chain transparency and traceability. *Finance & Accounting Research Journal*, 5(12), 479–501. https://doi.org/10.51594/FARJ.V5I12.1514

Alloun, W., & Calvio, C. (2024). Bio-driven sustainable extraction and AI-optimized recovery of functional compounds from plant waste: a comprehensive review. *Fermentation*, 10 (3), 126. https://doi.org/10.3390/FERMENTATION10030126

Al-Mamun, A., Hosen, A., Afroz Toma, M., & Anisur Rahman Mazumder, M. (2020). *Nutritional and sensory quality assessment of plain cake enriched with beetroot powder*. https://doi.org/10.26656/fr.2017.4(6).268

Alwazeer, D., Elnasanelkasim, M. A., Çiğdem, A., Engin, T., Kanmaz, H., Hayaloglu, A. A., Russell, G., & Hancock, J. T. (2023). Hydrogen incorporation into solvents can improve the extraction of phenolics, flavonoids, anthocyanins, and antioxidants: a case-study using red beetroot. *Industrial Crops and Products*, 202, 117005. https://doi.org/10.1016/J.INDCROP.2023.117005

Araujo-León, J. A., Sánchez-del Pino, I., Brito-Argáez, L. G., Peraza-Sánchez, S. R., Ortiz-Andrade, R., & Aguilar-Hernández, V. (2024). In silico mass spectrometric fragmentation and liquid chromatography with tandem mass spectrometry (LC-MS/MS) Betalainic Fingerprinting: identification of Betalains in Red Pitaya. *Molecules*, *29*(22), 5485. https://doi.org/10.3390/MOLECULES29225485/S1

Asgari, S., Pourjavadi, A., Licht, T. R., Boisen, A., & Ajalloueian, F. (2020). Polymeric carriers for enhanced delivery of probiotics. *Advanced Drug Delivery Reviews*, 161–162, 1–21. https://doi.org/10.1016/j.addr.2020.07.014

Baião, D. dos S., da Silva, D. V. T., & Paschoalin, V. M. F. (2020). Beetroot, a remarkable vegetable: its nitrate and phytochemical contents can be adjusted in novel formulations to benefit health and support cardiovascular disease therapies. *Antioxidants*, 9(10), 960. https://doi.org/10.3390/ANTIOX9100960

Bian, C., Ji, L., Xu, W., Dong, S., & Pan, N. (2024). Research progress on bioactive substances of beets and their functions. *Molecules*, 29 (19), 4756. https://doi.org/10.3390/MOLECULES29194756

Bordiga, M., Xu, B., Giovannelli, L., Sartzi, M. I., Drettas, D., Stramarkou, M., & Krokida, M. (2024). A comprehensive review of the latest trends in spray freeze drying and comparative insights with conventional technologies. *Pharmaceutics*, *16*(12), 1533. https://doi.org/10.3390/PHARMACEUTICS16121533

Borjan, D., Šeregelj, V., Andrejč, D. C., Pezo, L., Šaponjac, V. T., Knez, Ž., Vulić, J., & Marevci, M. K. (2022). Green techniques for preparation of red beetroot extracts with enhanced biological potential. *Antioxidants*, 11(5), 805. https://doi.org/10.3390/ANTIOX11050805

Calín-Sánchez, Á., Lipan, L., Cano-Lamadrid, M., Kharaghani, A., Masztalerz, K., Carbonell-Barrachina, Á. A., & Figiel, A. (2020). Comparison of traditional and novel drying techniques and its effect on quality of fruits, vegetables and aromatic herbs. *Foods*, *9*(9), 1261. https://doi.org/10.3390/FOODS9091261

Chaari, M., Akermi, S., Elhadef, K., Ennouri, K., Hlima, H. Ben, Fourati, M., Chakchouk-Mtibaa, A., Sarka, T., Shariati, M. A., Mellouli, L., & Smaoui, S. (2023). From modeling and optimizing extraction of peels beetroot (Beta vulgaris L.) betalains to in silico probing of their antibacterial multitarget mechanisms. *Biomass Conversion and Biorefinery*, 13(16), 14731–14754. https://doi.org/10.1007/S13399-022-03140-6

Chatzimitakos, T., Athanasiadis, V., Kalompatsios, D., Mantiniotou, M., Bozinou, E., & Lalas, S. I. (2023). Pulsed electric field applications for the extraction of bioactive compounds from food waste and by-products: a critical review. *Biomass*, 3(4), 367–401. https://doi.org/10.3390/BIOMASS3040022

Chen, L., Zhu, Y., Hu, Z., Wu, S., & Jin, C. (2021). Beetroot as a functional food with huge health benefits: antioxidant, antitumor, physical function, and chronic metabolomics activity. *Food Science & Nutrition*, 9(11), 6406. https://doi.org/10.1002/FSN3.2577

Da Silva, D. V. T., Baião, D. dos S., Magalhães, A., Almeida, N. F., Conte, C. A., & Paschoalin, V. M. F. (2023). Combining conventional organic solvent extraction, ultrasound-assisted extraction, and chromatographic techniques to obtain pure betanin from beetroot for clinical purposes. *Antioxidants*, *12*(10), 1823. https://doi.org/10.3390/ANTIOX12101823/S1

Devadiga, D., & Ahipa, T. N. (2020). Betanin: a red-violet pigment – chemistry and applications. *Chemistry and Technology of Natural and Synthetic Dyes and Pigments*. https://doi.org/10.5772/INTECHOPEN.88939

Dhiman, A., Suhag, R., Chauhan, D. S., Thakur, D., Chhikara, S., & Prabhakar, P. K. (2021). Status of beetroot processing and processed products: thermal and emerging technologies intervention. *Trends in Food Science & Technology*, 114, 443–458. https://doi.org/10.1016/J.TIFS.2021.05.042

Domínguez, R., Cuenca, E., Maté-Muñoz, J. L., García-Fernández, P., Serra-Paya, N., Estevan, M. C. L., Herreros, P. V., & Garnacho-Castaño, M. V. (2017). Effects of beetroot juice supplementation on cardiorespiratory endurance in athletes. a systematic review. *Nutrients*, 9(1), 43. https://doi.org/10.3390/NU9010043

Domínguez, R., Munekata, P. E. S., Pateiro, M., Maggiolino, A., Bohrer, B., & Lorenzo, J. M. (2020). Red beetroot. A potential source of natural additives for the meat industry. *Applied Sciences*, 10(23), 8340. https://doi.org/10.3390/APP10238340

Domínguez, R., Pateiro, M., Munekata, P. E. S., McClements, D. J., & Lorenzo, J. M. (2021). Encapsulation of bioactive phytochemicals in plant-based matrices and application as additives in meat and meat products. *Molecules*, 26(13), 3984. https://doi.org/10.3390/MOLECULES26133984

Dirim, S. N., Talih, M., & Koç, G. Ç. (2019). Drying characteristics of red beet (Beta vulgaris esculentacruenta) puree in a microwave oven. *Dergipark*, 44(4), 654–671. https://doi.org/10.15237/gida.GD19028

Ebrahimi, P., Mihaylova, D., Marangon, C. M., Grigoletto, L., & Lante, A. (2022). Impact of sample pretreatment and extraction methods on the bioactive compounds of sugar beet (Beta vulgaris L.) leaves. *Molecules*, 27(22), 8110. https://doi.org/10.3390/MOLECULES27228110

Ekrem Parlak, M., Irmak Sahin, O., Neslihan Dundar, A., Türker Saricaoglu, F., Smaoui, S., Goksen, G., Koirala, P., Al-Asmari, F., & Prakash Nirmal, N. (2024). Natural colorant incorporated biopolymers-based pH-sensing films for indicating the food product quality and safety. *Food Chemistry*, *439*. https://doi.org/10.1016/j.foodchem.2023.138160

Escribano, J., Cabanes, J., Jiménez-Atiénzar, M., Ibañez-Tremolada, M., Gómez-Pando, L. R., García-Carmona, F., & Gandía-Herrero, F. (2017). Characterization of betalains, saponins and antioxidant power in differently colored quinoa (Chenopodium quinoa) varieties. *Food Chemistry*, 234, 285–294. https://doi.org/10.1016/J.FOODCHEM.2017.04.187

Eyshi, S., Ghareaghajlou, N., Afshar Mogaddam, M. R., & Ghasempour, Z. (2024). Red beet betalains extraction process: a comprehensive review of methods, applications, and physicochemical properties. *Food Science & Nutrition*, 12(11). https://doi.org/10.1002/FSN3.4458

Fan, L., Fan, W., Mei, Y., Liu, L., Li, L., Wang, Z., & Yang, L. (2022). Mechanochemical assisted extraction as a green approach in preparation of bioactive components extraction from natural products – a review. *Trends in Food Science and Technology*, 129, 98–110. https://doi.org/10.1016/j.tifs.2022.09.009

Flores-Mancha, M. A., Ruíz-Gutiérrez, M. G., Sánchez-Vega, R., Santellano-Estrada, E., & Chávez-Martínez, A. (2020). Characterization of betabel extract (Beta vulgaris) encapsulated with maltodextrin and inulin. *Molecules, 25*(23). https://doi.org/10.3390/molecules25235498

Fox, B., Bellini, G., & Pellegrini, L. (2014). Drying. In: Fermentation and Biochemical Engineering Handbook: Principles, Process Design, and Equipment (Third Edition, pp. 283–305). https://doi.org/10.1016/B978-1-4557-2553-3.00014-3

Giampaoli, O., Sciubba, F., Conta, G., Capuani, G., Tomassini, A., Giorgi, G., Brasili, E., Aureli, W., & Miccheli, A. (2021). Red beetroot's NMR-based metabolomics: phytochemical profile related to development time and production year. *Foods, 10*(8). https://doi.org/10.3390/FOODS10081887

Gokhale, S. V., & Lele, S. S. (2014). Betalain content and antioxidant activity of beta vulgaris: effect of hot air convective drying and storage. *Journal of Food Processing and Preservation*, 38(1), 585–590. https://doi.org/10.1111/JFPP.12006

Harsha Sarma, H., Paul, A., Kakoti, M., Goswami, A., Talukdar, N., Kumar Dutta, S., Hazarika, P., & Borkotoky, B. (2023). The aerial genesis: seed bombing for nurturing earth and ecosystem renaissance – a review. *The Pharma Innovation Journal*, 9, 1719–1723. https://www.researchgate.net/publication/374059483

Hernández-Aguirre, O. A., Muro, C., Hernández-Acosta, E., Alvarado, Y., & Díaz-Nava, M. D. C. (2021). Extraction and stabilization of betalains from beetroot (Beta vulgaris) wastes using deep eutectic solvents. *Molecules*, 26(21), 6342. https://doi.org/10.3390/MOLECULES26216342

Jolvis Pou, K. R. (2021). Applications of high pressure technology in food processing. *International Journal of Food Studies*, 10(1), 248–281. https://doi.org/10.7455/IJFS/10.1.2021.A10

Karnwal, A., Kumar, G., Singh, R., Selvaraj, M., Malik, T., & Al Tawaha, A. R. M. (2025). Natural biopolymers in edible coatings: applications in food preservation. *Food Chemistry: X*, 25, 102171. https://doi.org/10.1016/J.FOCHX.2025.102171

Ko, G. P., Jo, H., Kim, J., Kim, J. S., Boo, K. H., & Kim, C. S. (2024). Enterotype-specific effects of red beetroot (Beta vulgaris L.) powder and betanin on human gut microbiota: a preliminary study based on in vitro fecal fermentation model. *Life, 14*(11), 1391. https://doi.org/10.3390/LIFE14111391/S1

Koubaier, H. B. H., Snoussi, A., Essaidi, I., Chaabouni, M. M., Thonart, P., & Bouzouita, N. (2014). Betalain and phenolic compositions, antioxidant activity of Tunisian red beet (Beta vulgaris L. conditiva) roots and stems extracts. *International Journal of Food Properties*, 17(9), 1934–1945. https://doi.org/10.1080/10942912.2013.772196

Lata, E. (2019, June 19). *(PDF) Thin-layer chromatographic investigation of the red beetroot dyes.* https://www.researchgate.net/publication/333662138_Thin-layer_chromatographic_investigation_of_the_red_beetroot_dyes

Lawag, I. L., Nolden, E. S., Schaper, A. A. M., Lim, L. Y., & Locher, C. (2023). A modified folin-ciocalteu assay for the determination of total phenolics content in honey. *Applied Sciences*, 13(4), 2135. https://doi.org/10.3390/APP13042135

Liu, Y., Sabadash, S., & Duan, Z. (2021). Effect of microwave-assisted drying methods on the physicochemical properties of beetroots. *IOP Conference Series: Earth and Environmental Science*, 792(1). https://doi.org/10.1088/1755-1315/792/1/012012

Łubek-Nguyen, A., Ziemichód, W., & Olech, M. (2022). Application of enzyme-assisted extraction for the recovery of natural bioactive compounds for nutraceutical and pharmaceutical applications. *Applied Sciences*, 12(7), 3232. https://doi.org/10.3390/APP12073232

Mahanti, N. K., Chakraborty, S. K., Sudhakar, A., Verma, D. K., Shankar, S., Thakur, M., Singh, S., Tripathy, S., Gupta, A. K., & Srivastav, P. P. (2021). Refractance WindowTM-Drying vs. other drying methods and effect of different process parameters on quality of foods: a comprehensive review of trends and technological developments. *Future Foods*, 3, 100024. https://doi.org/10.1016/J.FUFO.2021.100024

Mallick, K. K., & Winnett, J. (2014). 3D bioceramic foams for bone tissue engineering. *Bone Substitute Biomaterials*, 118–141. https://doi.org/10.1533/9780857099037.2.118

Martau, G. A., Mihai, M., & Vodnar, D. C. (2019). The use of chitosan, alginate, and pectin in the biomedical and food sector – biocompatibility, bioadhesiveness, and biodegradability. *Polymers, 11*(11), 1837. https://doi.org/10.3390/POLYM11111837

Martins, R., Barbosa, A., Advinha, B., Sales, H., Pontes, R., & Nunes, J. (2023). Green Extraction techniques of bioactive compounds: a state-of-the-art review. *Processes*, 11(8), 2255. https://doi.org/10.3390/PR11082255

Mehta, N., Kumar, P., Verma, A. K., Umaraw, P., Kumar, Y., Malav, O. P., Sazili, A. Q., Domínguez, R., & Lorenzo, J. M. (2022). Microencapsulation as a noble technique for the application of bioactive compounds in the food industry: a comprehensive review. *Applied Sciences*, *12*(3), 1424. https://doi.org/10.3390/APP12031424

Mitrevski, J., Pantelić, N., Dodevska, M. S., Kojić, J. S., Vulić, J. J., Zlatanović, S., Gorjanović, S., Laličić-Petronijević, J., Marjanović, S., & Antić, V. V. (2023). Effect of beetroot powder incorporation on functional properties and shelf life of biscuits. *Foods*, 12(2), 322. https://doi.org/10.3390/FOODS12020322/S1

Montini, L., Crocoll, C., Gleadow, R. M., Motawia, M. S., Janfelt, C., & Bjarnholt, N. (2020). Matrix-assisted laser desorption/ionization-mass spectrometry imaging of metabolites during sorghum germination. *Plant Physiology*, 183(3), 925–942. https://doi.org/10.1104/PP.19.01357

Mukherjee, P. K. (2019). High-performance liquid chromatography for analysis of herbal drugs. In: *Quality Control and Evaluation of Herbal Drugs* (pp. 421–458). https://doi.org/10.1016/B978-0-12-813374-3.00010-7

Neikov, O. D. (2009). Processing of powders and processing equipment. In: *Handbook of Non-Ferrous Metal Powders: Technologies and Applications* (pp. 225–264). https://doi.org/10.1016/B978-1-85617-422-0.00011-2

Nirmal, N. P., Medhe, S., Dahal, M., Koirala, P., Nirmal, S., Al-Asmari, F., & Xu, B. (2024). Betalains protect various body organs through antioxidant and anti-inflammatory pathways. *Food Science and Human Wellness*, 13(3), 1109–1117. https://doi.org/10.26599/FSHW.2022.9250093

Nistor, O. V., Seremet (Ceclu), L., Andronoiu, D. G., Rudi, L., & Botez, E. (2017). Influence of different drying methods on the physicochemical properties of red beetroot (Beta vulgaris L. var. Cylindra). *Food Chemistry*, 236, 59–67. https://doi.org/10.1016/J.FOODCHEM.2017.04.129

Nowak, D., & Jakubczyk, E. (2020). The freeze-drying of foods – the characteristic of the process course and the effect of its parameters on the physical properties of food materials. *Foods*, *9*(10), 1488. https://doi.org/10.3390/FOODS9101488

Paulauskienė, A., Šileikienė, D., Karklelienė, R., Tarasevičienė, Ž., & Česonienė, L. (2023). Quality research of the beetroots (Beta vulgaris L., ssp. vulgaris var. conditiva Alef.) grown in different farming systems applying chemical and holistic research methods. *Sustainability*, 15(9), 7102. https://doi.org/10.3390/SU15097102

Pérez, M., Dominguez-López, I., & Lamuela-Raventós, R. M. (2023). The chemistry behind the folin-ciocalteu method for the estimation of (Poly)phenol content in food: total phenolic intake in a mediterranean dietary pattern. *Journal of Agricultural and Food Chemistry*, 71(46), 17543–17553. https://doi.org/10.1021/ACS.JAFC.3C04022/ASSET/IMAGES/MEDIUM/JF3C04022_0008.GIF

Poornima, K., Priya, B., & Poornima, K. (2016). Cite this article as: K. Poornima, B. Priya, Extraction and encapsulation of beetroot (Beta Vulgaris.L) extract using spray dryer. *Asian Journal of Pharmthiaceutical Technology & Innovation*, 20, 54–57. www.asianpharmtech.com

Postružnik, V., Stajčić, S., Borjan, D., Ćetković, G., Knez, Ž., Knez Marevci, M., & Vulić, J. (2024). Impact of storage conditions on stability of bioactive compounds and bioactivity of beetroot extract and encapsulates. *Processes*, *12*(7), 1345. https://doi.org/10.3390/PR12071345

Priyoatmojo, D., Dewi, R. T., Septama, A. W., Angelina, M., Yuliani, T., & Wahyono, T. (2025). Edible layer-by-layer electrostatic adsorption nano and microcapsules based on plant proteins as natural bioactive delivery systems: a review. *Food Chemistry Advances*, 6, 100861. https://doi.org/10.1016/J.FOCHA.2024.100861

Punia Bangar, S., Sharma, N., Sanwal, N., Lorenzo, J. M., & Sahu, J. K. (2022). Bioactive potential of beetroot (Beta vulgaris). *Food Research International*, 158, 111556. https://doi.org/10.1016/J.FOODRES.2022.111556

Ratti, C. (2001). Hot air and freeze-drying of high-value foods: a review. *Journal of Food Engineering*, 49(4), 311–319. https://doi.org/10.1016/S0260-8774(00)00228-4

Razzak, M. A., Sharif, M. K., Naz, T., Rauf, M. A., Shahid, F., Shahzad, R., Saleem, R., Aslam, T., & Inam, A. (2024). Evaluating the bioactive compounds of beetroot and their pharmacological activities in promoting health. *European Journal of Health Sciences*, 10(1), 13–30. https://doi.org/10.47672/ejhs.1802

Rezagholizade-shirvan, A., Soltani, M., Shokri, S., Radfar, R., Arab, M., & Shamloo, E. (2024). Bioactive compound encapsulation: characteristics, applications in food systems, and implications for human health. *Food Chemistry:* X, 24, 101953. https://doi.org/10.1016/J.FOCHX.2024.101953

Rotich, V., Wangila, P., & Cherutoi, J. (2022). FT-IR analysis of beta vulgaris peels and pomace dye extracts and surface analysis of optimally dyed-mordanted cellulosic fabrics. *Journal of Chemistry*, 2022(1), 2233414. https://doi.org/10.1155/2022/2233414

Sadowska-Bartosz, I., & Bartosz, G. (2021). Biological properties and applications of betalains. *Molecules*, 26(9), 2520. https://doi.org/10.3390/MOLECULES26092520

Safety of the proposed extension of use of beetroot red (E 162) in foods for special medical purposes in young children. (2016). *EFSA Journal*, *14*(5). https://doi.org/10.2903/j.efsa.2016.4487

Sawicki, T., Juśkiewicz, J., & Wiczkowski, W. (2017). Using the SPE and micro-HPLC-MS/MS method for the analysis of betalains in rat plasma after red beet administration. *Molecules*, *22*(12). https://doi.org/10.3390/MOLECULES22122137

Shofinita, D., Fawwaz, M., & Achmadi, A. B. (2023). Betalain extracts: drying techniques, encapsulation, and application in food industry. *Food Frontiers*, 4(2), 576–623. https://doi.org/10.1002/FFT2.227

Shraim, A. M., Ahmed, T. A., Rahman, M. M., & Hijji, Y. M. (2021). Determination of total flavonoid content by aluminum chloride assay: a critical evaluation. *LWT*, 150, 111932. https://doi.org/10.1016/J.LWT.2021.111932

Stoica, F., Râpeanu, G., Rațu, R. N., Stănciuc, N., Croitoru, C., Țopa, D., & Jităreanu, G. (2025). Red beetroot and its by-products: a comprehensive review of phytochemicals, extraction methods, health benefits, and applications. *Agriculture*, 15(3), 270. https://doi.org/10.3390/AGRICULTURE15030270

Sun, L., Xin, F., & Alper, H. S. (2021). Bio-synthesis of food additives and colorants-a growing trend in future food. *Biotechnology Advances*, *47*. https://doi.org/10.1016/j.biotechadv.2020.107694

Tagnamas, Z., Idlimam, A., & Lamharrar, A. (2023). Predictive models of beetroot solar drying process through machine learning algorithms. *Renewable Energy*, *219*. https://doi.org/10.1016/J.RENENE.2023.119522

Theba, T., Nayi, P., & Ravani, A. (2024). Beetroot-based blended juice: process development, physico-chemical analysis and optimization of novel health drink. *Food Chemistry Advances*, 4, 100607. https://doi.org/10.1016/J.FOCHA.2024.100607

Thiruvengadam, M., Chung, I. M., Samynathan, R., Chandar, S. R. H., Venkidasamy, B., Sarkar, T., Rebezov, M., Gorelik, O., Shariati, M. A., & Simal-Gandara, J. (2024). A comprehensive review of beetroot (Beta vulgaris L.) bioactive components in the food and pharmaceutical industries. *Critical Reviews in Food Science and Nutrition*, 64(3), 708–739. https://doi.org/10.1080/10408398.2022.2108367

Trindade, L. R. da, Baião, D. dos S., da Silva, D. V. T., Almeida, C. C., Pauli, F. P., Ferreira, V. F., Conte-Junior, C. A., & Paschoalin, V. M. F. (2023). Microencapsulated and ready-to-eat beetroot soup: a stable and attractive formulation enriched in nitrate, betalains and minerals. *Foods*, *12*(7), 1497. https://doi.org/10.3390/FOODS12071497/S1

Trych, U., Buniowska-Olejnik, M., & Marszałek, K. (2022). Bioaccessibility of betalains in beetroot (Beta vulgaris L.) juice under different high-pressure techniques. *Molecules*, 27(20), 7093. https://doi.org/10.3390/MOLECULES27207093/S1

Tumbas Š aponjac, V. T., Čanadanović-Brunet, J. M., Ćetković, G. S., Jakišić, M. V., Vulić, J. J., Stajčić, S. S., & Šeregelj, V. N. (2020). Optimisation of beetroot juice encapsulation by freeze-drying. *Polish Journal of Food and Nutrition Sciences*, 70(1), 25–34. https://doi.org/10.31883/PJFNS/115153

Usman, M., Nakagawa, M., & Cheng, S. (2023). Emerging trends in green extraction techniques for bioactive natural products. *Processes*, 11(12), 3444. https://doi.org/10.3390/PR11123444

Vaessen, E. M. J., Timmermans, R. A. H., Tempelaars, M. H., Schutyser, M. A. I., & den Besten, H. M. W. (2019). Reversibility of membrane permeabilization upon pulsed electric field treatment in Lactobacillus plantarum WCFS1. *Scientific Reports*, 9(1), 1–11. https://doi.org/10.1038/s41598-019-56299-w

Vulić, J., Čanadanović-Brunet, J., Ćetković, G., Tumbas, V., Djilas, S., Četojević-Simin, D., & Čanadanović, V. (2012). Antioxidant and cell growth activities of beet root pomace extracts. *Journal of Functional Foods*, 4(3), 670–678. https://doi.org/10.1016/J.JFF.2012.04.008

Yao, Y., Pang, Y. X., Manickam, S., Lester, E., Wu, T., & Pang, C. H. (2022). A review study on recent advances in solar drying: mechanisms, challenges and perspectives. *Solar Energy Materials and Solar Cells*, 248. https://doi.org/10.1016/j.solmat.2022.111979

Yildirim, N., & Umdu, E. S. (2022). Geothermal energy in developing countries–The dilemma between renewable and nonrenewable. In: *Renewable Energy and Sustainability: Prospects in the Developing Economies* (pp. 227–262). https://doi.org/10.1016/B978-0-323-88668-0.00003-6

Yousefi, S., Kavyanirad, M., Aminifar, M., Weisany, W., & Mousavi Khaneghah, A. (2022). Yogurt fortification by microencapsulation of beetroot extract (Beta vulgaris L.) using maltodextrin, gum arabic, and whey protein isolate. *Food Science & Nutrition*, 10(6), 1875. https://doi.org/10.1002/FSN3.2804

Zin, M. M., Majrxi, E., & Banvolgyi, S. (2020). Conventional extraction of betalain compounds from beetroot peels with aqueous ethanol solvent. *Acta Alimentaria*, 49(2), 163–169. https://doi.org/10.1556/066.2020.49.2.5

7

Pharmacological Potential of Beetroot and Beetroot Juice

Abdul Mateen, Swasti Mudgal, and Aishwary Dubey

7.1 Introduction

Beetroot (*Beta vulgaris*) and beetroot juice have gained increasing attention due to their promising pharmacological properties, which are attributed to their complex profile of bioactive compounds such as Nitrate (NO_3^-), betalains, polyphenols, and various vitamins and minerals (Clifford et al., 2015; Lidder and Webb, 2013). One of the key pharmacological mechanisms of beetroot is its ability to improve bioavailability and reduce oxidative stress, both of which play critical roles in the pathophysiology of numerous chronic diseases. Many studies have highlighted the positive effects of beetroot on cardiovascular health, inflammation, oxidative damage, and cancer prevention (Clifford et al., 2015; Wootton-Beard et al., 2011). Dietary NO_3^- found in beetroot is converted into nitric oxide (NO) in the body, which enhances vasodilation and improves endothelial function, ultimately contributing to better cardiovascular health by lowering blood pressure and enhancing blood flow (Jones, 2014; Kapil et al., 2015). In addition to their vascular benefits, beetroot's antioxidant properties, primarily attributed to betalains, help neutralize reactive oxygen species (ROS), providing a defense against oxidative damage. This antioxidant action can potentially lower the risk of developing cardiovascular diseases (CVDs), neurodegenerative disorders, and certain cancers (Georgiev et al., 2010). Beyond cardiovascular and antioxidant effects, beetroot has been studied for its ergogenic potential, particularly in enhancing athletic performance. Research indicates that beetroot juice supplementation can improve endurance and reduce the oxygen cost of exercise, which is largely due to increased NO bioavailability that optimizes muscle oxygenation and reduces fatigue during high-intensity aerobic activities (Bailey et al., 2009). Furthermore, the anti-inflammatory and immunomodulatory effects of beetroot compounds make it relevant in the management of various inflammatory diseases, metabolic syndrome, and gastrointestinal disorders (Milton-Laskibar et al., 2021). The mechanisms behind these effects involve the modulation of inflammatory pathways, the regulation of oxidative stress markers, and the enhancement of mitochondrial efficiency, which are critical for maintaining metabolic homeostasis and preventing chronic disease (Pérez-Gregorio et al., 2011). Recent innovations in beetroot supplementation focus on improving the bioavailability and stability of its bioactive compounds through advanced technologies like nanoencapsulation, which

DOI: 10.1201/9781003527084-7

ensures that the health benefits are maximized in the human body (Thompson et al., 2015). Additionally, the development of beetroot-based functional foods and nutraceuticals, such as juices, powders, and energy gels, is expanding their therapeutic applications in both clinical settings and sports nutrition. In conclusion, beetroot and its derivatives offer a multifaceted pharmacological profile, with demonstrated benefits in cardiovascular health, exercise performance, and antioxidant protection. As research progresses, new formulations and delivery systems will further enhance the therapeutic efficacy of beetroot, making it a valuable component in both preventive health and disease management. Continued exploration of its molecular mechanisms, along with robust clinical trials, will help solidify beetroot's role in evidence-based nutrition and wellness strategies. The future of beetroot as a functional food and nutraceutical remains promising, with the potential for widespread use in both medical and sports contexts.

7.2 Bioavailability of Beetroot Compounds

Food elements must be bio-accessible in order to be considered nutritious. Bio-accessibility is the matrix of an active ingredient that is sufficiently provided from the food compounds in the gastrointestinal tract to be absorbed by the intestinal tract and used by the body's cells (Wootton-Beard and Ryan, 2011). The main compounds of beetroot are betalains and NO_3^-, each of which has unique absorption properties. However, solubility, interactions with other dietary components, molecular changes, various cellular transporters, metabolism, and the gut microbiota can all affect how well these compounds are absorbed, which can alter their bioavailability (Rein et al., 2013; Ting et al., 2014). The scientific research has examined the bioavailability of inorganic NO_3^- and betalains, the primary bioactive ingredients of beetroot. Inorganic dietary NO_3^- has a well-established bioavailability, with reports of about 100% absorption after digestion (Van Velzen et al., 2008). The average amount of inorganic NO_3^- in beet juice is 5–8 mmol. A complicated cycle transforms from NO_3^- to NO. In short, the stomach and small intestine absorb NO_3^- quickly and fully (Florin et al., 1990). After consuming raw beetroot and beetroot juice, Joris and Mensink (2013) investigated the plasma NO activity. When beetroot juice and raw beetroot were consumed, plasma NO activity was considerably higher than when a placebo was used. A prior study that found a sharp increase in NO activity 1–3 hours after consuming beetroot juice supports our results. These observations are significant when taken as a whole since a rise in the endogenous NO pool is linked to a number of physiological consequences that may be advantageous for health (Joris and Mensink, 2013). On the other hand, betacyanin bioavailability was examined by Frank et al. (2005). Six healthy volunteers received 500 mL of beetroot juice, and for 24 hours, they found betacyanins in their urine at levels that were around 0.3% of the amount they had consumed. The findings suggest limited bioavailability; however, betacyanins are not easily excreted through the renal system (Frank et al., 2005). The urine excretion alone to assess bioavailability is debated, as it ignores biliary and circulatory clearance, potentially underestimating true bioavailability (Rein et al., 2013). Furthermore, it is yet unclear how much betalains are metabolized and chemically changed into secondary metabolites, although this should be taken into account when assessing their absorption into the body (Frank et al., 2005). Wiczkowski et al. (2018) examined the bioavailability of betacyanins in persons who consumed various red beetroot preparations. Twelve volunteers willingly consumed beetroot juice and crunchy beetroot slices that contained betanin and isobetanin. They identified and analyzed betalain using high-performance liquid chromatography. Betacyanins and associated aglycones were found in the urine samples of those who drank beetroot juice and crunchy slices, according to the analysis. The urine sample of those who drank beetroot juice showed the highest excretion rate of betacyanins, 64 nmol/hour, after the first two hours. After two to four hours, the excretion rate (66 nmol/hour) for those consuming beetroot slices was noted. After consuming both products, they found that the overall excretion rate of betacyanine varied between 0.12% and 0.58% and was about 0.3%. Similar findings have been published by Kanner et al. (2001), who found that within 12 hours of eating 300 mL of beetroot juice, volunteers' urine contained 0.5% to 0.9% of the consumed betacyanins. This suggests that people may effectively absorb betacyanins, however, in low quantities. Tesoriere et al. (2013) researched the bioavailability of the betalains using a different method. Tesoriere and other researchers used Caco-2 cell monolayers to simulate an in vitro membrane system of the human gastrointestinal epithelium. Using this method, scientists were able to ascertain the bioavailability of betalains and explore their potential for absorption via a functioning intestinal barrier. Betanin and indicaxanthin, two betalains, were shown to be readily absorbed by the Caco-2 cell monolayer simulator of the intestinal lining, primarily in their unmetabolized form via paracellular transport. The final observation is especially significant since it shows that betalains may be assimilated into the bloodstream unaltered, maintaining their significant biological activity and molecular composition (Ting et al., 2014). Betanin may also be absorbed through transcellular transport, according to some studies. The concentration of nutrients reaching the bloodstream is influenced by factors like metabolism, gut microflora interactions, and protease degradation. Thus, in vitro studies, even those simulating the human gastrointestinal tract, may not fully reflect in *vivo* conditions (Rein et al., 2013; Ting et al., 2014). Beetroot contains antioxidants such as epicatechin, rutin, and caffeic acid (Georgiev et al., 2010), which are absorbed to varying extents (Manach et al., 2005). While total phenolic bioavailability has been studied, individual phenolic absorption remains unclear. Netzel et al. (2005) found that a 500 mL dose of beetroot juice increased urinary phenolic excretion by 97%, suggesting effective absorption and antioxidant potential. However, individual component bioavailability is yet to be determined. Further research is needed to understand metabolic pathways influencing beetroot's bioactive compounds. When combined, these findings would improve our knowledge of the botanical compounds bioavailability of beetroot and clarify its possibilities as an individual health-promoting assistance.

7.3 Oxidative Stress

Oxidative stress arises from an imbalance between the creation of ROS and the body's antioxidant defenses, contributing to chronic diseases like CVDs, neurological disorders, diabetes, and cancer. Beetroot's betalains and NO_3^- help counteract oxidative stress (Kanner et al., 2001). Betalains, particularly betanin, act as potent antioxidants by scavenging free radicals, inhibiting lipid peroxidation, and protecting cell membranes (Kanner et al., 2001; Reddy et al., 2005; Tesoriere et al., 2008). They also prevent LDL oxidation, reducing atherosclerosis risk. Two betalain metabolites, betanin and betanidin, were demonstrated in a study by Kanner et al. (2001) to lessen lipid membrane oxidation caused by H^2O^2-activated metmyoglobin and unbound iron (AA-Fe) as well as linoleate deterioration caused by cytochrome C oxidase. The most potent inhibitor of lipid peroxidation was betanin, the most prevalent (300–600 mg/kg) betalain in beetroot. According to Kanner et al. (2001), betanin's remarkable electron-donating ability and capacity to neutralize highly reactive radicals that target cell membranes seemed to be the source of its strong antioxidant activity. For instance, betalains have been shown to inhibit the oxidation of low-density lipoprotein (LDL) cholesterol, which is essential for preventing atherosclerosis. In vitro studies confirm the capability of betalains to neutralize a variety of free radicals (Wootton-Beard et al., 2011). Additionally, beetroot enhances antioxidant enzyme activity, including SOD, catalase, and glutathione peroxidase (Clifford et al., 2015). This enhancement contributes to reduced oxidative stress markers and tissue protection. For instance, a study by Clifford et al. (2015) found that beetroot juice consumption significantly decreased oxidative stress markers in hypertensive patients while improving overall vascular function. Nonetheless, as previously mentioned, betalains are not the only antioxidants present in beetroot. Caffeic acid, Rutin, and epicatechin are among the many relatively high bioactive phenolics found in beetroot that are also well-known for being potent antioxidants (Georgiev et al., 2010; Frank et al. 2005; Manach et al., 2005). Beetroot is a rich source of NO_3^-; also, NO_3^- converted into NO and played a pivotal role in maintaining vascular health and regulating oxidative stress. NO has been linked to increased expression of various antioxidant enzymes, thereby fostering a protective mechanism against oxidative stress (Vanhatalo et al., 2010; Lundberg et al., 2011; Tesoriere et al., 2008). NO influences mitochondrial activity, reducing ROS production by inhibiting enzymes within the electron transport chain. This action not only decreases oxidative damage but also enhances cellular energy efficiency (Bailey et al., 2009). Intense physical activity often leads to elevated ROS production. While moderate levels of ROS are necessary for muscle adaptation, excessive oxidative stress can result in muscle fatigue and delayed recovery. Beetroot juice supplementation has been shown to alleviate these effects. Research suggests that beetroot juice can reduce biomarkers of oxidative stress, such as malondialdehyde (MDA) and protein carbonyls, in athletes. Furthermore, the increased availability of NO improves blood flow and oxygen delivery, enhancing exercise performance and reducing muscle soreness (Clifford et al., 2015). Beetroot's preventive properties against exercise-induced oxidative damage have implications for both amateur and elite athletes, as well as individuals with cardiovascular or respiratory conditions (Gilchrist et al., 2011).

7.4 Health Benefits of Beetroot

7.4.1 Cardiovascular Health

Since insulin resistance, blood pressure, and high cholesterol levels are the main metabolic conditions that constitute risk factors for CVD, beetroot's cardioprotective action involves having a protective impact on human blood pressure, glucose levels, and lipid profiles (Volino-Souza et al., 2020). The impact of NO_3^- and other phytochemicals in several beetroot formulations to promote heart health has also been studied by Baiao et al. (2017). In the acidic conditions of the stomach, beetroot's high NO_3^- content naturally converts to the beneficial compound NO. By entering the circulation and forcing the endothelium to relax and dilate, this NO can control excessive blood pressure (Bahrami et al., 2021). According to this belief, Jakubcik et al. (2021) made an effort to comprehend the dose-related impact of NO_3^- and nitrite (NO_2^-) derived from beetroot on the cardiovascular assessments and plasma levels. The scientists noticed that consuming NO_3^- significantly raised plasma NO_3^- and NO_2^- levels, while lowering mean arterial pressure (MAP). Consumption of NO_3^- rich beetroot juice on day 1 resulted in a comparable positive chnages in MAP and blood pressure, but there was no apparent improvement until day 28 (Stanaway et al., 2019a). In another study, authors reported that older persons who had 150 mL of beetroot juice acutely observed a higher rise NO_3^- level in plasma and an improvement in diastolic blood pressure than younger ones (Stanaway et al., 2019b) (Figure 7.1).

A study found that 97% of pregnant women considered dietary NO_3^- supplementation via beetroot juice a suitable intervention (Ormesher et al., 2018). Researchers highlighted oral bacterial NO_3^- reductase as key in converting NO_3^- to nitrite. However, Tropea et al. (2020) reported that beetroot-derived NO_3^- did not significantly lower blood pressure or improve vascular function in eNOS knockout mice. Dyslipidemia, linked to CVD, involves high cholesterol, triglycerides, and LDL-C, along with low HDL-C. Sawicki et al. (2018)

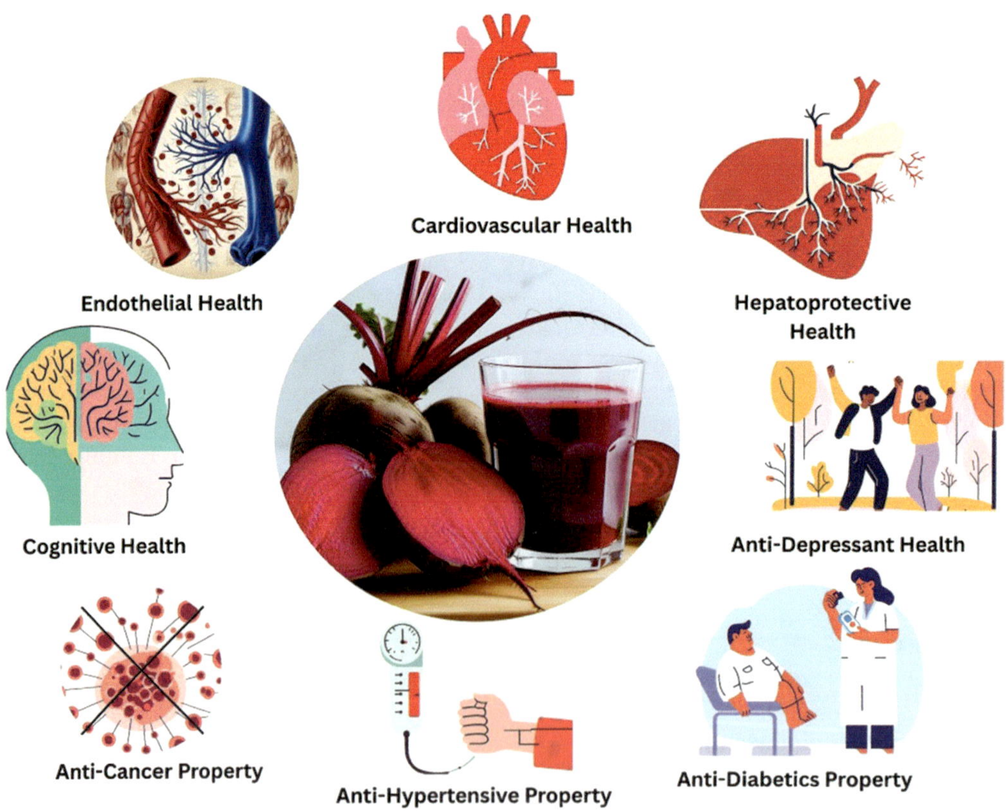

FIGURE 7.1
The summary points of the health benefits of beetroot and its juice.

noted that regular beetroot intake leads to stable betalain presence in bodily fluids and may enhance lipid profiles and endothelial function. Similarly, Rahimi et al. (2019) found that coronary artery disease patients consuming beet supplements with 50 mg of betalains (16.2 mg betanin) for two weeks had minimal betanin in plasma (<1 µg) but 120 µg in urine. The study suggested betalains may help mitigate atherogenic risk factors like hyperglycemia, hypertension, dyslipidemia, and hyperhomocysteinemia. Although many studies highlight beetroot's anti-oxidant, anti-inflammatory, anticancer, hepatoprotective, and cardioprotective properties, along with its benefits, remain largely limited to natural coloring and functional beverages.

7.4.2 Endothelial Health

Studies highlight beetroot's role in preserving endothelial function, counteracting stress, sedentary lifestyles, and high-calorie diets (Volino-Souza et al., 2020). As a rich source of NO_3^-, beetroot converts to nitrite and subsequently NO (Hobbs et al., 2013a; Hobbs, 2012). Nitrite reduction to NO is catalyzed by proteins like deoxymyoglobin, xanthine oxidoreductase, and antioxidants such as vitamin C (Hobbs et al., 2013a; Kim-Shapiro and Gladwin, 2014). NO is essential for endothelial health, regulating vascular homeostasis, thrombotic activity, platelet function, and vascular tone through vasodilators (NO, prostacyclin) and vasoconstrictors (endothelin-1, thromboxane) (Davignon and Ganz, 2004; Hobbs et al., 2012). Reduced NO availability, especially with aging, is a key factor in endothelial dysfunction, a major risk for cardiovascular diseases like hypertension and atherosclerosis (Lidder and Webb, 2013; Joris and Mensink, 2013). Beetroot, a natural NO donor, has been explored as a dietary approach to support endothelial function, with Webb et al. (2008) investigating its effects in healthy individuals. The flow-mediated dilatation (FMD) technique was used to assess the dilatation of the brachial artery (BA) before and after a 20-minute ischemia treatment in order to assess the endothelial function of the BA. The 60% reduction from before to after BAFMD response demonstrated that the ischemia technique was successful in causing endothelial dysfunction. The BAFMD response remained at pre-ischemic levels when individuals consumed a single 500 mL dose of beetroot juice (23 mmol NO_3^-) two hours prior, suggesting it preserved endothelial function. Hobbs et al. (2013b) expanded on this by examining the acute effects of beetroot-enriched bread (100 g beetroot, 1.1 mmol NO_3^-) on arterial stiffness and microvascular function in healthy young men. While it did not affect arterial stiffness, measured by pulse wave velocity and elevation index, it improved microvascular vasodilation, as shown by laser Doppler imaging (LDI). Six hours after consuming the beetroot-enriched bread, endothelium-independent vasodilation (perfusion units) was approximately 343% higher than that of the control bread. Crucially, this work demonstrated that substantial improvements in intravascular function may be augmented by even a modest NO_3^- load (1.1 mmol). Research on elderly people also revealed similar vascular effects. In order to figure out if adding beetroot juice will reduce postprandial deficits in BAFMD, Joris and Mensink (2013) used older volunteers (~61 years old) who appeared healthy but were slightly obese. In a randomized crossover design, the BAFMD response decreased by approximately 1.6% in the control condition, while only approximately 0.4% was impaired following beetroot juice (140 mL, NO_3^-; 500 mg). This suggests that beetroot has a beneficial effect on postprandial endothelial function. It has been more challenging to replicate the previously reported results in volunteers who are at a higher risk of endothelial dysfunction. Kenjale et al. (2011) reported that a 500 mL dose of beetroot juice did not affect endothelial function in PAD patients, as measured by BAFMD response (% arterial dilatation) to a hyperemic stimulus. In another study, Gilchrist et al. (2013) provided type 2 diabetic subjects with either 250 mL of beetroot juice daily (7.5 mmol NO_3^-) or a NO_3^- depleted placebo, nutritionally identical except for nitrate content, for 14 days. Following the treatment period, there were no changes in microvascular vasodilation as determined by LDI between the control and beetroot juice groups, and the BAFMD response was comparable between the two groups (4.94% vs. 4.97% change in vessel diameter, respectively). However, considering that the majority of participants were already taking vasoactive drugs for their diseases, it may not be unexpected that these trials showed little impact compared to those with healthy groups. It is possible that these drugs reduced or confused whatever possible vascular response that NO_3^- provided. Pre-clinical patients, or those who do not yet require prescription drugs that impair vascular function, could respond more profoundly to beetroot supplementation. A further interpretation for the study's results is that a dilatory response may have been mediated by bioactive components (such as betalains and caffeic acid) other than NO_3^-, which are probably present in both the NO_3^--rich beetroot juice and the NO_3^--depleted placebo. In fact, there is proof that flavonoids and other antioxidants have dilatatory effects on people (Frombaum et al., 2012). If betalains or other beetroot antioxidants had similar effects, no difference in BAFMD or microvascular function would be expected between the two conditions. Nevertheless, this remains hypothetical, as their dilatory effects and impact on endothelial function have not been studied. Additionally, conditions like type 2 diabetes are marked by chronic inflammation and excessive RONS (Donath and Shoelson, 2011). Because RONS like superoxide (O_2^-) immediately react with NO, reducing its bioavailability, this might restrict the effectiveness of NO_3^- intake (Gilchrist et al., 2013). Therefore, in the context of oxidative stress, any advantages of the excess NO produced by consuming dietary NO_3^- may be diminished. High dosages of beetroot juice can help these diseases combat this by increasing the quantity of NO_3^- and NO production that is consumed as well as by strengthening the antioxidant defense (i.e., betalains) against RONS. This strategy might improve the bioavailability of NO through the NO_3^-. More investigation is needed to justify the role of beetroot and its product intake in endothelial dysfunction.

7.4.3 Hepatoprotective Health

Beetroot intake is now generally recognized as a modern dietary strategy for providing protective benefits toward hepatic inflammation and oxidative stress (Sinaga et al., 2020) and as a potential treatment for liver damage (Butt et al., 2019). Beetroot extracts offer significant liver protection by alleviating oxidative stress, inflammation, fat accumulation, and lipid imbalances in diabetic animal models (Al-Harbi et al., 2021). Studies suggest that preventing NAFLD may involve boosting endogenous antioxidants, along with the presence of anti-apoptotic Bcl2 and PPARα. Regular beetroot intake has been associated with improved lipid metabolism and a slower increase in liver enzyme levels in individuals with NAFLD (Srivastava et al., 2019; Afzali et al., 2020). Comparable protective effects were noted with beetroot stem and leaf extract (Lorizola et al., 2018) and ultrafiltered beetroot juice in high-fat diet-induced obesity (Babarykin et al., 2019). A potential hepatic protective benefit was seen in individuals with NAFLD after taking organic beetroot powder supplements for three months (Srivastava et al., 2019). Administration of beetroot leaf extract to thioacetamide-induced hepatotoxic rats improved liver enzyme levels, highlighting the enhanced hepatoprotective potential of the study's self-nano-emulsifying drug delivery system (SNEDDS). The SNEDDS formulation also contributed to reductions in inflammatory markers (IL-6, TNF-α) and oxidative stress indicators (MDA, SOD, and glutathione) (Kassem et al., 2020). The antioxidants found in beetroots may be connected to this hepatoprotective action. The impact of beetroot extracts in reducing hepatic steatosis and damage to the liver caused by chemically stimulated liver damage in a laboratory animal model is another widely utilized paradigm to investigate the hepatoprotective benefits of beetroot. Dietary beetroot supplements support liver health by reducing hepatic stress, enhancing antioxidant status, and lowering inflammation and lipid peroxidation in rats with CCl_4^- and chlorpyrifos-induced hepatotoxicity (Butt et al., 2019; Olumese and Oboh, 2020). Additionally, beetroot extracts exhibit hepatoprotective effects in rats exposed to chronic nanotoxicity from silver nanoparticles (Albrahim and Alonazi, 2020). The scientists demonstrated that aqueous beetroot extract protected against lipid peroxidation, glutathione depletion, and decreased catalase and superoxide dismutase activity.

7.4.4 Anti-Depressant and Cognitive Health

Reduced cerebral blood flow is one of the major pathogenic processes that accompanies the decline of cognitive performance with aging (Bondonno et al., 2014; Bond et al., 2013). Numerous neurological disorders associated with diminished cognitive function, such as brain damage, clinical dementia, and Alzheimer's disease, have been correlated with an age-related drop in cerebral perfusion (De la Torre and Stefano, 2000). A disturbance in neurovascular function, which is partially caused by decreased NO activity, is one of the main causes and risk factors for the beginning and progression of cerebral hypo-perfusion (De la Torre and Stefano, 2000). Supplements derived from plants can help with mental disorders and cognitive abilities (Lundberg et al., 2011). Consuming raw beetroot has been shown to enhance cognitive performance in type II diabetics, as seen by improvements in Toulouse-Pieron and Digit learning test scores (an increase of 3.09 and 6.45 points from baseline) (Kim-Shapiro and Gladwin 2014). Reduced NO production can disrupt regular brain metabolism of energy (i.e., glucose supply) and neuronal activity (i.e., cellular communication), potentially leading to neurodegeneration and cognitive impairments over time (Bondonno et al., 2014; De la Torre, and Stefano, 2000). It is also possible that a NO producer, such as beetroot, might enhance the flow of blood to the brain and counteract impairments in cognitive performance. Beetroot supplementation for three days has not changed the cognitive abilities of elderly people (60–70 years old) (Lundberg et al., 2011). Anxiety and depression are two psychological disorders that can be triggered by stress. Significant anxiolytic and antidepressant effects were seen in stressed mice pretreated with doses of 100 and 200 mg beetroot leaf ethanolic extract/kg (Sulakhiya et al., 2016). The effect of dietary NO_3^- on cerebral blood flow was investigated in two recent human investigations. After giving a group of older persons (~75 years) either a NO_3^--depleted meal (~0.09 mmol) or a high NO_3^- diet (~12 mmol) that included beetroot juice for 24 hours, Presley et al. (2011) assessed cerebral perfusion. According to magnetic resonance imaging (MRI), a meal rich in NO_3^-s caused a significant and preferential rise in frontal cortical perfusion, a part of the brain involved in critical cognitive functions including executive function, working memory, and task-switching, was triggered by the high NO_3^- diet, according to MRI. Red beets have demonstrated therapeutic promise in the treatment of mental illnesses associated with stress and anxiety. Sulakhiya et al. (2016), investigated how beetroot leaf extract affected mice's anxiety, stress, and sadness. Anxiolytic, antidepressant, and excellent antioxidant properties were demonstrated in stressed mice. When combined with uridine (derived from sugar beet), omega-3 can help prevent depression by altering mood and lowering intense emotions, which causes the body to relax (Frank et al., 2005). A high NO_3^- diet (about 12 mmol) that includes beetroot juice has been shown in recent human studies to enhance perfusion in the frontal cortex, an area involved in cognitive functions. Although long-term clinical trials have not yet been carried out, these investigations demonstrated the impact of beetroot supplementation in age-related cognitive illnesses (Clifford et al., 2015). Another research examined the effects of a beetroot juice supplement (140 mL/day of NO_3^-: 9.6 mmol) on cognitive performance over the course of three days in healthy older individuals (~63 years). When compared to the control group, there were no discernible differences in cognitive ability (Kelly et al., 2013). As a result, cognitive function may not be demonstrably enhanced by an acute beetroot meal.

7.4.5 Anti-Hypertensive Property

About one billion people worldwide suffer from hypertension, which is the main cause of cardiovascular issues. A diet rich in dietary fiber and NO_3^- help in lowering blood pressure and oxidize low-density lipoprotein. Beetroot naturally contains a lot of inorganic NO_3^-, and it can be an excellent supplement to help prevent hypertension (Tawa et al., 2022). Asgary et al. (2016) investigated the experiment to find out how beetroot juice and cooked beetroot affected individuals with high blood pressure. In this, 24 hypertensive individuals between the ages of 25 and 68 have been divided into two groups. One group took cooked beetroot, whereas the other group drank juice. Consumption of 250 g of cooked beetroot or 250 mL of beetroot juice daily for 14 days led to reductions in systolic and diastolic blood pressure to 127 and 78.12 mmHg, respectively. A double-blind, placebo-controlled crossover study administering 150 mL of beetroot juice to different age groups found that older participants experienced a significant decrease in diastolic blood pressure, while younger participants showed a reduction in systolic pressure. These findings suggest that beetroot's antihypertensive effects may be more pronounced in older adults (Stanaway et al., 2019a, 2019b). Beetroot gel made without any chemicals was utilized in a study by Da Silva et al. (2016) to examine its positive effects on plasma NO and blood pressure. Five participants (four men and one woman) consumed 100 g of beetroot gel to test the sample. These participants' NO production was assessed in relation to the effects of acute inorganic NO_3^- supplementation. According to the results, beetroot gel intake decreased the coronary heart rate by 7 beats per minute, decreased systolic blood pressure by 6.2 mm Hg and diastolic blood pressure by 5.2 mm Hg, and enhanced plasma nitrite by three times after 60 minutes of ingestion. In addition to drinking juice, beetroot administration in the form of gel has shown encouraging results in terms of decreasing blood pressure. Silva et al. (2016) found that healthy individuals' systolic and diastolic blood pressure dropped by 6.2 and 5.2 mmHg, respectively, after just 60 minutes of gel intake. Since hypotensive pathways may entail diuretic action, natural supplements may have both diuretic and antihypertensive effects simultaneously (Prando et al., 2016). In mouse models, beetroot has demonstrated notable diuretic action and release of salt at dosages of 0.5 and 1 g/kg. A dosage of 0.5 g/kg had a diuretic effect that was equivalent to that of furosemide (Coles and Clifton, 2012). Both systolic and diastolic blood pressure have been shown positive effect when beetroot is consumed as a juice supplement or in bread products (Bailey et al., 2009; Hobbs et al., 2013a, 2013b; Hobbs et al., 2012; Jajja et al., 2014; Webb et al., 2008)

7.4.6 Anti-Diabetes Property

Type 2 diabetes mellitus, the most prevalent form of diabetes, arises from insulin resistance and insufficient insulin production, causing organ damage and obesity (Kumar et al., 2014). Kumar et al. (2014) compared beetroot extract to the antidiabetic drug glibenclamide in diabetic rats, finding that beetroot ethanolic extract significantly lowered blood sugar, serum lipids, and urinary sugar with ketone bodies, indicating potential antidiabetic properties (Kumar et al., 2016). Furthermore, red beet juice, high in neobetanin (1.3 g/225 mL), notably reduced postprandial insulin (0–60 minutes) and glucose response (0–30 minutes) in humans (Wootton-Beard et al., 2014).

7.4.7 Anti-Cancer Properties

Beetroot is a nutrient-dense plant that is used as an alternative medicine for biologically based therapies for cancer prevention and treatment (Sárváry and Sárváry, 2019). The advantages of betalains have been the subject of several reports. Two groups of cancer patients consumed beetroot in a trial conducted by Kapadia et al. (2011). It was discovered that the beetroot extract had a cytotoxic effect in human breast cancer cells that were positive for the estrogen receptor and androgen-independent human PC3. Furthermore, when red beetroot extract and an anticancer medication (doxorubicin) were combined, a synergistic antiproliferative effect was seen in breast and prostate cancer cell lines (Kapadia et al., 2013). According to a case study, an 80-year-old man's small cell lung cancer naturally regressed after two years of frequent beetroot juice consumption without medication or treatment. Despite the positive impact of beetroot diet and surgical stress on tumor regression, the exact mechanism remains unidentified (Song et al., 2022). Blázovics and Sárdi (2018) propose that the cytotoxic properties of beetroot arise from antioxidant and scavenging compounds (such as betanins, betacyanins, betaxanthins, vulgaxanthine, polyphenols, and ascorbic acid) and mineral constituents (including Al, B, Ba, Ca, Cu, Fe, K, Mg, Mn, Na, and Zn), which affect the redox equilibrium and signal transduction mechanisms. According to research, beetroot's cytotoxicity and anti-cancer properties mostly arise from its capacity to suppress cell division, encourage apoptosis, and stimulate autophagy. Nowacki et al. (2015) found that betanin-enriched beetroot extract caused breast cancer cells to undergo the cell death process and anticancer properties. Oral administration of water containing 78 mg/mL beetroot extract prevented N-nitroso methyl benzylamine-induced esophageal cancer in mice. Additionally, beet leaf extract promoted apoptosis while reducing inflammation and angiogenesis in murine tumor cells (Lechner et al., 2010). Several studies have explored beetroot's anti-cancer properties (Fu et al., 2020; Tan and Hamid, 2021; Chen et al., 2021). Studies indicate that beetroot and its leaves impede the division, migration, and colony formation of prostate cancer cells, with apigenin identified as the primary active compound in beet leaves (Mancini et al., 2021). Apigenin inhibits cancer progression through protein kinase B in hepatic cells (Yang et al., 2018) and increases cisplatin cytotoxicity in multiple cancer cell lines via p53-mediated apoptosis (Liu et al., 2017). Red beetroot extract demonstrated synergistic effects with doxorubicin in breast, pancreatic, and prostate cancer cell lines (Kapadia et al., 2013).

7.5 Biological Activity of Beetroot

7.5.1 Antioxidant Activity

Synthetic antioxidants have been linked to adverse side effects like carcinogenesis, CVD, and high blood pressure. On the other hand, plant extracts comprise natural antioxidants that could successfully substitute synthetic antioxidants and act as therapeutic agents to combat problems caused by oxidative stress. Many researchers have documented beetroot to have significantly high antioxidant content (Besbas and Takácsné, 2024; Wootton-Beard and Ryan, 2011; Younas et al., 2021; Reddy et. al., 2005; Vulić et al., 2014). Betalains are a main pigment found in beetroot, which comprises two main type of pigments: betacyanin and betaxanthin (Carrillo et al., 2017; Czapski et al., 2009; Wootton-Beard et al., 2011; Rodriguez-Amaya, 2019). Along with betanin, which accounts for approximately 50% of beetroot's overall antioxidant capacity (Koss-Mikołajczyk et al., 2019), it is also known to have high bioactivity of phenolics and polyphenols. Additionally, it contains a small quantity of vitamin E and C, all of which have been proven to contain significant antioxidant activity (Kazimierczak et al., 2014; Wruss et al., 2015; Sentkowska and Pyrzyn ́ska, 2018). An average betalain content of 120 mg/100 g (fresh weight) is present in beetroot, with the ratio of betacyanin to betaxanthin being 1:3 (Nemzer et al., 2011). Betacyanin levels in beetroots vary from 0.30 to 2.2 mg/g (fresh weight) and can even exceed 3 mg/g (Sadowska-Bartosz and Bartosz, 2021). The quantity of betaxanthines also varies significantly, ranging from 0.2 to 1.4 mg/g (Sadowska-Bartosz and Bartosz, 2021). According to studies, beetroot contains 1,046–1,364 mg GAE/100 g of total polyphenols, 133–233 mg CE/100 g of flavonoids, and 217–250 mg of vitamin C per 100 g. Platosz et al. (2020) investigated the antioxidant effects of fermented beet juice and observed that fermented beetroot juice possesses a higher total polyphenol level than fresh beetroot juice. Before and after an in vitro digestion procedure with simulated gastric (GAS) and duodenal (DUO) phases, Nahla et al. (2018) assessed the total antioxidant capacity (TAC) and total polyphenol (TP) content of a 70-mL beetroot juice shot, measured in gallic acid equivalents (GAE). The juice had a high TAC (697.9 ± 1.6 μmol/70 mL) and TP content (68.4 ± 0.3 mg GAE/70 mL). TAC increased threefold after gastric digestion (2,361.2 ± 20.9 μmol/70 mL) and remained high after duodenal digestion (1,740.3 ± 21.1 μmol/70 mL). TP levels rose fivefold after gastric digestion (341.6 ± 4.8 mg GAE/70 mL) and retained a 3.3-fold increase post-duodenal digestion (223.2 ± 5.4 mg GAE/70 mL). Other fruit and vegetable juices, including tomato and carrot, exhibited lower TAC and antioxidant bioavailability than beetroot (Wootton-Beard et al., 2011). Furthermore, it can be noted that beetroot scores quite high in the antioxidant level among vegetables, comparable to that of spinach, Brussels sprouts, or kale (Chu et al., 2002). Studies also found that beetroot juice exhibited a high antioxidant capacity with 53.34% effectiveness in scavenging free radicals, indicating its potential in preventing non-communicable diseases (Amelia et al., 2024) (Figure 7.2).

7.5.2 Antimicrobial Activity

Beetroot has been extensively studied for its antimicrobial and antibacterial properties against various gram-positive and gram-negative bacteria while showing no inhibitory effects on fungi and yeasts (Kumar and Brooks, 2018). These properties may contribute to the management of chronic diseases (Besbas and Takácsné, 2024). Platosz et al. (2023) researched the antimicrobial activity of fermented beetroot juice against pathogens such as Listeria monocytogenes and showed its potential for food safety and health applications. These properties are attributed to bioactive compounds like phenolic acids, flavonoids, betaxanthin, and betacyanin, contributing to their therapeutic applications in health and disease

FIGURE 7.2
Pictorial presentation of the biological activities of beetroot and its juice.

management. Researchers found that gram-negative species are considerably less susceptible to gram-positive species, which include B. cereus, S. epidermidis, S. aureus, and Str. pyogenes to red beet extracts. The high sensitivity of gram-positive bacteria was considered to be caused by polyphenolic compounds (such as betalains) in beetroot extract, which might easily alter their simple structure of cell wall, function, and accessibility. Bacillus cereus is extremely vulnerable among gram-positive organisms (20.33 inhibition zone of extract). On the contrary, pomace extract was ineffective toward wild bacterial species (Bacillus sp., Enterococcus faecalis, and Listeria monocytogenes), mold, and yeast (Vulić et al., 2013). Due to their double-layered, very hydrophilic lipopolysaccharide molecules and unique periplasm, Gram-negative bacteria are more susceptible. The extraction efficacy of polyphenolic compounds in beetroot correlates with microbial inhibitory activity, as different extracts contain substances that may inhibit or promote bacterial growth and their metabolites. A small inhibitory dosage of methanolic and 50% methanolic extracted pulp (4 mg/mL) demonstrates high effectiveness. The aqueous extract of peels and pulp exhibits a minimum inhibitory concentration (MIC) of 8 mg/mL for all tested bacterial and fungal strains. In contrast, the methanolic peel extract demonstrates greater efficacy against E. coli, with a MIC of 4 mg/mL, compared to other microorganisms (Salamatullah et al., 2021).

7.3.3 Anti-Inflammatory Activity

Molecular docking studies reveal strong negative interactions with cytokine and NO synthase active sites, significantly inhibiting microglial cell activation (Ahmadi et al., 2020). Beetroot demonstrates a pro-apoptotic effect on neutrophils after one day of stimulation at concentrations of 0.1%–10%. Neutrophils from obese patients showed a concentration-dependent inhibition of oxidative metabolism with beetroot chips and juice (Zielinska-Przyjemska et al., 2012). Beetroot may suppress transcriptional factor activation, affecting RNA synthesis for genes linked to inflammatory, cytoprotective, and antioxidant responses (Liu et al., 2019; Baird and Yamamoto, 2020). Both human and animal studies confirm beetroot's anti-inflammatory properties (Shepherd et al., 2019). In vitro research on human neutrophils suggests betanin reduces free radical production and regulates DNA damage and apoptosis (Zielinska-Przyjemska et al., 2012). In *vivo* studies using a carrageenan-induced rat paw edema model found that a water-soluble beetroot extract (500–1,000 mg/kg) exhibited significant anti-inflammatory effects comparable to indomethacin, especially at 1,000 mg/kg (Jain et al., 2011). A further experiment revealed that betanin influences the inhibitory mechanisms of the inflammatory enzymes lipoxygenase (LOX) and xanthine oxidase (XO). White and red beetroot and Swiss chard variants often inhibit LOX via an uncompetitive inhibition mechanism, while XO is inhibited by uncompetitive methods (Zein et al., 2015).

7.5.4 Radioprotective Activity

According to the research that is currently available, red beet betalains exhibit promising radioprotective properties. According to an in *vivo* (mice) study, red beet betalains exhibited radioprotective action in mice exposed to gamma radiation. Thymus index, spleen cells, and white blood cell counts were all decreased by gamma-ray irradiation. In a dose-dependent manner, betalains partially restored the spleen and thymus indices and mitigated the radiation-induced decrease in white blood cells (Lu et al., 2009).

7.5.5 Hypoglycemic Activity

Several research studies demonstrate that plant extracts have hypoglycemic action, block α-amylase, and have anti-hyperglycemic effects on β-glucosidase enzymes (Raikos et al., 2016; Poovitha, and Parani, 2016). According to a research, certain beetroot cultivars are more successful at inhibiting α-amylase than β-glucosidase (16% for red beetroot and 19% for white beetroot); however, both kinds are less effective than prickly pear (32%–50% inhibition) (Koss-Mikolajczyk et al., 2019). Consuming beetroot products has been demonstrated to help diabetic patients; in particular, eight weeks of consuming 100 g of raw beetroot improved type II diabetics' glucose metabolism and cognitive abilities (Aliahmadi et al., 2021). A single-blind cross-over experiment involving 16 healthy volunteers revealed that consuming 225 mL of commercially available neobetanin-rich beetroot drink (containing 1.3 g neobetanin/225 mL drink) improved postprandial blood sugar and insulin response when compared with a control drink that was adjusted with macronutrient compounds (Wootton-Beard et al., 2014). After 28 days of administering 10 mg/kg (Day 7: 348.50 mg/dL fasting blood glucose–Day 28: 259.00 mg/dL FBG at day 28) and 20 mg/kg betanin (Day 7: 355.00 mg/dL FBG–Day 28: 235.00 mg/dL FBG), another investigation on streptozotocin-induced diabetic rats revealed a substantial hypoglycemic effect (Abedimanesh et al., 2021).

7.6 Innovations and Latest Trends

Beetroot and beetroot juice have attracted considerable interest for their pharmacological potential, thanks to their abundance of bioactive compounds like NO_3^- and betalains, which offer numerous health benefits. Researchers are now delving into additional pharmacological properties and health advantages of these key bioactive compounds, as well as investigating other bioactive substances. Beetroots are rich in antioxidants, particularly betacyanins, which help reduce oxidative stress and inflammation linked to conditions like arthritis and diabetes (Khan et al., 2024). They also contain betaine, which supports liver health and is used in treating NAFLD (Liu et al., 2024). Research suggests betacyanins may have anti-cancer properties by inhibiting cancer cell growth and inducing apoptosis (Chung et al., 2022). Additionally, beetroot-derived NO_3^- is linked to improved cognitive function and is being explored for neurodegenerative diseases like Alzheimer's and Parkinson's (Phukan et al., 2023).

Retzl et al. (2020) isolated a peptide from beetroot. This peptide inhibits an enzyme that degrades messenger molecules in the body. The beetroot peptide, characterized by its stable molecular structure and pharmacological properties, demonstrates potential as a drug candidate for the treatment of inflammatory diseases, such as neurodegenerative and autoimmune conditions. The peptide, located in beetroot roots, is part of a class of molecules utilized by plants for chemical defense against pests, including bacteria, viruses, and insects. The beetroot peptide specifically hinders prolyloligopeptidase (POP), an enzyme that plays a role in the degradation of protein hormones and the regulation of inflammatory responses. POP represents a crucial drug target in the context of neurodegenerative and inflammatory diseases, including Alzheimer's disease and multiple sclerosis. Future research on the group of plant peptides termed 'knottins' may facilitate the creation of drug candidates for these diseases.

Nazari et al. (2024) investigated the therapeutic potential of beetroot extract, prepared with 20% ethanol, in mitigating inflammation and oxidative stress to prevent muscle atrophy following short-term immobilization in mice. The administration of beetroot extract led to increased muscle weight, cross-sectional area, and grip strength in mice with immobilization-induced muscle atrophy. It also reduced inflammation and oxidative stress biomarkers in atrophic muscle fibers. The beetroot extract-treated group exhibited higher NO_3^- levels and lower Troponin I (TnI) concentrations, indicating enhanced muscle structure and function.

Saad et al. (2023) found that beetroot betalains reduced gastric mucus and ulcer index (UI) while protecting against ethanol-induced gastric hemorrhagic damage. Supplementation preserved gastric mucus and significantly increased mucosal pepsin, hyaluronidase, and lipoxygenase activities, along with lipid peroxidation levels. The rising interest in beetroot's nutritional and pharmacological potential, along with systematic extraction, isolation, and characterization of its bioactive compounds, could support its broader application in health food and medicine.

7.7 Summary and Perspectives

Beetroots have been traditionally used in culinary practices across Asia and Europe and are known for their health-promoting properties. They are employed in traditional remedies for the treatment of kidney and liver disorders owing to their advantageous effects. Bioactive phytoconstituents in beetroot confer notable medicinal value, exhibiting a range of biological activities including antioxidant, antimicrobial (particularly antibacterial), anti-inflammatory, hepatoprotective, anxiolytic, antidiabetic, hypolipidemic, and antihypertensive effects.

This chapter examines the pharmacological potential of beetroot and its extract as documented in the literature. The antioxidant properties of phenolic compounds represent a significant area of biological activity that has been extensively researched. Antioxidants additionally play roles in anti-inflammatory, hypolipidemic, and hepatoprotective activities. Beetroot serves as a significant source of inorganic NO_3^-, demonstrating efficacy in lowering hypertension.

Phytochemicals in beetroots demonstrate complementary mechanisms in cancer prevention, potentially through the scavenging of oxidants, regulation of gene expression related to cell proliferation, and modulation of detoxification enzymes. Betalain pigments, responsible for the coloration of beetroots, exhibit activity against several conditions, such as hyperglycemia and inflammation. Stress may result in various psychotic complications, and beetroot supplementation has demonstrated efficacy in alleviating anxiety and depression. Crude extracts from plants encompass various phytochemicals that may exhibit antimicrobial properties; however, information regarding the antifungal activity of beetroot remains scarce.

The intake of beetroot significantly impacts a healthy lifestyle, since several components demonstrate varied biological actions. The mechanisms of action of numerous phytoconstituents are insufficiently explored, necessitating additional research to optimize their efficacy as nutraceutical supplements.

References

Abedimanesh, N., Asghari, S., Mohammadnejad, K., Daneshvar, Z., Rahmani, S., Shokoohi, S., ... and Motlagh, B. (2021). The anti-diabetic effects of betanin in streptozotocin-induced diabetic rats through modulating AMPK/SIRT1/NF-κB signaling pathway. *Nutrition and Metabolism*, 18, 1–13.

Afzali, N., Shahkarami, V., Vazirian, M., and Ebadi, S. S. (2020). Effect of beta vulgaris extract on liver enzymes in patients with non-alcoholic fatty liver disease: a randomized clinical trial. *Hepatitis Monthly*, *20*, e102125.

Ahmadi, H., Nayeri, Z., Minuchehr, Z., Sabouni, F., and Mohammadi, M. (2020). Betanin purification from red beetroots and evaluation of its anti-oxidant and anti-inflammatory activity on LPS-activated microglial cells. *PloS One*, *15*(5), e0233088.

Al-Harbi, L. N., Alshammari, G. M., Al-Dossari, A. M., Subash-Babu, P., Binobead, M. A., Alhussain, M. H., AlSedairy, S. A., Al-Nouri, D. M., and Shamlan, G. (2021). Beta vulgaris L. (Beetroot) methanolic extract prevents hepatic steatosis and liver damage in T2DM rats by hypoglycemic, insulin-sensitizing, antioxidant effects, and upregulation of PPARα. *Biology*, *10*(12), 1306.

Albrahim, T., and Alonazi, M. A. (2020). Role of beetroot (Beta vulgaris) juice on chronic nanotoxicity of silver nanoparticle-induced hepatotoxicity in male rats. *International Journal of Nanomedicine*, 15, 3471.

Aliahmadi, M., Amiri, F., Bahrami, L. S., Hosseini, A. F., Abiri, B., and Vafa, M. (2021). Effects of raw red beetroot consumption on metabolic markers and cognitive function in type 2 diabetes patients. *Journal of Diabetes and Metabolic Disorders*, *20*, 673–682.

Amelia, R. T., Ardyanto, T. D., and Sari, Y. (2024). Investigation of antioxidant activity from beetroot juice (Beta vulgaris L) as a healthy drink for the prevention of non-communicable disease. *AcTion: Aceh Nutrition Journal*, 9(1), 82–90.

Asgary, S., Afshani, M. R., Sahebkar, A., Keshvari, M., Taheri, M., Jahanian, E., Rafieian-Kopaei, M., Malekian, F., and Sarrafzadegan, N. (2016). Improvement of hypertension, endothelial function and systemic inflammation following short-term supplementation with red beet (Beta vulgaris L.) juice: a randomized crossover pilot study. *Journal of Human Hypertension*, 30(10), 627–632. https://doi.org/10.1038/jhh.2016.34

Babarykin, D., Smirnova, G., Markovs, J., Vasiljeva, S., Basova, N., Simanis, R., and Viksna, L. (2019). Therapeutic effect of fractionated by ultrafiltration red beetroot (*Beta vulgaris* L.) juice in rats with food-induced fatty liver. *European Journal of Biological Research*, 9(1), 1–9.

Bahrami, L. S., Arabi, S. M., Feizy, Z., and Rezvani, R. (2021). The effect of beetroot inorganic nitrate supplementation on cardiovascular risk factors: a systematic review and meta-regression of randomized controlled trials. *Nitric Oxide*, 115, 8–22.

Baiao, D. D. S., da Silva, D. V., Del Aguila, E. M., and Paschoalin, V. M. F. (2017). Nutritional, bioactive and physicochemical characteristics of different beetroot formulations. *Food Additives*, 6(6), 21–43.

Bailey, S. J., Winyard, P., Vanhatalo, A., Blackwell, J. R., DiMenna, F. J., Wilkerson, D. P., Tarr, J., Benjamin, N., and Jones, A. M., (2009). Dietary nitrate supple- mentation reduces the O_2 cost of low-intensity exercise and enhances tolerance to high-intensity exercise in humans. *Journal of Applied Physiology* 107, 1144–1155.

Baird, L., and M. Yamamoto. (2020). The molecular mechanisms regulating the KEAP1-NRF2 pathway. *Molecular and Cellular Biology 40* (13), e00099-20. https://doi.org/10.1128/MCB.00099-20

Besbas, M. E. A., and Takácsné, M. H. (2024). Evaluation of bioactive substances in different parts of the root in beetroot (Beta vulgaris ssp. esculenta var. rubra). *Acta Agraria Debreceniensis, 1*, 33–38.

Blázovics, A., and Sárdi, É. (2018). Methodological repertoire development to study the effect of dietary supplementation in cancer therapy. *Microchemical Journal, 136*, 121–127.

Bond Jr, V., Curry, B. H., Adams, R. G., Asadi, M. S., Millis, R. M., and Haddad, G. E. (2013). Effects of dietary nitrates on systemic and cerebrovascular hemodynamics. *Cardiology Research and Practice, 2013*(1), 435629.

Bondonno, C. P., Downey, L. A., Croft, K. D., Scholey, A., Stough, C., Yang, X., … and Hodgson, J. M. (2014). The acute effect of flavonoid-rich apples and nitrate-rich spinach on cognitive performance and mood in healthy men and women. *Food and Function*, 5(5), 849–858.

Butt, M. S., Randhawa, M. A., and Shahid, M. (2019). Hepatoprotective effects of red beetroot-based beverages against CCl4-induced hepatic stress in Sprague Dawley rats. *Journal of Food Biochemistry,* 43(12), e13057.

Carrillo, C., Rey, R., Hendrickx, M., del Mar Cavia, M., and Alonso-Torre, S. (2017). Antioxidant capacity of beetroot: traditional vs novel approaches. *Plant Foods for Human Nutrition, 72*, 266–273.

Chen, L., Zhu, Y., Hu, Z., Wu, S., and Jin, C. (2021). Beetroot as a functional food with huge health benefits: antioxidant, antitumor, physical function, and chronic metabolomics activity. *Food Science and Nutrition,* 9(11), 6406–6420.

Chu, Y. F., Sun, J. I. E., Wu, X., and Liu, R. H. (2002). Antioxidant and antiproliferative activities of common vegetables. *Journal of Agricultural and Food Chemistry, 50*(23), 6910–6916.

Chung, M. Y., Hwang, J. T., Lee, J., and Choi, H. K. (2022). The anti-cancer effects of red-pigmented foods: biomarker modulation and mechanisms underlying cancer progression. *Applied Sciences, 12*(5), 2584.

Clifford, T., Howatson, G., West, D. J., Stevenson, E. J. (2015). The potential benefits of red beetroot supplementation in health and disease. *Nutrients*, 7, 2801–2822.

Coles, L. T., and Clifton, P. M. (2012). Effect of beetroot juice on lowering blood pressure in free-living, disease-free adults: a randomized, placebo-controlled trial. *Nutrition Journal*, 11, 1–6.

Czapski, J., Mikołajczyk, K., and Kaczmarek, M. (2009). Relationship between antioxidant capacity of red beet juice and contents of its betalain pigments. *Polish Journal of Food and Nutrition Sciences, 59*(2), 119–122.

Da Silva, D. V. T., De Oliveira Silva, F., Perrone, D., Pierucci, A. P. T. R., Conte-Junior, C. A., Da Silveira Alvares, T., Del Aguila, E. M., and Paschoalin, V. M. F. (2016). Physicochemical, nutritional, and sensory analyses of a nitrateenriched beetroot gel and its effects on plasmatic nitric oxide and blood pressure. *Food and Nutrition Research*, 60. https://doi.org/10.3402/fnr.v60.29909

Davignon, J., and Ganz, P. (2004). Role of endothelial dysfunction in atherosclerosis. *Circulation, 109*(23_suppl_1), III–27.

De la Torre, J. C., and Stefano, G. B. (2000). Evidence that Alzheimer's disease is a microvascular disorder: the role of constitutive nitric oxide. *Brain Research Reviews, 34*(3), 119–136.

Donath, M. Y., and Shoelson, S. E. (2011). Type 2 diabetes as an inflammatory disease. *Nature Reviews Immunology, 11*(2), 98–107.

Florin, T. H. J., Neale, G., and Cummings, J. H. (1990). The effect of dietary nitrate on nitrate and nitrite excretion in man. *British Journal of Nutrition*, 64(2), 387–397.

Frank, T., Stintzing, F. C., Carle, R., Bitsch, I., Quaas, D., Straß, G., … and Netzel, M. (2005). Urinary pharmacokinetics of betalains following consumption of red beet juice in healthy humans. *Pharmacological Research, 52*(4), 290–297.

Frombaum, M., Le Clanche, S., Bonnefont-Rousselot, D., and Borderie, D. (2012). Antioxidant effects of resveratrol and other stilbene derivatives on oxidative stress and NO bioavailability: potential benefits to cardiovascular diseases. *Biochimie*, 94(2), 269–276.

Fu, Y., Shi, J., Xie, S. Y., Zhang, T. Y., Soladoye, O. P., and Aluko, R. E. (2020). Red beetroot betalains: perspectives on extraction, processing, and potential health benefits. *Journal of Agricultural and Food Chemistry*, 68(42), 11595–11611.

Georgiev, V. G., Weber, J., Kneschke, E. M., Denev, P. N., Bley, T., and Pavlov, A. I. (2010). Antioxidant activity and phenolic content of betalain extracts from intact plants and hairy root cultures of the red beetroot Beta vulgaris cv. Detroit dark red. *Plant Foods for Human Nutrition*, 65, 105–111.

Gilchrist, M., Shore, A. C., and Benjamin, N. (2011). Inorganic nitrate and nitrite and control of blood pressure. *Cardiovascular Research*, *89*(3), 492–498.

Gilchrist, M., Winyard, P. G., Aizawa, K., Anning, C., Shore, A., and Benjamin, N. (2013). Effect of dietary nitrate on blood pressure, endothelial function, and insulin sensitivity in type 2 diabetes. *Free Radical Biology and Medicine*, *60*, 89–97.

Hobbs, D. A., George, T. W., and Lovegrove, J. A., 2013a. The effects of dietary nitrate on blood pressure and endothelial function: a review of human intervention studies. *Nutrition Research Reviews*, 26, 210–222.

Hobbs, D. A., Kaffa, N., George, T. W., Methven, L., and Lovegrove, J. A., 2012. Blood pressure-lowering effects of beetroot juice and novel beetroot-enriched bread products in normotensive male subjects. *British Journal of Nutrition*, 108, 2066–2074.

Hobbs, D. A., Goulding, M. G., Nguyen, A., Malaver, T., Walker, C. F., George, T. W., ... and Lovegrove, J. A. (2013b). Acute ingestion of beetroot bread increases endothelium-independent vasodilation and lowers diastolic blood pressure in healthy men: a randomized controlled trial. *The Journal of nutrition*, 143(9), 1399–1405.

Jain, D., Singh, K. K., Patil, R. T., and Gupta, M. K. (2011). Physico-chemical and sensory quality of extruded foods developed from rice and defatted soyflour/chickpea splits supplemented with dried beetroot. *Journal of Agricultural Engineering*, *48*(4), 17–23.

Jajja, A., Sutyarjoko, A., Lara, J., Rennie, K., Brandt, K., Qadir, O., Siervo, M., 2014. Beetroot supplementation lowers daily systolic blood pressure in older, overweight subjects. *Nutrition Research* 34, 868–875.

Jakubcik, E. M., Rutherfurd-Markwick, K., Chabert, M., Wong, M., and Ali, A. (2021). Pharmacokinetics of nitrate and nitrite following beetroot juice drink consumption. *Nutrients*, 13(2), 281.

Jones, A. M. (2014). Dietary nitrate supplementation and exercise performance. *Sports Medicine*, 44(Suppl 1), 35–45.

Joris, P. J., and Mensink, R. P. (2013). Beetroot juice improves in overweight and slightly obese men post-prandial endothelial function after consumption of a mixed meal. *Atherosclerosis*, *231*(1), 78–83.

Kanner, J., Harel, S., and Granit, R. (2001). Betalains a new class of dietary cationized antioxidants. *Journal of Agricultural and Food chemistry*, 49(11), 5178–5185.

Kapadia, G. J., Rao, G. S., Ramachandran, C., Iida, A., Suzuki, N., and Tokuda, H. (2013). Synergistic cytotoxicity of red beetroot (Beta vulgaris L.) extract with doxorubicin in human pancreatic, breast and prostate cancer cell lines. *Journal of Complementary and Integrative Medicine*, *10*(1), 113–122.

Kapadia, G. J., Azuine, M. A., Rao, G. S., Arai, T., Iida, A., Tokuda, H., 2011. Cytotoxic effect of the red beetroot (*Beta vulgaris* L.) extract compared to doxorubicin (Adriamycin) in the human prostate (PC-3) and breast (MCF-7) cancer cell lines. *Anti-Cancer Agents in Medicinal Chemistry-Anti-Cancer Agents)*, *11*(3), 280–284.

Kapil, V., Khambata, R. S., Robertson, A., Caulfield, M. J., and Ahluwalia, A. (2015). Dietary nitrate provides sustained blood pressure lowering in hypertensive patients: a randomized, phase 2, double-blind, placebo-controlled study. *Hypertension*, 65(2), 320-327.

Kassem, A. A., Abd El-Alim, S. H., Salman, A. M., Mohammed, M. A., Hassan, N. S., and El-Gengaihi, S. E. (2020). Improved hepatoprotective activity of Beta vulgaris L. leaf extract loaded self-nanoemulsifying drug delivery system (SNEDDS): in vitro and in vivo evaluation. *Drug Development and Industrial Pharmacy*, *46*(10), 1589-1603.

Kazimierczak, R., Hallmann, E., Lipowski, J., Drela, N., Kowalik, A., Püssa, T., ... and Rembiałkowska, E. (2014). Beetroot (Beta vulgaris L.) and naturally fermented beetroot juices from organic and conventional production: metabolomics, antioxidant levels and anticancer activity. *Journal of the Science of Food and Agriculture*, *94*(13), 2618–2629.

Kelly, J., Fulford, J., Vanhatalo, A., Blackwell, J. R., French, O., Bailey, S. J., … and Jones, A. M. (2013). Effects of short-term dietary nitrate supplementation on blood pressure, O2 uptake kinetics, and muscle and cognitive function in older adults. *American Journal of Physiology-Regulatory, Integrative and Comparative Physiology*, *304*(2), R73–R83.

Kenjale, A. A., Ham, K. L., Stabler, T., Robbins, J. L., Johnson, J. L., VanBruggen, M., … and Allen, J. D. (2011). Dietary nitrate supplementation enhances exercise performance in peripheral arterial disease. *Journal of Applied Physiology*, *110*(6), 1582–1591.

Khan, S. A., Alsulami, M. N., Alsehimi, A. A., Alzahrani, M. S., Mosule, D. A., and Albohiri, H. H. (2024). Beta vulgaris betalains mitigate parasitemia and brain oxidative stress induced by plasmodium berghei in mice. *Pharmaceuticals*, *17*(8), 1064.

Kim-Shapiro, D. B., and Gladwin, M. T. (2014). Mechanisms of nitrite bioactivation. *Nitric Oxide*, 38, 58–68.

Koss-Mikołajczyk, I., Kusznierewicz, B., Wiczkowski, W., Sawicki, T., and Bartoszek, A. (2019). The comparison of betalain composition and chosen biological activities for differently pigmented prickly pear (Opuntia ficus-indica) and beetroot (Beta vulgaris) varieties. *International Journal of Food Sciences and Nutrition*, 70(4), 442–452.

Kumar, P. S., Bhaumik, A., Chopra, M., and Devi, K. N. 2016. Evaluation of anti diabetic activity of ethanolic extract of beet root (EEBT-*Beta vulgaris*) against streptozocin induced diabetic rats. *Journal of Drug Discoveries & Therapeutics* 4. https://www.semanticscholar.org/paper/Evaluation-of-Anti-diabetic-activity-of-Ethanolic-Kumar-Bhaumik/00262bac3467c6d4737275c9ff50e274099f61a1

Kumar, S., and Brooks, M. S. L. (2018). Use of red beet (Beta vulgaris L.) for antimicrobial applications – a critical review. *Food and Bioprocess Technology*, *11*, 17–42.

Kumar, V., Abbas, A. K., Fausto, N., Aster, J. C., 2014. *Robbins and Cotran Pathologic Basis of Disease*. Philadelphia, PA: Elsevier Health Sciences.

Lechner, J. F., Wang, L.-S., Rocha, C. M., Larue, B., Henry, C., McIntyre, C. M., Riedl, K. M., Schwartz, S. J., Stoner, G. D., 2010. Drinking water with red beetroot food color antagonizes esophageal carcinogenesis in *N*-nitrosomethylbenzylamine-treated rats. *Journal of Medicinal Food* 13, 733–739.

Lidder, S., and Webb, A. J. (2013). Vascular effects of dietary nitrate (as found in green leafy vegetables and beetroot) via the nitrate-nitrite-nitric oxide pathway. *British Journal of Clinical Pharmacology*, *75*(3), 677–696.

Liu, J., Liu, Y., Chen, Y., Liu, Y., Huang, C., Luo, Y., and Wang, X. (2024). Betaine alleviates nonalcoholic fatty liver disease (NAFLD) via a manner involving BHMT/FTO/m6A/PGC1α signaling. *The Journal of Nutritional Biochemistry*, *134*, 109738.

Liu, R., Ji, P., Liu, B., Qiao, H., Wang, X., Zhou, L., Deng, T., and Ba, Y. (2017). Apigenin enhances the cisplatin cytotoxic effect through p53–modulated apoptosis. *Oncology Letters*, 13(2), 1024–1030.

Liu, X., J. Huang, J. Li, Q. Mao, and J. He. (2019). Effects of liraglutide combined with insulin on oxidative stress and serum MCP-1 and NF-kB levels in type 2 diabetes. *Journal of the College of Physicians and Surgeons Pakistan*, *29*(3), 218–221. Available from: https://jcpsp.pk/archive/2019/Mar2019/06.pdf

Lorizola, I. M., Furlan, C. P., Portovedo, M., Milanski, M., Botelho, P. B., Bezerra, R., Sumere, B. R., Rostagno, M. A., and Capitani, C. D. (2018). Beet stalks and leaves (Beta vulgaris L.) protect against high-fat diet-induced oxidative damage in the liver in mice. *Nutrients*, 10(7), 872.

Lu, X., Wang, Y., and Zhang, Z. (2009). Radioprotective activity of betalains from red beets in mice exposed to gamma irradiation. *European Journal of Pharmacology*, *615*(1–3), 223–227.

Lundberg, J. O., Carlström, M., Larsen, F. J., and Weitzberg, E. (2011). Roles of dietary inorganic nitrate in cardiovascular health and disease. *Cardiovascular Research*, 89(3), 525–532.

Manach, C., Williamson, G., Morand, C., Scalbert, A., and Rémésy, C. (2005). Bioavailability and bioefficacy of polyphenols in humans. I. Review of 97 bioavailability studies. *The American Journal of Clinical Nutrition*, 81(1), 230S–242S.

Mancini, M. C. S., Ponte, L. G. S., Silva, C. H. R., Fagundes, I., Pavan, I. C. B., Romeiro, S. A., da Silva, L. G. S., Morelli, A. P., Rostagno, M. A., Simabuco, F. M., and Bezerra, R. M. N. (2021). Beetroot and leaf extracts present protective effects against prostate cancer cells, inhibiting cell proliferation, migration, and growth signaling pathways. *Phytotherapy Research*, 35(9), 5241–5258.

Milton-Laskibar, I., Martínez, J. A., and Portillo, M. P. (2021). Current knowledge on beetroot bioactive compounds: role of nitrate and betalains in health and disease. *Foods, 10*(6), 1314.
Nazari, S. E., Khalili-Tanha, N., Babaei, F., Jafarzadeh Esfahani, A., Khalili-Tanha, G., Asgharzadeh, F., Khojasteh-Leylakoohi, F., Asghari, S. Z., Hadjzadeh, M.-A.-R., Hassanian, S. M., Ferns, G., Avan, A., Rezvani, R., and Khazaei, M. (2024). The therapeutic potential of red beetroot (Beta vulgaris L.) intake on muscle atrophy in immobilized mouse skeletal muscles. *Food Science and Nutrition*, 12, 9085–9093. https://doi.org/10.1002/fsn3.4335
Nemzer, B., Pietrzkowski, Z., Spórna, A., Stalica, P., Thresher, W., Michałowski, T., and Wybraniec, S. (2011). Betalainic and nutritional profiles of pigment-enriched red beet root (*Beta Vulgaris L.*) dried extracts. *Food Chemistry, 127*(1), 42–53. https://doi.org/10.1016/j.foodchem.2010.12.081
Netzel, M., Stintzing, F. C., Quaas, D., Strass, G., Carle, R., Bitsch, R., ... and Frank, T. (2005). Renal excretion of antioxidative constituents from red beet in humans. *Food Research International, 38*(8–9), 1051–1058.
Nowacki, L., Vigneron, P., Rotellini, L., Cazzola, H., Merlier, F., Prost, E., ... and Vayssade, M. (2015). Betanin-enriched red beetroot (Beta vulgaris L.) extract induces apoptosis and autophagic cell death in MCF-7 cells. *Phytotherapy Research*, 29(12), 1964–1973.
Olumese, F. E., and Oboh, H. A. (2020). Aqueous beetroot juice extract improves renal function and some biochemical parameters in carbon tetrachloride-induced toxicity in Sprague Dawley rats. *Annals of Tropical Pathology*, 11(2), 171.
Ormesher, L., Myers, J. E., Chmiel, C., Wareing, M., Greenwood, S. L., Tropea, T., Lundberg, J. O., Weitzberg, E., Nihlen, C., Sibley, C. P., and Cottrell, E. C. (2018). Effects of dietary nitrate supplementation, from beetroot juice, on blood pressure in hypertensive pregnant women: a randomised, double-blind, placebo-controlled feasibility trial. *Nitric Oxide*, 80, 37–44.
Pérez-Gregorio, M. R., Regueiro, J., Alonso-González, E., Pastrana-Castro, L. M., and Simal-Gándara, J. (2011). Influence of alcoholic fermentation process on antioxidant activity and phenolic levels from mulberries (Morus nigra L.). *LWT-Food Science and Technology, 44*(8), 1793–1801.
Phukan, B. C., Roy, R., Paul, R., Mazumder, M. K., Nath, J., Bhattacharya, P., and Borah, A. (2023). Traversing through the cell signaling pathways of neuroprotection by betanin: therapeutic relevance to Alzheimer's disease and Parkinson's disease. *Metabolic Brain Disease, 38*(3), 805–817.
Platosz, N., Sawicki, T., and Wiczkowski, W. (2020). Profile of phenolic acids and flavonoids of red beet and its fermentation products. Does long-term consumption of fermented beetroot juice affect phenolics profile in human blood plasma and urine?. *Polish Journal of Food and Nutrition Sciences, 70*(1), 55–65.
Poovitha, S., and Parani, M. (2016). In vitro and in vivo α-amylase and α-glucosidase inhibiting activities of the protein extracts from two varieties of bitter gourd (Momordica charantia L.). *BMC Complementary and Alternative Medicine, 16*, 1–8.
Prando, T. B. L., Barboza, L. N., de Oliveira Araújo, V., Gasparotto, F. M., de Souza, L. M., Lourenço, E. L. B., and Junior, A. G. (2016). Involvement of bradykinin B2 and muscarinic receptors in the prolonged diuretic and antihypertensive properties of Echinodorus grandiflorus (Cham. and Schltdl.) Micheli. *Phytomedicine, 23*(11), 1249–1258.
Presley, T. D., Morgan, A. R., Bechtold, E., Clodfelter, W., Dove, R. W., Jennings, J. M., ... and Miller, G. D. (2011). Acute effect of a high nitrate diet on brain perfusion in older adults. *Nitric Oxide*, 24(1), 34–42.
Rahimi, P., Mesbah-Namin, S. A., Ostadrahimi, A., Abedimanesh, S., Separham, A., and Jafarabadi, M. A. (2019). Effects of betalains on atherogenic risk factors in patients with atherosclerotic cardiovascular disease. *Food and Function*, 10(12), 8286–8297
Raikos, V., McDonagh, A., Ranawana, V., and Duthie, G. (2016). Processed beetroot (Beta vulgaris L.) as a natural antioxidant in mayonnaise: effects on physical stability, texture and sensory attributes. *Food Science and Human Wellness, 5*(4), 191–198.
Reddy, M. K., Alexander-Lindo, R. L., and Nair, M. G. (2005). Relative inhibition of lipid peroxidation, cyclooxygenase enzymes, and human tumor cell proliferation by natural food colors. *Journal of Agricultural and Food Chemistry, 53*(23), 9268–9273.
Rein, M. J., Renouf, M., Cruz-Hernandez, C., Actis-Goretta, L., Thakkar, S. K., and da Silva Pinto, M. (2013). Bioavailability of bioactive food compounds: a challenging journey to bioefficacy. *British Journal of Clinical Pharmacology, 75*(3), 588–602.

Retzl, B., Hellinger, R., Muratspahić, E., Pinto, M. E., Bolzani, V. S., and Gruber, C. W. (2020). Discovery of a beetroot protease inhibitor to identify and classify plant-derived cystine knot peptides. *Journal of Natural Products*, *83*(11), 3305–3314.

Rodriguez-Amaya, D. B. (2019). Update on natural food pigments – a mini-review on carotenoids, anthocyanins, and betalains. *Food Research International*, *124*, 200–205. http://doi.org/10.1016/j.foodres.2018.05.028.

Saad, F., Al-Shaikh, T. M., Zouidi, F., Taher, M. A., Saidi, S. A., and Hamden, K. (2023). Betalain-enriched beetroots exhibit antiulcer and anti-inflammatory potentials. *Journal of Food Processing and Preservation*, *2023*(1), 9522830.

Sadowska-Bartosz, I., and Bartosz, G. (2021). Biological properties and applications of betalains. *Molecules*, *26*(9), 2520.

Salamatullah, A. M., Hayat, K., Alkaltham, M. S., Ahmed, M. A., Arzoo, S., Husain, F. M., … and Al-Harbi, L. N. (2021). Bioactive and antimicrobial properties of oven-dried beetroot (pulp and peel) using different solvents. *Processes*, *9*(4), 588.

Sawicki, T., Topolska, J., Romaszko, E., and Wiczkowski, W. (2018). Profile and content of betalains in plasma and urine of volunteers after long-term exposure to fermented red beet juice. *Journal of Agricultural and Food Chemistry*, 66(16), 4155–4163.

Sentkowska, A., and Pyrzyńska, K. (2018). Zwitterionic hydrophilic interaction liquid chromatography coupled to mass spectrometry for analysis of beetroot juice and antioxidant interactions between its bioactive compounds. *LWT*, *93*, 641–648.

Shepherd, A. I., Costello, J. T., Bailey, S. J., Bishop, N., Wadley, A. J., Young-Min, S., … and Eglin, C. M. (2019). "Beet" the cold: beetroot juice supplementation improves peripheral blood flow, endothelial function, and anti-inflammatory status in individuals with Raynaud's phenomenon. *Journal of Applied Physiology*, *127*(5): 1478–1490.

Silva, D. V. T. D., Silva, F. D. O., Perrone, D., Pierucci, A. P. T. R., Conte-Junior, C. A., Alvares, T. D. S., … and Paschoalin, V. M. F. (2016). Physicochemical, nutritional, and sensory analyses of a nitrate-enriched beetroot gel and its effects on plasmatic nitric oxide and blood pressure. *Food and Nutrition Research*, *60*(1), 29909.

Sinaga, F. A., Hasibuan, R., Risfandi, M., and Sinaga, R. N. (2020). Hepatoprotective effects of beetroot juice at maximum physical activity. *Journal of Physics: Conference Series*, 1462(1), 012007).

Song, X., Hao, X., Song, B., Zhao, X., Wu, Z., Wang, X., … and Zhang, X. (2022). The oxidative damage and morphological changes of sugar beet (Beta vulgaris L.) leaves at seedlings stage exposed to boron deficiency in hydroponics. *Sugar Tech*, *24*(2), 1–10.

Srivastava, S., Siddiqi, Z., Singh, T., and Bala, L. (2019). Beetroot supplementation on non-alcoholic fatty liver disease patients. *Current Research in Nutrition and Food Science Journal*, 7(1), 96–101.

Stanaway, L., Rutherfurd-Markwick, K., Page, R., Wong, M., Jirangrat, W., Teh, K. H., and Ali, A. (2019a). Acute supplementation with nitrate-rich beetroot juice causes a greater increase in plasma nitrite and reduction in blood pressure of older compared to younger adults. *Nutrients*, 11(7), 1683.

Stanaway, L., Rutherfurd-Markwick, K., Page, R., Wong, M., Jones, B., and Ali, A. (2019b). Effects of chronic supplementation with nitrate-rich beetroot juice on cardiovascular responses in healthy adults. *Multidisciplinary Digital Publishing Institute Proceedings*, 37(1), 27.

Sulakhiya, K., Patel, V. K., Saxena, R., Dashore, J., Srivastava, A. K., and Rathore, M. (2016). Effect of Beta vulgaris Linn. leaves extract on anxiety-and depressive-like behavior and oxidative stress in mice after acute restraint stress. *Pharmacognosy Research*, 8(1), 1.

Tan, M. L., and Hamid, S. B. S. (2021). Beetroot as a potential functional food for cancer chemoprevention, a narrative review. *Journal of Cancer Prevention*, 26(1), 1.

Tawa, M., Nagano, J., Kitama, J., Abe, S., Fujita, A., Nakagawa, K., and Ohkita, M. (2022). Ameliorative effects of beetroot juice supplementation on monocrotaline-induced pulmonary hypertension in rats. *Future Pharmacology*, 2(4), 547–557.

Tesoriere, L., Fazzari, M., Angileri, F., Gentile, C., and Livrea, M. A. (2008). In vitro digestion of betalainic foods. Stability and bioaccessibility of betaxanthins and betacyanins and antioxidative potential of food digesta. *Journal of Agricultural and Food Chemistry*, *56*(22), 10487–10492.

Tesoriere, L., Gentile, C., Angileri, F., Attanzio, A., Tutone, M., Allegra, M. Y., and Livrea, M. A. (2013). Trans-epithelial transport of the betalain pigments indicaxanthin and betanin across Caco-2 cell monolayers and influence of food matrix. *European Journal of Nutrition*, 52, 1077–1087.

Thompson, C., Wylie, L. J., Fulford, J., Kelly, J., Black, M. I., McDonagh, S. T., … and Jones, A. M. (2015). Dietary nitrate improves sprint performance and cognitive function during prolonged intermittent exercise. *European Journal of Applied Physiology*, *115*, 1825–1834.

Ting, Y., Jiang, Y., Ho, C. T., and Huang, Q. (2014). Common delivery systems for enhancing in vivo bioavailability and biological efficacy of nutraceuticals. *Journal of Functional Foods*, *7*, 112–128.

Tropea, T., Renshall, L. J., Nihlen, C., Weitzberg, E., Lundberg, J. O., David, A. L., Tsatsaris, V., Stuckey, D. J., Wareing, M., Greenwood, S. L., and Cottrell, E. C. (2020). Beetroot juice lowers blood pressure and improves endothelial function in pregnant eNOS$^{-/-}$ mice: importance of nitrate-independent effects. *The Journal of Physiology*, 598(18), 4079–4092.

van Velzen, A. G., Sips, A. J., Schothorst, R. C., Lambers, A. C., and Meulenbelt, J. (2008). The oral bioavailability of nitrate from nitrate-rich vegetables in humans. *Toxicology Letters*, 181(3), 177–181.

Vanhatalo, A., Bailey, S.J., Blackwell, J. R., DiMenna, F.J., Pavey, T.G., Wilkerson, D.P., Benjamin, N., Winyard, P.G., and Jones, A.M. (2010). Acute and chronic effects of dietary nitrate supplementation on blood pressure and the physiological responses to moderate-intensity and incremental exercise. *American Journal of Physiology*, *299*, R1121–1131.

Volino-Souza, M., Oliveira, G. V. D., Conte-Junior, C. A., and Alvares, T. S. (2020). Covid-19 quarantine: impact of lifestyle behaviors changes on endothelial function and possible protective effect of beetroot juice. *Frontiers in Nutrition*, 7, 204.

Vulić, J. J., Ćebović, T. N., Čanadanović, V. M., Ćetković, G. S., Djilas, S. M., Čanadanović-Brunet, J. M., … and Tumbas, V. T. (2013). Antiradical, antimicrobial and cytotoxic activities of commercial beetroot pomace. *Food and Function*, 4(5), 713–721.

Webb, A. J., Patel, N., Loukogeorgakis, S., Okorie, M., Aboud, Z., Misra, S., Rashid, R., Miall, P., Deanfield, J., Benjamin, N., 2008. Acute blood pressure lowering, vasoprotective, and antiplatelet properties of dietary nitrate via bioconversion to nitrite. *Hypertension*, 51, 784–790.

Wiczkowski, W., Romaszko, E., Szawara-Nowak, D., and Piskula, M. K. (2018). The impact of the matrix of red beet products and interindividual variability on betacyanins bioavailability in humans. *Food Research International*, 108, 530–538.

Wootton-Beard, P. C., and Ryan, L. (2011). A beetroot juice shot is a significant and convenient source of bioaccessible antioxidants. *Journal of Functional Foods*, *3*(4), 329–334.

Wootton-Beard, P. C., Brandt, K., Fell, D., Warner, S., and Ryan, L. (2014). Effects of a beetroot juice with high neobetanin content on the early-phase insulin response in healthy volunteers. *Journal of Nutritional Science*, *3*, e9.

Wruss, J., Waldenberger, G., Huemer, S., Uygun, P., Lanzerstorfer, P., Müller, U., … and Weghuber, J. (2015). Compositional characteristics of commercial beetroot products and beetroot juice prepared from seven beetroot varieties grown in Upper Austria. *Journal of Food Composition and Analysis*, *42*, 46–55.

Yang, J., Pi, C., and Wang, G. (2018). Inhibition of PI3K/Akt/mTOR pathway by apigenin induces apoptosis and autophagy in hepatocellular carcinoma cells. *Biomedicine and Pharmacotherapy*, 103, 699–707.

Younas, U., Iqbal, S., Bashir, R., Sajjad, N., Saeed, Z., Pervaiz, M., … and Iqbal, M. (2021). An eco-friendly approach for the extraction of antioxidant components from Artemisia annua leaves using response surface methodology. *Polish Journal of Environmental Studies*, 30, 4827–4833.

Zein, H., Hashish, A. E., and Ismaiel, G. (2015). The antioxidant and anticancer activities of Swiss chard and red beetroot leaves. *Current Science International*, *4*(4), 491–498.

Zielińska-Przyjemska, M., Olejnik, A., Dobrowolska-Zachwieja, A., and Grajek, W. (2009). In vitro effects of beetroot juice and chips on oxidative metabolism and apoptosis in neutrophils from obese individuals. *Phytotherapy Research: An International Journal Devoted to Pharmacological and Toxicological Evaluation of Natural Product Derivatives*, 23(1), 49–55.

Zielinska-Przyjemska, M., Olejnik, A., Kostrzewa, A., Łuczak, M., Jagodziński, P. P., and Baer-Dubowska, W. (2012). The beetroot component betanin modulates ROS production, DNA damage and apoptosis in human polymorphonuclear neutrophils. *Phytotherapy Research*, 26(6), 845–852.

8

Valorization of Beetroot and Its Waste: An Unexplored Potential

Moufida Chaari, Elahesadat Hosseini, Zenebe Tadesse Tsegay, Theodoros Varzakas, and Slim Smaoui

8.1 Introduction

Agricultural waste has become a significant global concern due to its environmental, economic, and social implications (Bellemare et al., 2017; Gustavsson et al., 2011). The rapid growth of the agri-food sector has led to an increase in waste generation, particularly during production, processing, and distribution (Mann and Sooch, 2020; Mirabella et al., 2014; Parfitt et al., 2010; Ray, 2022; Sooch et al., 2022). For instance, fruit and vegetable industries contribute 25%–30% of total agricultural waste, often in the form of peels, seeds, pomaces, and other by-products (Ajila et al., 2010). If not properly managed, these waste materials break down quickly, leading to the release of unpleasant odors, greenhouse gases, and harmful contaminants, which significantly contribute to environmental degradation (Rechkemmer, 2001; Thiviya et al., 2022). Consequently, fruits and vegetables are recognized as major contributors to food waste on a large scale. Beyond their nutritional benefits, fruits and vegetables are also valued as rich sources of various phytochemicals. These bioactive compounds, whether acting individually or synergistically, are known to support human health and are often highly nutrient-dense (Bangal and Ray, 2022; Sooch et al., 2022). Among the various vegetables grown globally, beetroot (Beta vulgaris) exemplifies the challenges posed by agricultural waste. The European Union leads in global beetroot production, accounting for approximately 70% of the total output. In the United Kingdom alone, beetroot juice production generates waste amounting to 35%–40% of the initial biomass. Beetroot is particularly valued for its functional properties, owing to its rich content of water-soluble nitrogenous pigments known as betalains. These pigments are widely used, including as natural food colorants authorized in the European Union (E162: beet red) (Koul et al., 2002; Roy et al., 2004; Sawicki and Wiczkowski, 2018). The betalains, which include betacyanins (red-violet) and betaxanthins (yellow-orange), offer a range of nutritional and health benefits (Panghal et al., 2017).

In addition to betalains, beetroot is a source of flavonoids, carotenoids, vitamins, and minerals. Research has consistently highlighted beetroot as a significant source of health-promoting phytochemicals (Clifford et al., 2015; Mirmiran et al., 2020). Despite its remarkable nutritional and functional properties, the processing of beetroot results in substantial amounts of waste, such as peels and pomace. While these by-products are

DOI: 10.1201/9781003527084-8

biodegradable, they are often discarded, ignoring their significant potential as sources of valuable compounds like betalains, phenolic acids, and dietary fibers. Sustainable valorization of these residues offers an opportunity to transform them into high-value products, contributing to the advancement of circular economy principles and supporting sustainable development goals. This chapter delves into innovative applications for beetroot and its by-products, focusing on their use as natural colorants, functional ingredients, and materials for food coatings. Additionally, it explores the latest advancements in valorization technologies, providing insights into eco-friendly practices that convert beetroot waste into both economic and environmental assets.

8.2 Beetroot as a Natural Coloring Pigment

It is a truism that betalains, plant secondary metabolites, are the principal compounds existing in beetroot (Chaari et al., 2023b, 2024). As anthocyanins, betalains have comparable chemical features, biological potential, and color ranges. However, these two families betalains and anthocyanins) don't jointly coincide in plants (Mannino et al., 2021; Schweiggert, 2023). In accordance with their color and chemical structure, betalains are nitrogenous hydrophilic pigments and can be divided into two main groups (Abedi-Firoozjah et al., 2023; Martins et al., 2024; Sadowska-Bartosz and Bartosz, 2021). In this line, betalain family members are classified on betacyanin pigments (red-violet) or betaxanthin (yellow-orange) pigments (Coy-Barrera, 2020; Fernández-López et al., 2020). In the literature, there are inconsistencies concerning the percentages among beetroot betacyanins (Bc) and betaxanthins (Bx); notwithstanding this, it appears clear that the level of Bc in beetroot is much larger than that of Bx. Some investigations concluded that the BC captured till 90% of the entire betalains (Czyzowska et al., 2020; Nikan and Manayi, 2019), while others established that the Bc levels ranged between 55% and 73% (Desseva et al., 2020). It is obvious that the difference in these concentrations controls the coloring root, and the elevated Bc provides its characteristic red-purple coloring (Vieira Teixeira da Silva et al., 2019). In nature, 93 diverse betalains (33 Bx and 60 Bc) have been discovered. Regarding Bx, the descending order is: vulgaxanthin I, vulgaxanthin II, indicaxanthin, miraxanthin, and portulaxanthin (Dias et al., 2020; Kumar and Brooks, 2018; Nowacka et al., 2019). On the other hand, the highest concentration of Bc was represented by isobetanin, neobetanin, and probetanin.

In dried beetroot extracts, Nemzer et al. (2011) noticed a total of 20 Bc and eleven Bx. These authors reported that the Bc was mainly illustrated by 15-decarboxy-betanin, 17-decarboxy-isobetanin, 17-decarboxy-neobetanin, and betanidin. Regarding Bx, the isoleucine-betaxanthin, γ-aminobutyric acid-betaxanthin, and isovulgaxanthin IV are the major detected compounds. Later, Sawicki and Wiczkowski (2018) characterized 18 Bc and 12 Bx in diverse beetroot varieties. These authors found high levels of Bc in 2,17-bidecarboxy-neobetanin, 17-decarboxy-neobetanin, 2,17-bidecarboxy-betanin/isobetanin, 17-decarboxy-betanidin/isobetanidin, and 17-decarboxy-betanin and high levels of Bx in valine-betaxanthin, vulgaxanthin IV, 3-methoxytyramine-betaxanthin, γ-aminobutyric-acid-betaxanthin, and vulgaxanthin III (Bucur et al., 2016). As a practical natural substitute replacing artificial colorants in the food industry, betanin was accepted in Europe (E-162) and the USA (Food and Drug Administration, Title 1 of Code of Federal Regulations, 21 DFR 73, 40) (de Oliveira et al., 2021). However, its utilization as a food colorant agent is acclimatized by its susceptibility to degradation at different processes in the food industry (de Oliveira et al., 2021). Biochemically, and in agreement with their stability, we must note that betalains maintain their structure across a large pH extent (between pH 3 and 7) (Bucur et al., 2016; Jafari, 2017). This fact could make betalains proper in pH-neutral and low-acid foods (Villaño et al., 2016). In comparison with further natural substances like anthocyanins, betalains (i) were invariable over a broader temperature and pH range, (ii) were more water-soluble and obtained higher coloring strength and superior free radical ability, and (iii) had an advanced molar extinction coefficient than artificial dyes, which provides them an extremely significant coloring capacity (Domínguez et al., 2020). On the contrary, throughout food processing and storage, betalains are disposed to oxidation, enzymatic activity, light, metal ions, and heat (Martins et al., 2024). It should be noted that the betalain thermostability is the utmost between pH 4 and 5, and diverse investigations decided that Bc are more thermostable than Bx 5 (Martins et al., 2017). To increase the betalain content, the employment of non-conventional tools, for instance, microwave, could be used (Punia Bangar et al., 2023), and some chemical degradation reactions, including decarboxylation and isomerization, could cleave betalains through heat and/or acid-treating approaches. Nabi et al. (2023) confirmed that thermal processes could augment the level of isobetanin as a result of the epimerization of betanin to isobetanin.

Together with the coloring properties and in comparison with other compounds, the betalains' application as bioactive agents has not been extensively exploited. For instance, beetroot belongs to the group of 10 vegetables with the highest antioxidant potential (Desseva et al., 2020). This is related to the antioxidant potential of betanin, which is ~2-fold more active than anthocyanins, ten-fold more than tocopherol, and threefold more than catechin (Chaari et al., 2022; Chaari et al., 2023a; Thiruvengadam et al., 2024). Moreover, beetroot is renowned for its extraordinary free radical scavenging and antioxidant activities of betalains, which are similar to butylated hydroxytoluene (BHT): a chemical antioxidant (Fu et al., 2020).

8.3 Potential Applications of Beetroot Juice and Its Waste

Beetroot (*Beta vulgaris* L.) can be processed into many products like juice, concentrates, colorants, powder, and so forth. Moreover, by-products of beetroot such as leaf, peel, and pomace have been utilized for many potential applications. In particular, the leaf and peel extracts for antioxidant and the pomace for antimicrobial applications have been investigated (Abdo et al., 2022b; Al-Harbi et al., 2021; Costa et al., 2017; Tsegay and Mulaw, 2024).

Beetroot products have been consumed as a source of nutritional and health components. In particular, the minimally processed beetroot is a source of soluble and insoluble fiber, which is employed for pasta, cakes, and cookies enrichment. Mainly, the best betalain attribute of beetroot makes it a healthy diet, which accounts for protection against degenerative diseases (Costa et al., 2017). Figure 8.1 shows potential applications of beetroot and its by-products.

8.3.1 Health and Nutritional Potentials

Beetroot products are sources of many bioactive compounds like betalains, polyphenols, flavonoids, and other particular compounds such as 5-hydroxymethylfurfural, methyl pyruvate, furfural, and 2,3-dihydro-3,5-dihydroxy-6-methyl-4H-Pyran-4-one (Abdo

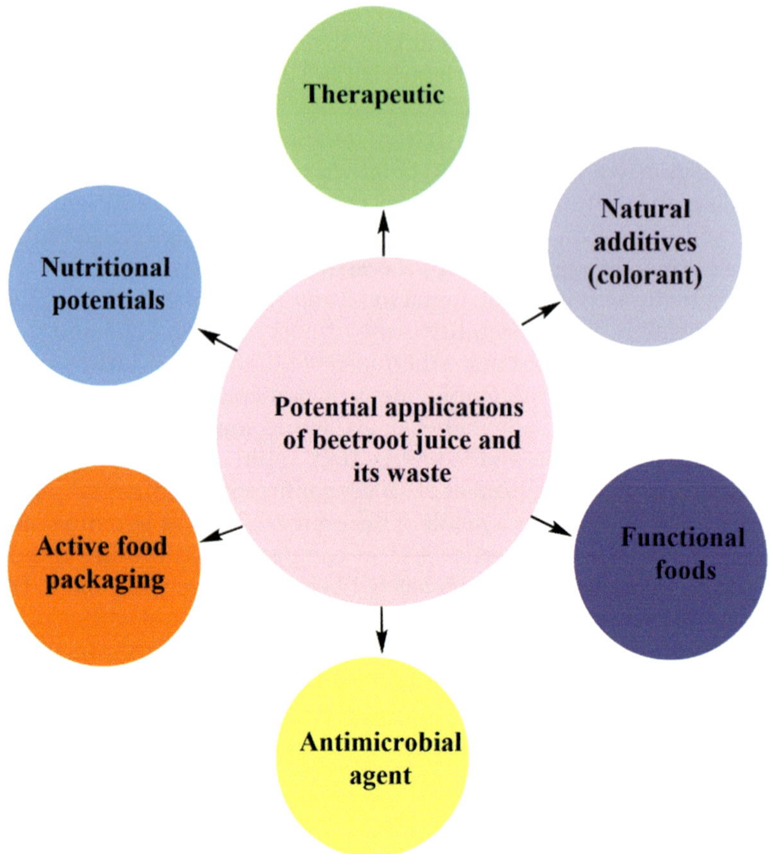

FIGURE 8.1
Potential applications of beetroot and its by-products.

et al., 2022b; Al-Harbi et al., 2021; Kim et al., 2020). These bioactive compounds act as chemopreventive (antioxidant and anti-inflammatory activities, anti-proliferative, cardioprotective activities, ameliorating cancer-related fatigue, hyperlipidemia, and obesity attenuation) as summarized in Table 8.1.

Beetroot, one of the minimally processed vegetable foods, is in high demand by consumers as a natural colorant to impart nutritional, functional, and health benefits in food, cosmetics, and drugs. Availability of betalains beyond the anthocyanins content in beetroots makes them suitable for the stability of a variety of foods and drinks (Costa et al., 2017). In particular, beetroot juice extracted by cold atmospheric pressure plasma helped to enhance the availability of dehydrogenated forms (betalains), bioavailable metals, and phenolic compounds, which inhibit proliferation of Caco-2 cell lines and exhibit non-cytotoxic effects for HIMEC non-malignant endothelial cells (Dzimitrowicz et al., 2021). Moreover, water extracts of beetroot contain mainly betalains, which have shown antioxidant activities against human colon adenocarcinoma and detoxification of enzymes (Koss-Mikołajczyk et al., 2019). Saad et al. (2023) studied the anti-inflammatory activity of red beetroot betalains extracted by a water/methanol/formic acid mixture, which contained mainly betanidin 5-glucoside, isobetanidin 5-glucoside, 2,17-bidecarboxy-neobetanin

TABLE 8.1
Health Applications of Beetroot (*Beta vulgaris* L.) Products

Health Effects	Processing Method	Main Bioactive Compounds	Experimental Model	References
Antioxidant	Beetroot ethanol extract (lyophilized)	Polyphenol and flavonoid	Rat with Cell-free system using DPPH NO inhibitory efficacy in mousemacrophage RAW 264.7 cell line	Kim et al. (2020)
	Beetroot leaf and stem juice fortified with orange juice	Flavonoids and betalains	High-fat diet in a rat model	Abdo et al. (2022b)
	Water extracts of beetroot	Betalains	Cell-free system using standard chemical tests Cellular antioxidant activity assay in HT29 (human colon adenocarcinoma) cells Activity of phase II detoxification enzymes: glutathione S-transferases (GST) and quinone oxidoreductase (NQO1)	Koss-Mikołajczyk et al. (2019)
	Beetroot peel methanol extract	5-hydroxymethylfurfural, methyl pyruvate, furfural, and 2,3-dihydro-3,5-dihydroxy-6-methyl-4H-Pyran-4-one	Cell proliferation assay (human umbilical vein endothelial cells (HUVECs)	Al-Harbi et al. (2021)
Cytoprotective	Beetroot juice by cold atmospheric pressure plasma	Phenolic, dehydrogenated forms (betalains) bioavailable metals	Human epithelial colorectal adenocarcinoma (Caco-2) and human non-malignant intestine microvascular endothelial cells (HIMEC)	Dzimitrowicz et al. (2021)
Anti-inflammatory	Red beetroot betalains extracted by a water/methanol/formic acid mixture	Betanidin 5-glucoside, isobetanidin 5-glucoside, 2,17-bidecarboxy-neobetanin (7 07%), 2-O-glucosyl-betanin, 17-decarboxy-betanidin, neobetanin	Inflammation induction in rat	Saad et al. (2023)
Anti-proliferative	Hydro-alcoholic extract	Betanin	Caco-2 and HT-29 colorectal cancer cells using 3-(4,5-dimethylthiazol-2-yl)-2,5-diphenyltetrazolium bromide (MTT) assay	Saber et al. (2023)
	Red beetroot hydro-alcoholic extract	Betanin	Caco-2 and HT-29 colorectal cancer cells, and KDR/293 normal epithelial cells, using 3-(4,5-dimethylthiazol-2-yl)-2,5-diphenyltetrazolium bromide (MTT) assay	Saber et al. (2021)
	Juice from young shoots of beetroot and roots at full maturity	Polyphenols	Human breast adenocarcinoma MCF-7 (hormone-dependent HTB-22TM) and MDA-MB-231 cell lines (hormone-independent HTB-26™), the human colon cancer cell line Caco-2 (HTB-37™), and the human normal breast MCF-12A (CRL-10782) cultured in medium containing fetal bovine serum (FBS), supplements, and antibiotic mixture	Piasna-Słupecka et al. (2023)

(*Continued*)

TABLE 8.1 ***(Continued)***

Health Applications of Beetroot (*Beta vulgaris* L.) Products

Health Effects	Processing Method	Main Bioactive Compounds	Experimental Model	References
Cardioprotective	Beet root juice powder	Nitrates and antioxidants	Adult male Wistar rats' cardiomyocytes	Das et al., (2016)
Ameliorating cancer-related fatigue	Beetroot juice as a dietary nitrate supplementation	Nitrates	Basal mitochondrial function and whole-body oxygen consumption in healthy volunteers	Larsen et al. (2011)
Hyperlipidemia reduction	Beet leaves and stalk juices	Polyphenols	Human patients from 20 to 59 years of age with dyslipidemia	Gomes et al. (2019)
Obesity attenuation	Beetroot leaf and stem juice fortified with orange juice	Flavonoids and betalains	High-fat diet in a rat model	Abdo et al. (2022b)
Hepatoprotective	Crude aqueous and ethanolic extracts of the root tissue of red and high-pigment beet	Vulgaxanthins I and II, and (iso)betanin	Rats	Lee et al. (2005)

(7.07%), 2-O-glucosyl-betanin, 17-decarboxy-betanidin, and neobetanin. The beetroot hydro-alcoholic extract contains betanin as its major betacyanin, which shows anticancer and pro-apoptotic effects, as reported by Saber et al. (2023). About 92 μg/mL and 107 μg/mL of beetroot hydro-alcoholic extract were found to be optimum IC_{50} to treat HT-29 and HT-29 colorectal cancer cells, respectively. Nirmal et al. (2024) summarized the studies on betalains that have been used as potential organ-protecting compounds, in particular, heart protection [25 and 100 mg/(kg·day)], liver protection [25 and 100 mg/(kg·day)], kidney protection [25 and 100 mg/(kg·day)], neuroprotection [50 and 100 mg/(kg·day)], lung protection (at 20, 60, and 180 mg/kg), gut protection (at 50 mg/kg and 1 g/kg), and repro protection (at different doses of red beetroot extract, 300 mg/kg beetroot extract), in which all the tests were performed in animal models (rats and mice). The potential therapeutic applications of betalains are depicted in Figure 8.2.

Different processing techniques and different parts of beetroot have been studied for their nutritional and health-related applications. For instance, juice from young shoots of beetroot and mature roots was found to possess potential antiproliferative and apoptotic effects on human breast cancer and normal cells when consumed with proper eating habits (Piasna-Słupecka et al., 2023). Commercial beetroot juice powder was analyzed for its protection against doxorubicin toxicity in cardiomyocytes. Moreover, it was found to enhance apoptosis in breast cancer cells, which was observed in adult male Wistar rats' cardiomyocytes. The main reason for the effective cardioprotective effect of beetroot juice was the presence of nitrate and antioxidants (Das et al., 2016). Apart from the main sources of beetroot juice, the by-products, such as leaves and stalks, were found to be rich in polyphenols, which play a significant role in preventing cardiovascular disease in humans. Mainly, two organic beet leaves and stalk juices contained polyphenols (32 or 77.5 mg EAG/100 mL), which reduced diastolic blood pressure (high-density lipoprotein cholesterol) (Gomes et al., 2019).

Secondary metabolites, commonly known as betalains, present in beetroot have been used as natural dyes in the food industry, which show anticancer activity (Ninfali and Angelino, 2013). Many functional foods and nutraceutical products have been developed from beetroot. "Beetroot has been employed as a natural food colorant and to enhance the nutritional and sensory qualities of foods. For instance, beetroot pulp (30%) fortified with wheat flour enhanced the color, functional, phytochemical, nutritional, and sensory characteristics of the final Asian noodle (Chhikara et al., 2019).

Beetroot peel and pomace are particular by-products. Red beetroot peel and pomace are rich sources of protein, carbohydrate, total dietary fiber, lipids, soluble sugar, and trace minerals. These significant nutritional qualities of beetroot by-products are applicable for enhancing the nutritional qualities of pasta, cakes, and cookies. Mainly, they can be used to replace synthetic pigments (Tsegay et al., 2024).

8.3.2 Natural Additives, Functional Ingredients (Colorant)

The inherent colorant capacity of betalain in beetroot is due to the presence of betanin, and it is approved as a food and pharmaceutical product that can be utilized as a natural red colorant (Vieira Teixeira da Silva et al., 2019). Beetroots are potential additives or preservatives in food processing. Mainly, beetroot processed by drying and freezing has a stable nature and is nontoxic, non-carcinogenic, and nonpoisonous. Moreover, active components like betacyanin, betaxanthin, minerals, amino acids, flavonoids, and phenolic acids are present in beetroot. This makes it suitable to develop functional food varieties (Thiruvengadam

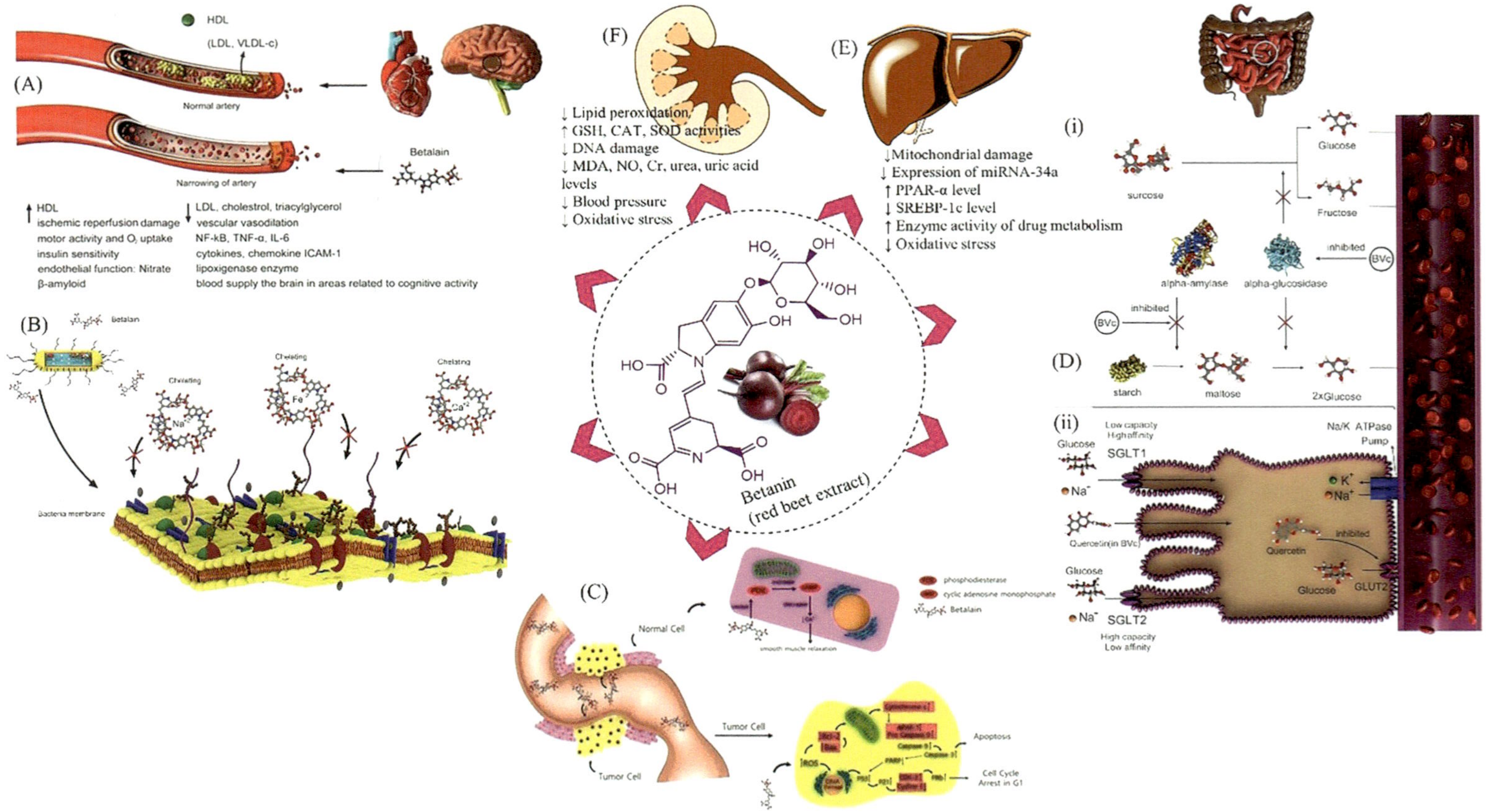

FIGURE 8.2
Betalains as therapeutic applications. (a) Reduction of blood low-density lipoprotein (LDL), increasing high-density lipoprotein (HDL), and vascular vasodilation by betalain. (b) Betalain prevents bacterial poisoning by chelating Fe^{+2}, $Ca^{+2,}$ and Mg^{+2} ions, in which bacterial growth is inhibited due to the loss of these ions. (c) Effects of betalain on normal cells, cancer cells, and induction of apoptosis. (d) Beetroot has a hypoglycemic function (i) controlling glucose in diabetes by alpha-amylase and alpha-glucosidase (ii) glucose transporter isoform 2 (GLUT2) inhibition. (e) Nephroprotective effect of betalains on human kidney organ (f) Hepatoprotective effect of betalains on human liver. The ↑ and ↓ arrows indicate upregulation and downregulation, respectively. (a, b, c, and d reproduced from Madadi et al. (2020) with permission from MDPI, copyright 2020.)

et al., 2024). Beetroot red (E 162) is considered a natural color since it contains a number of pigments, which have a long history in food consumption. This can be prepared as spray–dried powders, simple extracts, and/or refined extracts. However, due to many reasons like limited studies on toxicological qualities, acute and short-term toxicity studies are very limited to draw conclusions, studies on chronic toxicity and carcinogenicity are inadequate, and the toxicological database is inadequate to establish an acceptable daily intake (ADI) (EFSA Panel on Food Additives and Nutrient Sources added to Food (ANS), 2015). Beetroot betanin extract purified by semi-preparative HPLC-LC/MS has shown mild stability at cold temperatures and attractive stability up to −30°C. These properties make betanin effective as a natural antioxidant displaying the characteristics of a peroxy-radical scavenger beyond its colorant capacity (Vieira Teixeira da Silva et al., 2019). Artificial colorants are frequently utilized in food, pharmaceutical, and cosmetic industries as additives due to their easy availability, low cost, and higher stability (Samanta et al., 2024). However, beyond their accessibility and colorant capacity, natural betanin is the most preferable since it incorporates natural antioxidants and many health-related benefits. Costa et al. (2017) investigated minimally processed beetroot waste dried at 70°C, valorized for flour development. This flour has shown color stability for 20 days without changing microbiological or color change. The minimally processed beetroot waste can be valorized as an alternative source of natural and functional ingredients that have high antioxidant activity and betalain content.

8.3.3 Functional Food and Beverage

Functional beverages that are rich in protein and antioxidants have been developed by mixing beetroot with other food resources. For instance, a functional beverage was developed by mixing 5% of whey protein isolate and 2.5% concentrations of beetroot peel water extract, which was flavored by adding 5% strawberry puree. This mix was found to be optimum for a healthy body with a high-quality protein and a considerable capacity of antioxidants. This is recommended to promote muscle performance and prevent muscle damage (Abdo et al., 2022a). Besides, Stoica et al. (2024) studied the application of the peel of red beetroot powder in developing a whey fruit-based beverage. They reported that the powder had a high concentration of bioactive phytochemicals such as betalains and dietary fibers, which make it applicable as a natural food ingredient. The addition of peel of red beetroot powder to the whey fruit-based beverage enhanced the total polyphenols (1239.35 mg GAE/100 g dw), antioxidant activity (90.02%), and its nutritional value. They also reported that valorizing this type of by-products can promote sustainable food processing. Fermented milk gels were produced by fortifying stevia as an alternative non-caloric sweetener and red beetroot as a natural colorant. The addition of red beetroot enriched the bacterial growth of *L. acidophilus* and the acceptable sensory and color quality of the final fermented milk gel. The main reason for this enrichment is due to the presence of biologically metabolized compounds of red beetroot. This beetroot fortification is able to improve therapeutic dairy products that have positive synergistic interactions by providing a rich source of probiotic bacteria (Ozdemir and Ozcan, 2020). Račkauskienė et al. (2015) investigated the effect of dried beetroot juice on the formation of the Maillard reaction using a milk model system, considering the formation of (carboxymethyl)lysine and Nε-(2-furoylmethyl)-L-lysine (furosine). They reported that the beetroot juice reduced the formation of furosine than the (carboxymethyl)lysine. Such a novel source of functional ingredients is potent in anti-carcinogenic and antiglycation activities.

By-products of beetroot, like the stalk, have been valorized for functional food development. In particular, the extract of beetroot stalk was used as a functional natural colorant to improve yogurt beverages, which affected nutritional quality and its storage stability (Abdo et al., 2023). Ultrafiltered betalains extract from the beetroot (*Beta vulgaris*) bagasse was investigated for its potential application as active food packaging by incorporating it into zein films. Antioxidant activity and mechanical, morphological, and physicochemical properties of zein films were improved due to the addition of an ultrafiltered betalain extract (Rodríguez-Félix et al., 2022).

8.3.4 Antimicrobial Agent

The presence of high levels of betalains and other bioactive compounds in red beets is responsible for the antimicrobial quality. Mainly, the presence of betacyanins and betaxanthins makes beetroot a potential antimicrobial agent. The antimicrobial qualities of beetroots are being potential agents in food, consumer products, and pharmaceutical processing industries. Generally, gram-positive bacteria were found to be more sensitive to red beet extract than gram-negative bacteria (Kumar and Brooks, 2018).

Gong et al. (2022) studied the antibacterial application of beetroot (*Beta vulgaris*) extract against Listeria monocytogenes by applying it in cooked pork. They reported that 20 mg/mL of the beetroot extract was the minimum concentration to inhibit *Listeria monocytogenes*. This study implies that the beetroot extract could be a potential natural preservative in foods like cooked meat production, which inhibits the growth of bacteria like *L. monocytogenes*.

Vulić et al. (2013) investigated the phytochemical profile and the antiradical, antimicrobial, and cytotoxic activities of beetroot pomace extract industrial by-products. The extract contained phenolics (45.68 mg gallic acid equivalents g^{-1}), flavonoids (25.89 mg rutin equivalents g^{-1}), and betalains (4.09 mg betanin g^{-1}; 7.32 mg vulgaxanthin I g^{-1}). These bioactive compounds were potent antibacterial against Gram negative bacteria (*Salmonella typhimurium, and Citrobacter freundii*) and Gram positive bacteria (*Staphylococcus aureus, Staphylococcus sciuri, and Bacillus cereus*) Red beet pomace was found to be an antimicrobial agent against gram-positive bacteria (*S. epidermidis*=13.33 mm IZ, 0.75 MIC (mg/ml), *B. cereus* =20.33 mm IZ, 0.5 MIC (mg/ml), *S. aureus* =12 mm IZ)), gram-negative bacteria (*Escherichia coli*=13.33 to 18 mm IZ, 1.5 mg/ml MIC, *P. aeruginosa* =13.33 mm IZ, 4.5 mg/ml MIC, *C. freundii*=20.33 mm IZ, *E. cloacae*=12 mm IZ, *S., typhymurium*=12 mm IZ, *Shigella flexneri*=12 mm IZ, *K. pneumonia*=22 mm IZ), and against fungi (*Rhodotorula rubra* =0.5 mg/mL MIC *C. albicans*=1 mg/mL MIC) (Tsegay and Mulaw, 2024).

8.4 Functional Ingredient in Dairy and Food Products

Beetroot is a nutrient-dense root vegetable packed with proteins, carbohydrates, vitamins (including vitamin B complex and vitamin C), minerals, and fiber. It also contains a notable concentration of phenolic compounds, betalains, and antioxidants such as coumarins, carotenoids, sesquiterpenoids, triterpenes, and flavonoids (Figure 8.3) (Babarykin et al., 2019).

In addition to these compounds, beetroot is a nutritious vegetable packed with bioactive compounds such as alkaloids, tannins, and amino acids, including methionine, leucine, isoleucine, cysteine, histidine, and arginine (Babarykin et al., 2019; Lechner and Stoner, 2019). The exact chemical composition can fluctuate depending on the beet variety and the part of the plant (leaf, stem, root, or peel) (Sawicki et al., 2016). For example, carotenoids are more abundant in beetroot leaves than in tubers because they are found in chloroplasts, the structures responsible for photosynthesis (Lechner and Stoner, 2019).

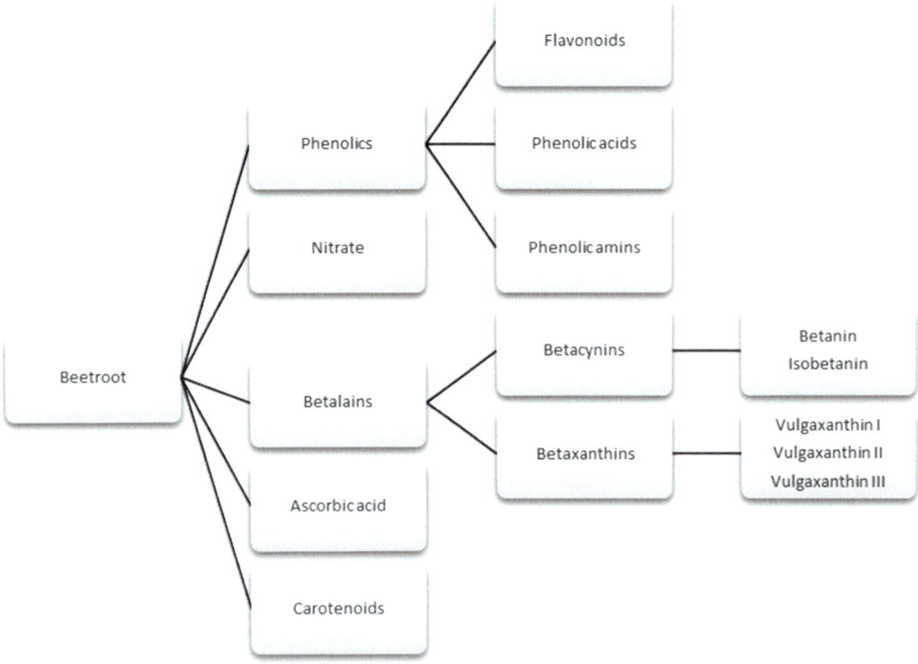

FIGURE 8.3
Beetroot's bioactive components. (Clifford et al., 2015.)

These compounds have been associated with numerous health benefits, including anti-carcinogenic, antihypertensive, anti-diabetic, anti-inflammatory, anti-oxidant, hepatoprotective, and anti-microbial properties and can also support vascular health and immune regulation (Figure 8.4) (Fu et al., 2020; Hadipour et al., 2020; Punia Bangar et al., 2023).

Beetroot's vibrant red color, primarily due to betalains (a water-soluble nitrogenous pigment derived from tyrosine, a type of amino acid), makes it a popular addition to many food products. Betalains come in two main colors: betaxanthins, which are yellow-orange, and betacyanins, which are red-violet. Beetroot can be processed into various forms and incorporated into different foods, from dairy and cereal to meat and beverage products. The specific properties of the final food product depend on the type and concentration of the beetroot component used.

8.4.1 Dairy Products

As a natural food colorant and antioxidant, beetroot offers a healthier alternative to artificial additives in dairy products. By incorporating beetroot (in various forms such as paste, juice, puree, powder, and extract) into dairy products, manufacturers can create more nutritious and appealing options for consumers (Table 8.2). In ice cream, beetroot addition can improve nutritional quality, viscosity, overrun, and melting rate. Beetroot-enriched ice cream contains higher levels of bioactive compounds such as betalains, B-carotene, total phenolic acids, and flavonoids, which makes it a healthier option with improved antioxidant properties (Atteallah et al., 2019; Evstigneeva et al., 2020; Ozturk et al., 2014; Punia Bangar et al., 2023; Roriz et al., 2018). Beetroot can boost the nutritional value, viscosity, overrun, melting rate, and overall quality of ice cream. However, the high concentration

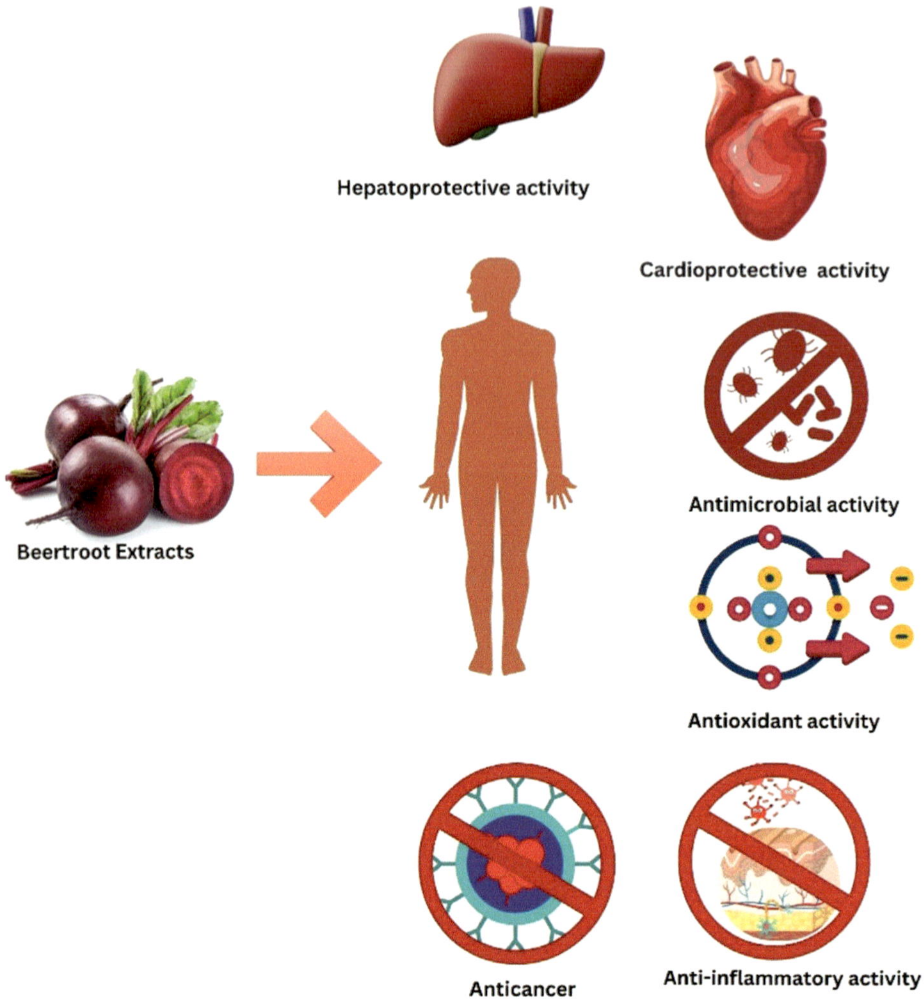

FIGURE 8.4
Health benefits of beetroot and its extracts.

of bioactive compounds in beetroot ice cream can sometimes affect its texture and stability. Adding stabilizers like carboxymethyl cellulose can help maintain the desired consistency (Chairuni et al., 2019). Beetroot can be used as a natural coloring ingredient in strawberry-flavored ice cream and flavored milk, offering a healthier alternative to artificial strawberry color. However, higher concentrations of beetroot can impart a malignant, dark color and a pronounced vegetable flavor, which may not be desirable for consumers (Manoharan et al., 2012; Punia Bangar et al., 2023).

Butter, cream cheese, and cheese contain beetroot powder, puree, and juice, which can act as natural coloring and antimicrobial agents. Beetroot's high antioxidant activity results in a longer shelf life for these products (Asadaii et al., 2020; Junqueira-Goncalves et al., 2011; Prudencio et al., 2008; Raikos et al., 2016; Sandhya and Priya, 2017), reducing peroxide value, lowering lipid oxidation, and suppressing the growth of bacteria and molds (Jovanović et al., 2021).

TABLE 8.2

The Use of Diverse Beetroot Forms in Dairy Production and Their Impact on Product Quality

Product	Form of Beetroot	Description	Reference
Ice cream	Beetroot paste	Including beetroot improved the viscosity, overrun, boosted sensory evaluations, and slowed the melting process of the ice cream. A concentration of 20% beetroot paste produced premium-quality beetroot ice cream	Chairuni et al. (2019)
	Beetroot juice	Incorporating beetroot juice influenced the sensory characteristics of the product. At 4% concentration, beetroot juice produced a sample comparable in acceptability to standard strawberry-flavored ice cream. Lower concentrations of beetroot juice resulted in a lighter color, while higher amounts negatively impacted the ice cream's appearance	Chairuni et al. (2019)
		Adding beetroot juice improved the ice cream's nutritional value, reduced overrun, and enhanced melting resistance. It also increased antioxidants and other nutrients, with 5% concentration yielding high sensory acceptability.	Ateteallah et al. (2019)
	Beetroot powder	While adding beetroot can improve the nutritional value, texture, and melting rate of ice cream, reducing sesame powder can also positively impact these properties. The optimal formulation, with 19% sesame and 16% beetroot flour, offers a higher content of iron and protein while maintaining a lower fat content.	Punia Bangar et al. (2023)
		The addition of beetroot juice to strawberry ice cream noticeably changed its taste, texture, and overall sensory qualities. Among the different inclusion levels of beetroot juice, 4% had the maximum scores.	Manoharan et al. (2012)
	Beetroot puree	Beetroot puree increases viscosity and declines overrun. Melting resistance peaks at 15% puree but falls with higher amounts. It also raises air bubble size and acidity. Ice cream containing 20% beetroot puree was rated the most favorably in terms of taste and overall sensory appeal.	Evstigneeva et al. (2020)
	Beetroot extract	Beetroot extracts reduced sucrose, fat, lactose, and energy in ice cream and increased redness, ash, and protein.	Roriz et al. (2018)
		Beetroot ice cream can raise hemoglobin levels compared to regular ice cream, while both types of ice cream increase total blood protein levels.	Bakara et al. (2022)
	Beetroot pigment	Betalain ice cream had better acidity, pH, fat, protein, microorganism count, and sensory scores than Red Dye 40 (synthetic pigment) ice cream. Betalain also preserved better over 70 days	Ali and Jameel (2023)
Mellorine	Beetroot Juice	Beetroot juice in mellorine reduced viscosity, consistency, and overrun while increasing antioxidant activity, phenolic content, and melting rate. Mellorine containing 5% beetroot juice was rated the most favorably in terms of taste and overall sensory appeal.	Ozturk et al. (2014)

(Continued)

TABLE 8.2 (*Continued*)

The Use of Diverse Beetroot Forms in Dairy Production and Their Impact on Product Quality

Product	Form of Beetroot	Description	Reference
Flavored Milk	Beetroot Juice	Flavored milk made with 2% beetroot juice is successful. A high-quality milk alternative can be created by combining 80% cow's milk, 20% soy milk, and 2% beetroot juice.	Punia Bangar et al. (2023)
	Beetroot extract	Beetroot-flavored milk, rich in antioxidants, offers therapeutic benefits for hypertension, cancer, obesity, heart health, and immunity.	Ashraf et al. (2022)
Curd	Beetroot peel	Beetroot peel extract boosted curd's antioxidant activity, shelf-life, and protein while reducing peroxide formation and increasing acidity in kefir acidified curd.	Vanlallawmzuali and Pushpa (2021)
Whey beverage	Beetroot extract	The best sensory scores were achieved with 20% beetroot extract in paneer whey, making it the most acceptable, cost-effective, and adaptable option for producing beetroot whey beverages.	Kamate and Padghan (2018)
Butter	Beetroot extract powder	Red beetroot extracts effectively controlled peroxide and acidity values in cow's milk butter, showing the best chemical preservation effects.	Asadaii et al. (2020)
Cream cheese	Beetroot puree	Beetroot puree reduced the fat content and boosted the protein in cream cheese while maintaining acceptable color and shelf stability for 30 days.	Sandhya and Priya (2017)
Cream cheese	beetroot juice extract	Beetroot extract performed similarly to carmine cochineal in flavor and overall appearance, with only a slight preference for carmine in color. This highlights the potential of beetroot extract as an effective natural food colorant.	Junqueira-Goncalves et al. (2011)
Cheese	beetroot betalains	Adding beetroot betalains to Petit suisse cheese resulted in satisfactory pigment stability and color retention over 40 days at 6°C, with adequate performance in maintaining color.	Prudencio et al. (2008)
Yogurt	Beetroot juice	Yogurt containing 4% beetroot juice was rated the most favorably in terms of taste and overall sensory appeal. Adding beetroot juice increased the protein, fat, and ash content without affecting the acidity. It also improved consumer preference for goat milk yogurt.	Damunupola et al. (2014)
	Beetroot extract powder	Encapsulated beetroot extract powder improves yogurt's fat, protein, and texture, while beetroot juice increases pH and antioxidant levels but causes more syneresis.	Flores-Mancha et al. (2021a, 2021b)
	Beetroot powder	Beetroot powder improved yogurt's pH, texture, and nutritional content, with 8% powder being most preferred, and enhanced bacterial counts during storage.	Yadav et al. (2016)
	Beetroot powder	Beetroot increased ash, protein, lipids, and sugars in yogurt, while ensuring safety and high consumer acceptance.	de Oliveira et al. (2020)
	Beetroot juice	Yogurt with 3% beetroot juice was found to have high antioxidant properties, a rich concentration of bioactive compounds, and was well-received by consumers.	Eveline and Nawangsih (2019)

(*Continued*)

TABLE 8.2 (*Continued*)

The Use of Diverse Beetroot Forms in Dairy Production and Their Impact on Product Quality

Product	Form of Beetroot	Description	Reference
	Beetroot extract powder	Adding encapsulated beetroot powder lowered acidity and ash content compared to the extract, without affecting yogurt quality, and increased product acceptability.	Antigo et al. (2020)
	Beetroot bits	Adding beetroot reduced yogurt's lactose, protein, and lactic acid content, with 2% beetroot bits producing the best quality yogurt.	Chairunnissa et al. (2017)
	Beetroot powder	Beetroot powder increased yogurt's acidity, improved protein, fat, and folic acid levels, and was most acceptable at 10%.	Nyamete and Mongi (2017)
	Beetroot juice	Beetroot juice improved vitamin C and mineral content in yogurt but decreased protein and fat. Yogurt with 10% beetroot juice was highly acceptable.	Elizabeth and Onyinyechi (2012)
	Beetroot powder	Beetroot powder boosted protein and total solids in yogurt, improved color, and enhanced overall acceptability. It increased viscosity and acidity but reduced fat, syneresis, and pH.	Shalaby and Hassenin (2020)
	Beetroot pomace powder	Yogurt with beetroot powder had high *St. thermophilus* viability, lower syneresis, and better palatability. It also inhibited colon cancer cells and had anti-*E. Coli* activity.	Jovanović et al. (2021)
	Beetroot powder	Beetroot powder and basil seed gum improved yogurt's texture, probiotic viability, and antioxidant activity. The best acceptability was achieved at 0.2% basil seed gum and 0.1% beetroot powder.	Ghasempour et al. (2020)
	Beetroot powder	Beetroot powder improved yogurt's acidity and water-holding capacity and reduced syneresis, enhancing consumer acceptability.	Adjei et al. (2024)
	Beetroot puree	Beetroot puree at an optimal concentration of 2.03% exhibited the best performance in yogurt, improving color, flavor, and consumer acceptability while maintaining pH and titratable acidity.	Abdo et al. (2023)
	Beetroot stalk water extract (BSE)	Beetroot stalk extract (BSE) improved raspberry yogurt's nutritional quality, increased betalain and phenolic content, and enhanced color stability and antioxidant activity. Over 14 days, yogurt with BSE retained a reddish hue and superior dietary profile.	Abdo et al. (2023)

Beetroot, in different forms, can boost the nutritional value, antioxidant properties, and betalain content of curd, whey beverages, and yogurt. This can also make these products more appealing to consumers (Antigo et al., 2020; Chairunnissa et al., 2017; Damunupola et al., 2014; Elizabeth and Onyinyechi, 2012; Eveline and Nawangsih, 2019; Flores-Mancha et al., 2021a, 2021b; Kamate and Padghan, 2018). Beetroot-enriched yogurt has been shown to have lower syneresis (water separation), improved acidity, firmer gel, and better consumer acceptance (Abdo et al., 2023; Adjei et al., 2024; Dabija et al., 2019; Nyamete and Mongi, 2017; Shalaby and Hassenin, 2020; Yadav et al., 2016). Beetroot can also increase the probiotic viability of these products by promoting the growth of *Lactobacillus bulgaricus* and *Streptococcus thermophilus*, ensuring microbiological safety (Ghasempour et al., 2020; Jovanović et al., 2021).

8.5 Cereal-Based Products

Cereal-based foods, while providing essential nutrients like carbohydrates and proteins, often fall short in providing phytochemicals, vitamins, minerals, and fiber. As consumers increasingly seek healthier options, there is a rising demand for cereal-based products that offer additional functional health benefits (Kamali Rousta et al., 2021). Beetroot, a functional ingredient, can be added to these products in various forms, such as powder, juice, pomace, or extract. This incorporation provides a nutritional boost to these daily consumed foods, addressing the dietary needs of a specific population (Abdo et al., 2021; Mitrevski et al., 2023). Beetroot powder, rich in nutrients, can improve the nutritional content of baked goods like bread, cakes, and cookies. Incorporating beetroot powder can increase the fiber, ash, and protein content of various foods while reducing their fat content (Daunaravičiūtė et al., 2020; Lucky et al., 2020; Nagib and Zidan, 2019). Adding beetroot to baked goods like cookies, extruded snacks, ready-to-eat snacks, breakfast cereals, pasta, and noodles can enhance their nutritional value and influence their texture and taste. For example, beetroot can make bread firmer and harder due to its high fiber content, while in cookies and biscuits, it can increase weight and density, alter hardness, and reduce the spread factor. These changes in texture and sensory characteristics may depend on the specific formulation of the baked good (Alshehry, 2019; Aulia and Sunarharum, 2020; Daunaravičiūtė et al., 2020; Ingle et al., 2017; Kohajdová et al., 2018; Mitrevski et al., 2023). Beetroot extract significantly enhances baked goods' bioactive properties, increasing phenolic compounds' levels. Beyond these nutritional benefits, beetroot-enriched baked goods offer functional health advantages such as cardioprotective, anti-anemic, blood pressure-lowering, and antioxidant effects and improved liver function enzymes (Amnah, 2013; Čakarević et al., 2021; Hidalgo et al., 2018; Mitrevski et al., 2023; Ranawana et al., 2016; Srivastava and Singh, 2018). These products also have higher essential minerals like Ca^+, Mg, K, zinc, and iron (Abdo et al., 2021). Consumers generally accept beetroot-enriched products, especially when beetroot is used as a natural red coloring agent, making them a healthier and more appealing choice for consumers (Ambedkar and Pradesh, 2019; Nagarajaiah et al., 2019).

8.6 Meat Products

Beetroot, a nutrient-dense vegetable, serves as a multifunctional ingredient in various meat products, from poultry to sausages. Its incorporation not only provides an appealing red hue but also enhances the overall quality of these products. This is attributed to beetroot's rich composition of bioactive compounds like betalains, carotenoids, and phenolic acids, which contribute to its antioxidant and antimicrobial properties (Aykın-Dinçer et al., 2021; Fernández-López et al., 2023; Marrone et al., 2021; Munekata et al., 2021; Pennisi et al., 2020).

The addition of beetroot to meat products has been shown to decrease pH levels and improve storage stability (Aykın-Dinçer et al., 2021; Fernández-López et al., 2023; Marrone et al., 2021; Pennisi et al., 2020).

Moreover, it can increase product yield and reduce cooking shrinkage (Rayees-ul-Islam and Khan, 2016; Singh et al., 2020). The characteristic red color of betalains, coupled with beetroot's nitrate content, enhances the redness of meat products, both internally and externally, and maintains this color throughout storage (Aykın-Dinçer et al., 2021; Bellucci et al., 2020; Ozaki et al., 2021; Packer et al., 2021; Sucu and Turp, 2018).

Beetroot is a substantial source of nitrates (Brzezińska-Rojek et al., 2023; Munekata et al., 2021), making it a potential natural alternative to synthetic nitrates and nitrites in meat curing. This aligns with growing consumer preferences for cleaner-label products. The bioactive compounds in beetroot contribute to the product's oxidative stability by inhibiting lipid oxidation and microbial growth (Aksu et al., 2020; Aykın-Dinçer et al., 2021; Bellucci et al., 2020; Chaari et al., 2023b; Marrone et al., 2021). Consequently, beetroot-incorporated meat products exhibit extended shelf life and improved overall acceptability.

8.7 Beverages and Gels

Beetroot, in various forms, can be incorporated into a range of beverages and gels to enhance their nutritional and functional properties. In probiotic milk gels, beetroot pulp can increase antioxidant properties while also promoting the growth of *Lactobacillus acidophilus* (Özdemir and Özcan, 2020). The functional properties of beetroot juice can be enhanced through gelation, resulting in a product with a higher concentration of bioactive compounds.

Consuming functional beetroot gels offers a nutritional strategy to increase the intake of bio-accessible nutrients, including nitrate, antioxidants, and potassium, which may benefit cardiovascular health and exercise performance (Morgado et al., 2016; Silva et al., 2016).

Milk-based drinks like fermented buttermilk, low-calorie lassi, rayeb milk, and yogurt can benefit from the addition of beetroot. Beetroot can lower acidity, improve antibacterial and antioxidant properties, increase total phenolic content, and support the growth of beneficial probiotic bacteria like Streptococcus thermophiles and Lactobacillus. This can also make these beverages more appealing to consumers (Ganguly et al., 2017; Hashem, 2018; Özdemir and Özcan, 2020; Pasca et al., 2016).

Beetroot can add value to fruit juices and non-alcoholic beverages, whether fermented or not. Beetroot juice, spray-dried powder, or betatalin extract can improve taste, consumer preference, and levels of essential nutrients like Ca, Fe, Na, Zn, antioxidants, total phenols, and flavonoids (Hashem, 2018; Kaimainen et al., 2015; Sharma et al., 2020; Yang et al., 2021).

8.8 Beetroot as Coating Material

The current progressions in emerging active and intelligent functional films resulted from their fortification with bioactive fractions from beetroot powder or extracts (Table 8.3). The beetroot extract and beetroot powder addition is an innovative method for producing edible, biodegradable, active, and smart packaging films. They serve as a natural additive in the films. Newly established films showed good structural, mechanical, biological, barrier, and functional features (Cejudo-Bastante et al., 2020; Otàlora Gonzàlez et al., 2020; Sucheta et al., 2019; Tran et al., 2017; Yao et al., 2021) (Tran et al., 2017; Sucheta et al., 2019; Cejudo-Bastante et al., 2020; Otàlora Gonzàlez et al., 2020; Drohsler et al., 2021). By virtue of their bioactive compounds, viz., Bc, Bx, and betanin, established films with beetroot

TABLE 8.3

The Application of Beetroot Powder/Extracts Forms in Films and Coatings and Their Quality Features

Beetroot Fraction and Concentrations	Wall Material	Quality Product	References
Beetroot powder at 0.02 g/mL	Pectin and corn flour at (60/40; 80/20 and 50/50)	• otal soluble matter and viscosity of the film solution. • Moisture, ↘ thickness, and water solubility. • tensile strength and ↘ of water vapor and permeability • a* and ↘ L* values. • ↗ of antioxidant activity	Sucheta et al. (2019)
Beetroot powder at levels ranged between 2% and 10%	Starch	• oisture uptake • static water contact angles • ntioxidant properties. • Young's modulus • → strain at break values and elongation properties	Tran et al. (2017)
Betacyanins from beetroot at 0, 20–2 mg/100 g of film solution	Starch/Polyvinyl alcohol	• L* value of the films, and ↗a* and b* • thickness, ↘water vapor permeability, ↗tensile strength elongation break. • he antioxidant activity of films. • Sensitivity to pH 3<12 buffer and ammonia • Film's color changes from red to purple when shrimp is soiled	Yao et al. (2021)
Beetroot extract at 0.1%–2.5%	Ethylene vinyl	• ilm thickness, solubility, and moisture • extensibility • stress at the rupture • *, ↗ a* and b* • water vapor permeability • ↗ wettability and contact angles	Cejudo-Bastante et al. (2020)
Beetroot extract at 1%–3%	Pectin	• L* and ↗ a* • the O2 and water vapor permeability • the tensile strength • A change of color from pink to yellow with a ↗in pH. • Practical application to track the freshness of fresh, chilled pork.	Otàlora Gonzàlez et al. (2020)
Blanched beetroot powder at 1.5%	Pectin	• thickness • water solubility and ↗wettability • * and ↗ a* • No color change with pH during storage (One month)	Torres Vargas et al. (2021)
Beetroot extract 0.5%–5%	Chitosan and starch	• thickness of the films and ↘ the density • Variation of color from ink to red wine • ransmittance and transparency and ↗opacity. • ↗ antioxidant and antimicrobial activity	Guo et al. (2021)
Beetroot extract at 5%	Polylactide and Polycaprolactone	• Young's modulus of the films • ↘of the total color change, whiteness index, and ↗for chroma.	Drohsler et al. (2021)

(Continued)

TABLE 8.3 ***(Continued)***

The Application of Beetroot Powder/Extracts Forms in Films and Coatings and Their Quality Features

Beetroot Fraction and Concentrations	Wall Material	Quality Product	References
Beetroot extract at 0.25%, 0.5%, and 1%	gelatin-sodium alginate	• f mechanical and physical properties • ↗ antioxidant and antimicrobial activity	Chaari et al. (2022)
Betacyanin from beetroot peels at 0.5%	CMC/flaxseed gum	• echanico-physical traits of films • high sensitivity to ammonia and pH change. • Good monitoring freshness ability	Chaari et al. (2024)

extracts displayed influential antimicrobial and antioxidant activities. On the other hand, due to their warmth to an extensive pH range, beetroot pigments can be employed in active and intelligent packaging as quality modification markers in the packaging of fresh food products, owing to their sensitivity to a wide pH range and gases. As an illustration, high protein foods such as meat, fish, and seafood, throughout their proteolytic spoilage, leading to the production of different gases such as ammonia, amines and volatile nitrogen compounds which upsurge their pH and this pH change aids to predict the change in the quality of the product (Chaari et al., 2024; Guo et al., 2021). Obviously, the employment of beetroot pigments in the biodegradable and functional edible films could decrease the application of synthetic as indicator to expect the fresh food quality.

8.9 Summary and Perspectives

By virtue of the presence of betalains, phenolic compounds, carotenoids, ascorbic acids, flavonoids, vitamins, minerals, and nitrates in it, beetroot is considered an excellent vegetable with powerful nutritive features, health-supporting properties, and functional importance. This material, used in various forms such as extract, powder, juice, and pulp, has been successfully incorporated into several food products and applied in active and intelligent packaging as well as edible coatings. Many supplements based on beetroot are currently obtainable and inexpensive, particularly when compared to synthetically industrial drugs. This intensified interest has encouraged widespread investigation into the nutraceutical and medicinal features of beetroot by scientists. Over the years, beetroot and its by-products have revealed aptitude in numerous applications, like functional foods and pharmaceuticals, which provide potential added value to the food/pharmaceutical industry, while decreasing waste. When employing beetroots as pigments in different matrices, the color stability is a major matter, and the upcoming investigation should focus on stabilizing the beetroot pigments throughout formulation, handling, and preservation. Once applied in the powder au lieu of juice or extract, beetroot pigments and activity could maintain their quality and integrity; consequently, beetroot extract/powder must be considered a possible approach to preserve the products' quality. In addition, the beetroot toxicity should also be evaluated at different stages of production.

References

Abdo, E. M., Allam, M. G., Gomaa, M. A. E., Shaltout, O. E., & Mansour, H. M. M. (2022a). Valorization of whey proteins and beetroot peels to develop a functional beverage high in proteins and antioxidants. *Frontiers in Nutrition, 9.* https://www.frontiersin.org/journals/nutrition/articles/10.3389/fnut.2022.984891

Abdo, E. M., et al. (2023). Beetroot stalk extract as a functional colorant for stirred yogurt beverages: effect on nutritional value and stability during storage. *Fermentation, 9*(10).878.

Abdo, E. M., Shaltout, O. E.-S., El-Sohaimy, S., Abdalla, A. E. M., & Zeitoun, A. M. (2021). Effect of functional beetroot pomace biscuit on phenylhydrazine induced anemia in albino rats: hematological and blood biochemical analysis. *Journal of Functional Foods*, 78, 104385. https://doi.org/10.1016/j.jff.2021.104385

Abdo, E. M., Shaltout, O. E., Ali, S., & Mansour, H. M. M. (2022b). A functional orange juice fortified with beetroot by-products attenuates hyperlipidemia and obesity induced by a high-fat diet. *Antioxidants, 11*(3).457.

Abedi-Firoozjah, R., et al. (2023). Betalains as promising natural colorants in smart/active food packaging. *Food Chemistry*, 424, 136408. https://doi.org/10.1016/j.foodchem.2023.136408

Adjei, M. L., Boakye, A. A., Deku, G., Pepra-Ameyaw, N. B., Simpah Anim Jnr, A., Oduro, I. N., & Ellis, W. O. (2024). Development of yoghurt incorporated with beetroot puree and its effect on the physicochemical properties and consumer acceptance. *Heliyon, 10*(3), e25492. https://doi.org/10.1016/j.heliyon.2024.e25492

Ajila, C. M., Aalami, M., Leelavathi, K., & Rao, U. J. S. P. (2010). Mango peel powder: a potential source of antioxidant and dietary fiber in macaroni preparations. *Innovative Food Science & Emerging Technologies*, 11(1), 219–224. https://doi.org/10.1016/j.ifset.2009.10.004

Aksu, M. İ., Erdemir, E., Turan, E., & Sat, İ. G. (2020). Effects of red beet extracts on protein and lipid oxidation, colour, microbial, sensory properties and storage stability of Turkish pastırma. *Journal of Stored Products Research*, 89, 101721. https://doi.org/10.1016/j.jspr.2020.101721

Al-Harbi, L. N., et al. (2021). Beta vulgaris rubra L. (Beetroot) peel methanol extract reduces oxidative stress and stimulates cell proliferation via increasing VEGF expression in H_2O_2 Induced Oxidative Stressed Human Umbilical Vein Endothelial Cells. *Genes, 12*(9), 1380.

Ali, R. T. M., & Jameel, Q. Y. (2023). Red beetroot betalains as a novel source of colorent in ice-cream as compared with Red Dye 40 (E129). *Functional Foods in Health and Disease*, 13(4), 225–239.

Alshehry, G. A. (2019). Utilization of beetroot as a natural antioxidant, pigment and antimicrobial in cupcake during the storage period. *International Journal of Engineering Research & Technology*, 8(10), 652–659.

Ambedkar, B. B., & Pradesh, U. (2019). Betalian pigments extracted from the red beet and its potential uses as natural dyes in new food product development. *Journal of Pharmacognosy and Phytochemistry*, 8(6), 991–993.

Amnah, M. (2013). Nutritional, sensory and biological study of biscuits fortified with red beet roots. *Life Science Journal*, 10(3), 1579–1584.

Antigo, J. L. D., da Silva, J. M., de Cássia Bergamasco, R., & Madrona, G. S. (2020). Microencapsulation of beet dye (Beta vulgaris L.) using maltodextrin and xanthan gum as encapsulant agents and application in yogurt. *Research, Society and Development*, 9(12), e14091210896–e14091210896.

Asadaii, H., Sani, A. M., Arianfar, A., & Salehi, E. A. (2020). Effect of tomato lycopene, turmeric and beetroot extract on microbial and chemical properties of cow's milk butter. *Journal of BioScience and Biotechnology*, 9(1), 59–64.

Ashraf, S., et al. (2022). Assessment of potential benefits of functional food characteristics of beetroot energy drink and flavored milk. BioMed Research International, *2022*(1), 1971018. https://doi.org/10.1155/2022/1971018

Ateteallah, H., Abd-Elkarim, N., & Hassan, N. A. (2019). Effect of adding beetroot juice and carrot pulps on rheological, chemical, nutritional and organoleptic properties of ice cream. *Journal of Food and Dairy Sciences*, 10(6), 175–179.

Aulia, F., & Sunarharum, W. B. (2020). Beetroot (Beta vulgaris L. var. rubra L.) flour proportion and oven temperature affect the physicochemical characteristics of beetroot cookies. *IOP Conference Series: Earth and Environmental Science*, *475*(1), 012040.

Aykın-Dinçer, E., Güngör, K. K., Çağlar, E., & Erbaş, M. (2021). The use of beetroot extract and extract powder in sausages as natural food colorant. *International Journal of Food Engineering*, 17(1), 75–82.

Babarykin, D., et al. (2019). Red beet (Beta vulgaris) impact on human health. *Journal of Biosciences and Medicines*, 7(3), 61–79.

Bakara, T. L., Rumida, U. S., & Siahaan, G. (2022). The effect of beetroot (Beta vulgaris L.) ice cream on haemoglobin and total protein levels on teenage girl. *Journal of Natural Science, Biology and Medicine*, 13(1), 66–72.

Bancal, V., & Ray, R. C. (2022). Overview of food loss and waste in fruits and vegetables: from issue to resources. In R. C. Ray (Ed.), *Fruits and Vegetable Wastes: Valorization to Bioproducts and Platform Chemicals* (pp. 3–29). Singapore: Springer Nature Singapore. https://doi.org/10.1007/978-981-16-9527-8_1

Bellemare, M. F., Çakir, M., Peterson, H. H., Novak, L., & Rudi, J. (2017). On the measurement of food waste. *American Journal of Agricultural Economics*, 99(5), 1148–1158. https://doi.org/10.1093/ajae/aax034

Bellucci, E. R. B., et al. (2020). Natural colorants improved the physicochemical and sensorial properties of frozen Brazilian sausage (linguiça) with reduced nitrite. *Scientia Agricola*, 78, e20190211.

Brzezińska-Rojek, J., et al. (2023). Antioxidant capacity, nitrite and nitrate content in beetroot-based dietary supplements. Foods, *12*(5), 1017.

Bucur, L., Taralunga, G., & Schroder, V. (2016). The betalains content and antioxidant capacity of red beet (Beta vulgaris L. subsp. vulgaris) root. *Farmacia*, 64(2), 198–201.

Čakarević, J., et al. (2021). Pumpkin oil cake protein as a new carrier for encapsulation incorporated in food matrix: effect of processing, storage and in vitro digestion on bioactivity. *International Journal of Food Science & Technology*, 56(7), 3400–3408. https://doi.org/10.1111/ijfs.14964

Cejudo-Bastante, M. J., Cejudo-Bastante, C., Cran, M. J., Heredia, F. J., & Bigger, S. W. (2020). Optical, structural, mechanical and thermal characterization of antioxidant ethylene vinyl alcohol copolymer films containing betalain-rich beetroot. *Food Packaging and Shelf Life*, 24, 100502. https://doi.org/10.1016/j.fpsl.2020.100502

Chaari, M., et al. (2022). Novel active food packaging films based on gelatin-sodium alginate containing beetroot peel extract. Antioxidants, *11*(11), 2095.

Chaari, M., et al. (2023a). Multiobjective response and chemometric approaches to enhance the phytochemicals and biological activities of beetroot leaves: an unexploited organic waste. *Biomass Conversion and Biorefinery*, 13(16), 15067–15081. https://doi.org/10.1007/s13399-022-03645-0

Chaari, M., et al. (2023b). Potentials of beetroot (Beta vulgaris L.) peel extract for quality enhancement of refrigerated beef meat. *Quality Assurance and Safety of Crops & Foods*, 15(4), 99–115.

Chaari, M., et al. (2024). Development of a novel colorimetric pH-indicator film based on CMC/flaxseed gum/betacyanin from beetroot peels: a powerful tool to monitor the beef meat freshness. *Sustainable Chemistry and Pharmacy*, 39, 101543. https://doi.org/10.1016/j.scp.2024.101543

Chairuni, A. R., Sari, P. M., & Rusnaini, R. (2019). Effect of beetroot extract (Beta vulgaris L) and CMC (carboxyl methyl cellulose) concentration on ice cream quality. *Serambi Journal of Agricultural Technology*, 1(2), 80–91.

Chairunnissa, H., Balia, R. L., Pratama, A., & Hadiat, D. (2017). Karakteristik Kimia Set Yoghurt dengan Bahan Baku Susu Tepung dengan Penambahan Jus Bit (Beta Vulgaris L.). *Jurnal Ilmu Ternak*, 17(1), 35–39.

Chhikara, N., Kushwaha, K., Jaglan, S., Sharma, P., & Panghal, A. (2019). Nutritional, physicochemical, and functional quality of beetroot (Beta vulgaris L.) incorporated Asian noodles. *Cereal Chemistry*, 96(1), 154–161. https://doi.org/10.1002/cche.10126

Clifford, T., Howatson, G., West, D. J., & Stevenson, E. J. (2015). The potential benefits of red beetroot supplementation in health and disease. *Nutrients*, 7(4), 2801–2822.

Costa, A. P. D., Hermes, V. S., Rios, A. D. O., & Flôres, S. H. (2017). Minimally processed beetroot waste as an alternative source to obtain functional ingredients. *Journal of Food Science and Technology*, 54(7), 2050–2058. https://doi.org/10.1007/s13197-017-2642-4

Coy-Barrera, E. (2020). Chapter 17- Analysis of betalains (betacyanins and betaxanthins). In A. Sanches Silva, S. F. Nabavi, M. Saeedi, & S. M. Nabavi (Eds.), *Recent Advances in Natural Products Analysis* (pp. 593–619). Elsevier. https://doi.org/10.1016/B978-0-12-816455-6.00017-2

Czyzowska, A., Siemianowska, K., Sniadowska, M., & Nowak, A. (2020). Bioactive compounds and microbial quality of stored fermented red beetroots and red beetroot juice. *Polish Journal of Food and Nutrition Sciences*, 70(1).35–44.

Dabija, A., Codină, G. G., Ropciuc, S., & Stroe, S. G. (2019). Studies regarding the production of a novel yogurt using some local plant raw materials. *Journal of Food Processing and Preservation*, 43(6), e13826. https://doi.org/10.1111/jfpp.13826

Damunupola, D., Weerathilake, W., & Sumanasekara, G. S. (2014). Evaluation of quality characteristics of goat milk yogurt incorporated with beetroot juice. *International Journal of Scientific and Research Publications*, 4(10), 1–5.

Das, S., Filippone, S. M., Williams, D. S., Das, A., & Kukreja, R. C. (2016). Beet root juice protects against doxorubicin toxicity in cardiomyocytes while enhancing apoptosis in breast cancer cells. *Molecular and Cellular Biochemistry*, 421(1), 89–101. https://doi.org/10.1007/s11010-016-2789-8

Daunaravičiūtė, M., Paulauskienė, A., Tarasevičienė, Ž., & Silkartaitė, B. (2020). Influence of vegetable additives on spelt wheat (Triticum spelta L.) bread quality. *Žemės ūkio mokslai*, *27*(2), pp. 69–76.

de Oliveira, J. M., Macêdo, L. T., Polari, I. D. L. B., de Oliveira, F. L. N., & Azerêdo, G. A. (2020). Functional potential of beet yogurt. *Research, Society and Development*, 9(9), e196997153.

de Oliveira, S. P. A., do Nascimento, H. M. A., Sampaio, K. B., & de Souza, E. L. (2021). A review on bioactive compounds of beet (Beta vulgaris L. subsp. vulgaris) with special emphasis on their beneficial effects on gut microbiota and gastrointestinal health. *Critical Reviews in Food Science and Nutrition*, 61(12), 2022–2033.

Desseva, I., Stoyanova, M., Petkova, N., & Mihaylova, D. (2020). Red beetroot juice phytochemicals bioaccessibility: an in vitro approach. *Polish Journal of Food and Nutrition Sciences*, 70(1).45–53

Dias, S., et al. (2020). Application of natural pigments in ordinary cooked ham. Molecules, *25*(9), 1556.

Domínguez, R., et al. (2020). Red beetroot. a potential source of natural additives for the meat industry. Applied Sciences, *10*(23), 8340.

Drohsler, P., Cisar, J., Sopik, T., Sedlarik, V., & Pummerova, M. (2021). Effect of an antioxidant based on red beetroot extract on the abiotic stability of polylactide and polycaprolactone. Molecules, *26*(17), 5190.

Dzimitrowicz, A., et al. (2021). Application of cold atmospheric pressure plasmas for high-throughput production of safe-to-consume beetroot juice with improved nutritional quality. *Food Chemistry*, 336, 127635. https://doi.org/10.1016/j.foodchem.2020.127635

EFSA Panel on Food Additives and Nutrient Sources added to Food (ANS) (2015). Scientific opinion on the re-evaluation of beetroot red (E 162) as a food additive. *EFSA Journal*, 13(12), 4318. https://doi.org/10.2903/j.efsa.2015.4318

Elizabeth, I., & Onyinyechi, N. G. (2012). Production and evaluaton of yoghurt flavoured with beetroot (Beta vulgaris L.). *Journal of Food Science and Engineering*, 2(10), 583.

Eveline, E., & Nawangsih, A. A. (2019). Variasi Rasio Sari Bit Merah (Beta vulgaris L.), Susu Skim, Dan Kultur Starter Terhadap Karakteristik Yoghurt Sari Bit Merah [The Variation of Red Beet (Beta vulgaris L.) Juice Ratio, Skim Milk, and Starter Culture to Red Beet Juice Yogurt]. *FaST-Jurnal Sains dan Teknologi (Journal of Science and Technology)*, 3(1), 29–44.

Evstigneeva, T., Iakovchenko, N., Kuzmicheva, N., & Skvortsova, N. (2020). Applying beetroot as food ingredient in ice-cream production. *Agronomy Research*, *18*(Special Issue 3), 1662–1672. https://doi.org/10.15159/AR.20.115

Fernández-López, J., et al. (2023). Beetroot juices as colorant in plant-based minced meat analogues: color, betalain composition and antioxidant activity as affected by juice type. *Food Bioscience*, 56, 103156. https://doi.org/10.1016/j.fbio.2023.103156

Fernández-López, J. A., Fernández-Lledó, V., & Angosto, J. M. (2020). New insights into red plant pigments: more than just natural colorants [10.1039/D0RA03514A]. *RSC Advances*, 10(41), 24669–24682. https://doi.org/10.1039/D0RA03514A

Flores-Mancha, M. A., et al. (2021a). Stirred yogurt added with beetroot extracts as an antioxidant source: rheological, sensory, and physicochemical characteristics. *Journal of Food Processing and Preservation*, 45(7), e15628. https://doi.org/10.1111/jfpp.15628

Flores-Mancha, M. A., Ruíz-Gutiérrez, M. G., Sánchez-Vega, R., Santellano-Estrada, E., & Chávez-Martínez, A. (2021b). Effect of encapsulated beet extracts (Beta vulgaris) added to yogurt on the physicochemical characteristics and antioxidant activity. Molecules, *26*(16).4768.

Fu, Y., et al. (2020). Red beetroot betalains: perspectives on extraction, processing, and potential health benefits. *Journal of Agricultural and Food Chemistry*, 68(42), 11595–11611. https://doi.org/10.1021/acs.jafc.0c04241

Ganguly, S., Chakraborty, C., & Bandyopadhyay, K. (2017). Development and characterization of biocolour (Beta vulgaris) enriched low calorie Lassi (Yoghurt based beverage). International Journal of Current Microbiology and Applied Sciences, 6(4), 2265–2270.

Ghasempour, Z., et al. (2020). Development of probiotic yogurt containing red beet extract and basil seed gum; techno-functional, microbial and sensorial characterization. *Biocatalysis and Agricultural Biotechnology*, 29, 101785. https://doi.org/10.1016/j.bcab.2020.101785

Gomes, A. P. O., et al. (2019). Organic beet leaves and stalk juice attenuates HDL-C reduction induced by high-fat meal in dyslipidemic patients: a pilot randomized controlled trial. *Nutrition*, 65, 68–73. https://doi.org/10.1016/j.nut.2019.03.004

Gong, S., Jiao, C., & Guo, L. (2022). Antibacterial mechanism of beetroot (Beta vulgaris) extract against Listeria monocytogenes through apoptosis-like death and its application in cooked pork. *LWT*, 165, 113711. https://doi.org/10.1016/j.lwt.2022.113711

Guo, Z., et al. (2021). Active-intelligent film based on pectin from watermelon peel containing beetroot extract to monitor the freshness of packaged chilled beef. *Food Hydrocolloids*, 119, 106751. https://doi.org/10.1016/j.foodhyd.2021.106751

Gustavsson, J., Cederberg, C., Sonesson, U., Van Otterdijk, R., & Meybeck, A. (2011). Global food losses and food waste-FAO report. In: *Food and Agriculture Organization (FAO) of the United Nations:* Rome, Italy (pp. 1–37).

Hadipour, E., Taleghani, A., Tayarani-Najaran, N., & Tayarani-Najaran, Z. (2020). Biological effects of red beetroot and betalains: a review. *Phytotherapy Research*, 34(8), 1847–1867. https://doi.org/10.1002/ptr.6653

Hashem, M. I. (2018). Supplementation of buttermilk with red beet root for producing fermented milk beverage. *Egyptian Journal of Agricultural Research*, 96(3), 1111–1125.

Hidalgo, A., Brandolini, A., Čanadanović-Brunet, J., Ćetković, G., & Tumbas Šaponjac, V. (2018). Microencapsulates and extracts from red beetroot pomace modify antioxidant capacity, heat damage and colour of pseudocereals-enriched einkorn water biscuits. *Food Chemistry*, 268, 40–48. https://doi.org/10.1016/j.foodchem.2018.06.062

Ingle, M., Thorat, S. S., Kotecha, P. M., & Nimbalkar, C. A. (2017). Nutritional assessment of beetroot (Beta vulgaris L.) powder cookies. *Asian Journal of Dairy & Food Research*, 36(3), 222–228.

Jafari, S. M. (2017). *Nanoencapsulation of Food Bioactive Ingredients: Principles and Applications*. Cambridge, MA, USA: Academic Press.

Jovanović, M., et al. (2021). Functionality and palatability of yogurt produced using beetroot pomace flour granulated with lactic acid bacteria. Foods, *10*(8), 1696.

Junqueira-Goncalves, M. P., et al. (2011). Irradiated beetroot extract as a colorant for cream cheese. *Radiation Physics and Chemistry*, 80(1), 114–118. https://doi.org/10.1016/j.radphyschem.2010.08.002

Kaimainen, M., Laaksonen, O., Järvenpää, E., Sandell, M., & Huopalahti, R. (2015). Consumer acceptance and stability of spray dried betanin in model juices. *Food Chemistry*, 187, 398–406. https://doi.org/10.1016/j.foodchem.2015.04.064

Kamali Rousta, L., et al. (2021). Optimization of novel multigrain pasta and evaluation of physicochemical properties: using D-optimal mixture design. *Food Science & Nutrition*, 9(10), 5546–5556. https://doi.org/10.1002/fsn3.2514

Kamate, R. D., & Padghan, P. V. (2018). Studies on sensory/organoleptic properties of beetroot whey beverage. *International Journal of Current Microbiolgy Applied Sciences*, 7, 3309–3315.

Kim, B.-H., Jung, S.-H., & Jung, S. (2020). Beet root (Beta vulgaris) protects lipopolysaccharide and alcohol-induced liver damage in rat. *Toxicological Research*, 36(3), 275–282. https://doi.org/10.1007/s43188-019-00030-4

Kohajdová, Z., Karovičová, J., Kuchtová, V., & Lauková, M. (2018). Utilisation of beetroot powder for bakery applications. *Chemical Papers*, 72(6), 1507–1515. https://doi.org/10.1007/s11696-018-0392-z

Koss-Mikołajczyk, I., Kusznierewicz, B., Wiczkowski, W., Sawicki, T., & Bartoszek, A. (2019). The comparison of betalain composition and chosen biological activities for differently pigmented prickly pear (Opuntia ficus-indica) and beetroot (Beta vulgaris) varieties. *International Journal of Food Sciences and Nutrition*, 70(4), 442–452. https://doi.org/10.1080/09637486.2018.1529148

Koul, V. K., et al. (2002). Spray drying of beet root juice using different carriers. *Indian Journal of Chemical Technology*, *9*(5), 442–445.

Kumar, S., & Brooks, M. S.-L. (2018). Use of red beet (Beta vulgaris L.) for antimicrobial applications – a critical review. *Food and Bioprocess Technology*, 11(1), 17–42. https://doi.org/10.1007/s11947-017-1942-z

Larsen, F. J., et al. (2011). Dietary inorganic nitrate improves mitochondrial efficiency in humans. *Cell Metabolism*, 13(2), 149–159. https://doi.org/10.1016/j.cmet.2011.01.004

Lechner, J. F., & Stoner, G. D. (2019). Red beetroot and betalains as cancer chemopreventative agents. Molecules, *24*(8), 1602.

Lee, C.-H., Wettasinghe, M., Bolling, B. W., Ji, L.-L., & Parkin, K. L. (2005). Betalains, phase II enzyme-inducing components from red beetroot (Beta vulgaris L.) *Nutrition and Cancer*, 53(1), 91–103. https://doi.org/10.1207/s15327914nc5301_11

Lucky, A. R., Al-Mamun, A., Hosen, A., Toma, M. A., & Mazumder, M. A. R. (2020). Nutritional and sensory quality assessment of plain cake enriched with beetroot powder. Food Res*earch*, 4(6), 2049–2053.

Madadi, E., et al. (2020). Therapeutic application of betalains: a review. Plants, *9*(9), 1219.

Mann, M. K., & Sooch, B. S. (2020). Emerging trends in food industry waste valorization for bioethanol production. In P. Verma (Ed.), *Biorefineries: A Step Towards Renewable and Clean Energy* (pp. 57–92). Springer, Singapore. https://doi.org/10.1007/978-981-15-9593-6_3

Mannino, G., et al. (2021). Proanthocyanidins and where to find them: a meta-analytic approach to investigate their chemistry, biosynthesis, distribution, and effect on human health. Antioxidants, *10*(8), 1229.

Manoharan, A., Ramasamy, D., Dhanalashmi, B., Gnanalashmi, K. S., & Thyagarajan, D. (2012). Organoleptic evaluation of beetroot juice as natural color for strawberry flavor ice cream. *Indian Journal of Medicine and Healthcare*, 1(1), 34–37.

Marrone, R., et al. (2021). Effect of beetroot (Beta vulgaris) extract on Black Angus burgers shelf life. *Italian Journal of Food Safety*, 10(1), 9031.

Martins, I. R., Martins, L. H. D. S., Chisté, R. C., Picone, C. S. F., & Joele, M. R. S. P. (2024). Betalains from vegetable peels: extraction methods, stability, and applications as natural food colorants. *Food Research International*, 195, 114956. https://doi.org/10.1016/j.foodres.2024.114956

Martins, N., Roriz, C. L., Morales, P., Barros, L., & Ferreira, I. C. F. R. (2017). Coloring attributes of betalains: a key emphasis on stability and future applications. *Food & Function*, 8(4), 1357–1372. https://doi.org/10.1039/C7FO00144D

Mirabella, N., Castellani, V., & Sala, S. (2014). Current options for the valorization of food manufacturing waste: a review. *Journal of Cleaner Production*, 65, 28–41. https://doi.org/10.1016/j.jclepro.2013.10.051

Mirmiran, P., Houshialsadat, Z., Gaeini, Z., Bahadoran, Z., & Azizi, F. (2020). Functional properties of beetroot (Beta vulgaris) in management of cardio-metabolic diseases. *Nutrition & Metabolism*, 17(1), 3. https://doi.org/10.1186/s12986-019-0421-0

Mitrevski, J., et al. (2023). Effect of beetroot powder incorporation on functional properties and shelf life of biscuits. Foods, *12*(2), 322.

Morgado, M., et al. (2016). Development of a beetroot-based nutritional gel containing high content of bioaccessible dietary nitrate and antioxidants. *International Journal of Food Sciences and Nutrition*, 67(2), 153–160. https://doi.org/10.3109/09637486.2016.1147531

Munekata, P. E. S., et al. (2021). Beta vulgaris as a natural nitrate source for meat products: a review. Foods, *10*(9), 2094.

Nabi, B. G., et al. (2023). Natural pigments: anthocyanins, carotenoids, chlorophylls, and betalains as colorants in food products. *Food Bioscience*, 52, 102403. https://doi.org/10.1016/j.fbio.2023.102403

Nagarajaiah, S. B., Ramakrishna, M. G., & Prakash, J. (2019). Nutritional, antioxidant and sensory properties of value added muffins. *Current Nutrition & Food Science*, 15(5), 454–466.

Nagib, R. M., & Zidan, N. S. (2019). Fortification of cake with sweet potato and beetroot flour as natural antioxidant during storage. *Alexandria Science Exchange Journal*, 40(October–December), 754–766. https://doi.org/10.21608/asejaiqjsae.2019.70264

Nemzer, B., et al. (2011). Betalainic and nutritional profiles of pigment-enriched red beet root (Beta vulgaris L.) dried extracts. *Food Chemistry*, 127(1), 42–53. https://doi.org/10.1016/j.foodchem.2010.12.081

Nikan, M., & Manayi, A. (2019). Chapter 3.5- Beta vulgaris L. In S. M. Nabavi & A. S. Silva (Eds.), *Nonvitamin and Nonmineral Nutritional Supplements* (pp. 153–158). Academic Press. https://doi.org/10.1016/B978-0-12-812491-8.00021-7

Ninfali, P., & Angelino, D. (2013). Nutritional and functional potential of Beta vulgaris cicla and rubra. *Fitoterapia*, 89, 188–199. https://doi.org/10.1016/j.fitote.2013.06.004

Nirmal, N. P., et al. (2024). Betalains protect various body organs through antioxidant and anti-inflammatory pathways. *Food Science and Human Wellness*, 13(3), 1109–1117. https://doi.org/10.26599/FSHW.2022.9250093

Nowacka, M., et al. (2019). The impact of pulsed electric field on the extraction of bioactive compounds from beetroot. Foods, *8*(7), 244.

Nyamete, F. A., & Mongi, R. J. (2017). Folate contents, nutritional quality and consumer acceptability of yogurt fortified with red beetroot extract. *Tanzania Journal of Agricultural Sciences*, 16(2), pp. 90–100.

Otàlora Gonzàlez, C. M., Flores, S. K., Basanta, M. F., & Gerschenson, L. N. (2020). Effect of beetroot (Beta vulgaris L. var conditiva) fiber filler and corona treatment on cassava starch films properties. *Food Packaging and Shelf Life*, 26, 100605. https://doi.org/10.1016/j.fpsl.2020.100605

Ozaki, M. M., et al. (2021). Beetroot and radish powders as natural nitrite source for fermented dry sausages. *Meat Science*, 171, 108275. https://doi.org/10.1016/j.meatsci.2020.108275

Ozdemir, T., & Ozcan, T. (2020). Effect of steviol glycosides as sugar substitute on the probiotic fermentation in milk gels enriched with red beetroot (Beta vulgaris L.) bioactive compounds. *LWT*, 134, 109851. https://doi.org/10.1016/j.lwt.2020.109851

Ozturk, G., Dogan, M., & Said Toker, O. (2014). Physicochemical, functional and sensory properties of mellorine enriched with different vegetable juices and TOPSIS approach to determine optimum juice concentration. *Food Bioscience*, 7, 45–55. https://doi.org/10.1016/j.fbio.2014.05.001

Packer, V. G., Selani, M. M., Spada, F. P., & Contreras-Castillo, C. J. (2021). Guava and beetroot residues extract as natural antioxidants in cooked chicken meat stored in aerobic and vacuum packaging. *Vacuum*, 1(7), 14.

Panghal, A., Virkar, K., Kumar, V., Dhull, S. B., Gat, Y., & Chhikara, N. (2017). Development of probiotic beetroot drink. *Current Research in Nutrition and Food Science Journal*, 5(3), 257–262. https://doi.org/10.12944/CRNFSJ.5.3.10

Parfitt, J., Barthel, M., & Macnaughton, S. (2010). Food waste within food supply chains: quantification and potential for change to 2050. *Philosophical Transactions of the Royal Society B: Biological Sciences*, *365*(1554), 3065–3081.

Pasca, F. P., Nurwantoro, N., & Pramono, Y. B. (2016). Total bakteri asam laktat, kadar asam laktat, dan warna yogurt drink dengan penambahan ekstrak bit (Beta vulgaris L.). Jurnal Aplikasi Teknologi Pangan, *5*(4), 154–156.

Pennisi, L., Verrocchi, E., Paludi, D., & Vergara, A. (2020). Effects of vegetable powders as nitrite alternative in Italian dry fermented sausage. *Italian Journal of Food Safety*, 9(2), 8422. https://doi.org/10.4081/ijfs.2020.8422

Piasna-Słupecka, E., et al. (2023). Young shoots of red beet and the root at full maturity inhibit proliferation and induce apoptosis in breast cancer cell lines. *International Journal of Molecular Sciences*, 24(8), 6889.

Prudencio, I. D., Prudêncio, E. S., Gris, E. F., Tomazi, T., & Bordignon-Luiz, M. T. (2008). Petit suisse manufactured with cheese whey retentate and application of betalains and anthocyanins. *LWT – Food Science and Technology*, 41(5), 905–910. https://doi.org/10.1016/j.lwt.2007.05.019

Punia Bangar, S., Singh, A., Chaudhary, V., Sharma, N., & Lorenzo, J. M. (2023). Beetroot as a novel ingredient for its versatile food applications. *Critical Reviews in Food Science and Nutrition*, 63(26), 8403–8427.

Račkauskienė, I., et al. (2015). Effects of beetroot (Beta vulgaris) preparations on the Maillard reaction products in milk and meat-protein model systems. *Food Research International*, 70, 31–39. https://doi.org/10.1016/j.foodres.2015.01.026

Raikos, V., McDonagh, A., Ranawana, V., & Duthie, G. (2016). Processed beetroot (Beta vulgaris L.) as a natural antioxidant in mayonnaise: effects on physical stability, texture and sensory attributes. *Food Science and Human Wellness*, 5(4), 191–198. https://doi.org/10.1016/j.fshw.2016.10.002

Ranawana, V., et al. (2016). Breads fortified with freeze-dried vegetables: quality and nutritional attributes. Part II: breads not containing oil as an ingredient. Foods, *5*(3), 62.

Ray, R. C. (2022). *Fruits and Vegetable Wastes: Valorization to Bioproducts and Platform Chemicals*. Springer Nature Singapore, Singapore. https://doi.org/10.1007/978-981-16-9527-8

Rayees-ul-Islam, R.-U.-I., & Khan, M. A. (2016). Effect of olive oil and freeze dried beetroot powder on quality characteristics of chicken meat patties by response surface methodology study. *International Journal of Processing and Post Harvest Technology*, 7(2), 220–227.

Rechkemmer, G. (2001). Funktionelle Lebensmittel-Zukunft der Ernahrung oder Marketing-Strategie? *Molkerei Zeitung Welt Der Milch*, 55(12/13), 332.

Rodríguez-Félix, F., et al. (2022). Physicochemical, structural, mechanical and antioxidant properties of zein films incorporated with no-ultrafiltered and ultrafiltered betalains extract from the beetroot (Beta vulgaris) bagasse with potential application as active food packaging. *Journal of Food Engineering*, 334, 111153. https://doi.org/10.1016/j.jfoodeng.2022.111153

Roriz, C. L., Barreira, J. C. M., Morales, P., Barros, L., & Ferreira, I. C. F. R. (2018). Gomphrena globosa L. as a novel source of food-grade betacyanins: incorporation in ice-cream and comparison with beet-root extracts and commercial betalains. *LWT*, 92, 101–107. https://doi.org/10.1016/j.lwt.2018.02.009

Roy, K., Gullapalli, S., Chaudhuri, U. R., & Chakraborty, R. (2004). The use of a natural colorant based on betalain in the manufacture of sweet products in India. *International Journal of Food Science & Technology*, 39(10), 1087–1091.

Saad, F., et al. (2023). Betalain-enriched beetroots exhibit antiulcer and anti-inflammatory potentials. *Journal of Food Processing and Preservation*, 2023(1), 9522830. https://doi.org/10.1155/2023/9522830

Saber, A., Abedimanesh, N., Somi, M.-H., & Khosroushahi, A. Y. (2021). Antiproliferative and apoptotic effects of red beetroot and betanin on human colorectal cancer cell lines. Preprint [Internet]. Available from: https://doi.org/10.21203/rs.3.rs-955140/v1

Saber, A., Abedimanesh, N., Somi, M.-H., Khosroushahi, A. Y., & Moradi, S. (2023). Anticancer properties of red beetroot hydro-alcoholic extract and its main constituent; betanin on colorectal cancer cell lines. *BMC Complementary Medicine and Therapies*, 23(1), 246. https://doi.org/10.1186/s12906-023-04077-7

Sadowska-Bartosz, I., & Bartosz, G. (2021). Biological properties and applications of betalains. Molecules, *26*(9), 2520.

Samanta, S., Sarkar, T., & Chakraborty, R. (2024). Multifunctional applications of natural colorants: preservative, functional ingredient, and sports supplements. *Biocatalysis and Agricultural Biotechnology*, 56, 103026. https://doi.org/10.1016/j.bcab.2024.103026

Sandhya, P. S., & Priya, S. (2017). Formulation of beetroot cream cheese spread. *International Journal of Information Research and Review*, 4(2), 3710–3712.

Sawicki, T., Bączek, N., & Wiczkowski, W. (2016). Betalain profile, content and antioxidant capacity of red beetroot dependent on the genotype and root part. *Journal of Functional Foods*, 27, 249–261. https://doi.org/10.1016/j.jff.2016.09.004

Sawicki, T., & Wiczkowski, W. (2018). The effects of boiling and fermentation on betalain profiles and antioxidant capacities of red beetroot products. *Food Chemistry*, 259, 292–303. https://doi.org/10.1016/j.foodchem.2018.03.143

Schweiggert, R. (2023). *Handbook on Natural Pigments in Food and Beverages: Industrial Applications for Improving Food Color*. Amsterdam, Netherlands: Elsevier. https://doi.org/10.1016/C2021-0-04386-8

Shalaby, H. S., & Hassenin, A. S. H. (2020). Effects of fortification stirred yoghurt with red beet powder (RBP) on Hypercholesterolemia Rats. *European Journal of Agriculture and Food Sciences*, 2(5), pp. 1–7. https://doi.org/10.24018/ejfood.2020.2.5.61

Sharma, A., Mazumdar, B., & Keshav, A. (2020). Formulation, standardization and characterization of novel sattu beverage enriched with beetroot juice. *Journal of Food Science and Technology*, 57(5), 1936–1943. https://doi.org/10.1007/s13197-019-04229-z

Silva, D. V. T. d., et al. (2016). Physicochemical, nutritional, and sensory analyses of a nitrate-enriched beetroot gel and its effects on plasmatic nitric oxide and blood pressure. *Food & Nutrition Research*, 60(1), 29909. https://doi.org/10.3402/fnr.v60.29909

Singh, P., Yadav, S., Pathera, A., & Sharma, D. (2020). Effect of vacuum tumbling and red beetroot juice incorporation on quality characteristics of marinated chicken breast and leg meats. *Nutrition & Food Science*, 50(1), 143–156.

Sooch, B. S., Mann, M. K., Sharma, P., & Ray, R. C. (2022). Phenolic and other bioactive compounds from vegetable and fruit waste: extraction methods and their possible utilization. In R. C. Ray (Ed.), *Fruits and Vegetable Wastes : Valorization to Bioproducts and Platform Chemicals* (pp. 81–114). Springer Nature Singapore. https://doi.org/10.1007/978-981-16-9527-8_4

Srivastava, S., & Singh, K. (2018). Nutritional differences found in two values added baked products of beetroot (Beta vulgaris). *International Journal of Science, Engineering and Management*, 3(4), 209–212.

Stoica, F., et al. (2024). Exploitation of red beet peel powder as a natural food ingredient in whey-fruit based beverage. *International Journal of Food Properties*, 27(1), 44–67. https://doi.org/10.1080/10942912.2024.2426660

Sucheta, Rai, S. K., Chaturvedi, K., & Yadav, S. K. (2019). Evaluation of structural integrity and functionality of commercial pectin based edible films incorporated with corn flour, beetroot, orange peel, muesli and rice flour. *Food Hydrocolloids*, 91, 127–135. https://doi.org/10.1016/j.foodhyd.2019.01.022

Sucu, C., & Turp, G. Y. (2018). The investigation of the use of beetroot powder in Turkish fermented beef sausage (sucuk) as nitrite alternative. *Meat Science*, 140, 158–166. https://doi.org/10.1016/j.meatsci.2018.03.012

Thiruvengadam, M., et al. (2024). A comprehensive review of beetroot (Beta vulgaris L.) bioactive components in the food and pharmaceutical industries. *Critical Reviews in Food Science and Nutrition*, 64(3), 708–739. https://doi.org/10.1080/10408398.2022.2108367

Thiviya, P., Gamage, A., Kapilan, R., Merah, O., & Madhujith, T. (2022). Single cell protein production using different fruit waste: a review. Separations, *9*(7), 178.

Torres Vargas, O. L., Galeano Loaiza, Y. V., & González, M. L. (2021). Effect of incorporating extracts from natural pigments in alginate/starch films. *Journal of Materials Research and Technology*, 13, 2239–2250. https://doi.org/10.1016/j.jmrt.2021.05.091

Tran, T. N., Athanassiou, A., Basit, A., & Bayer, I. S. (2017). Starch-based bio-elastomers functionalized with red beetroot natural antioxidant. *Food Chemistry*, 216, 324–333. https://doi.org/10.1016/j.foodchem.2016.08.055

Tsegay, Z. T., Gebreegziabher, S. T.-B., & Mulaw, G. (2024). Nutritional qualities and valorization trends of vegetable and fruit byproducts: a comprehensive review. *Journal of Food Quality*, 2024(1), 5518577. https://doi.org/10.1155/2024/5518577

Tsegay, Z. T., & Mulaw, G. (2024). Antimicrobial activities and mode of action of bioactive substances from vegetable and fruit byproducts as a current option for valorization. Waste and Biomass Valorization, *16*(1), 1–28. https://doi.org/10.1007/s12649-024-02587-0

Vanlallawmzuali, C., & Pushpa, N. (2021). Production of natural antioxidant rich food products (bread, tofu & curd)–an alternate to synthetic food products. *Annals of the Romanian Society for Cell Biology*, 25(6), 5683–5693.

Villaño, D., García Viguera, C., & Mena Parreno, P. M. (2016). Colors: health effects. In B. Caballero, P. M. Finglas & F. Toldrá (Eds.), *Encyclopedia of Food and Health* (pp. 265–272). Oxford, UK: Elsevier / Academic Press. https://doi.org/10.1016/B978-0-12-384947-2.00190-2

Vulić, J. J., et al. (2013). Antiradical, antimicrobial and cytotoxic activities of commercial beetroot pomace. *Food & Function*, 4(5), 713–721. https://doi.org/10.1039/C3FO30315B

Yadav, M., Masih, D., & Sonkar, C. (2016). Development and quality evaluation of beetroot powder incorporated yogurt. *International Journal of Science, Engineering and Technology*, 4(4), 582–586.

Yang, W., et al. (2021). Red beet (Beta vulgaris) betalains and grape (Vitis vinifera) anthocyanins as colorants in white currant juice–Effect of storage on degradation kinetics, color stability and sensory properties. *Food Chemistry*, 348, 128995. https://doi.org/10.1016/j.foodchem.2020.128995

Yao, X., et al. (2021). Development of active and smart packaging films based on starch, polyvinyl alcohol and betacyanins from different plant sources. *International Journal of Biological Macromolecules*, 183, 358–368. https://doi.org/10.1016/j.ijbiomac.2021.04.152

9

Physicochemical, Functional, and Microbiological Studies on Beetroot

Kanchan Suri, Anu Sharma, and Supriya Kumari

9.1 Introduction

Beetroot (*Beta vulgaris* L.) is a herbaceous biennial plant from the Chenopodiaceae family. Known for its fleshy root, this versatile crop is widely used in cooking, including as a vegetable, in salads, and for pickling and canning. It comes in various bulb colors, ranging from yellow to red, with multiple distinct varieties available (Ramappa et al., 2024). This root vegetable is highly valued for its rich content of carbohydrates, proteins, micronutrients, and various functional components with notable health benefits (Azam and Pandey, 2020). Red beetroot is abundant in polyphenols, including flavonoids, phenolic compounds, and primarily betalains. These compounds are associated with beneficial effects in preventing chronic diseases and reducing the toxic impact of various substances in the body (Paula et al., 2025). The manifold health advantages of these compounds have been extensively examined, rendering beetroot a valuable dietary inclusion due to its numerous nutritional and health benefits (Panghal et al., 2017). The beetroot, apart from consumption in its fresh form, is also a valuable vegetable used in the food industry to produce dried and frozen food, non-concentrated and concentrated juices, as well as natural colorants (betalains) used as additives in food manufacturing. Beetroot juice is used as a coloring agent in various dairy products such as processed cheese, yogurt, candy, ice cream, sweets, and other confectionery products, which require no thermal treatment because change may occur upon heating (Szekely et al., 2016)

In diet, beetroots can be administered through various formulations, offering a convenient and alternative dietary source such as fresh cut beetroot pieces, beetroot juice, chips, grated beetroot, ready to drink health drinks, or beetroot powder (Baião et al., 2017). Different parts of beetroot have different medicinal properties, such as anti-oxidant, anti-hypertensive, anti-microbial, anti-inflammatory, hepato-protective, anti-hyperglycemic, anti-cancer, and diuretic. Beetroot juice improves oxygenation to the brain, slowing the progression of dementia in older adults. Due to its high fiber content, it prevents constipation and promotes regularity for a healthy digestive tract (Tulp and Bohlin 2004). It helps to preserve brain function with nitrates that improve blood flow and beet, having the ability to increase the production of glutathione naturally in the body; this compound helps to prevent colon cancer (Clifford et al., 2015). Fresh beetroot can be pounded into a powder and dehydrated at low temperatures (50°C for 6 hours) to preserve its nutritional qualities for an extended period of time. It should then be stored in tightly sealed containers.

DOI: 10.1201/9781003527084-9

Beetroot can be used in this way to improve the nutritional value and health-promoting qualities of a variety of food products (Sahni and Shere, 2017). Recent years have seen a significant rise in the demand for readily consumable functional foods, which has prompted producers to create novel, nutrient-dense goods with positive health effects. This can prevent or delay the onset of numerous diseases brought on by inadequate nutrition, including diabetes, cardiovascular disease, and some types of cancer, in addition to lowering obesity (Betoret, 2011). Unquestionably, functional food with high fiber content and bioactive ingredients improves digestion and speeds up metabolism, both of which are critical for a healthy functioning of the entire body (Betoret, 2011). This chapter provides a concise review of various physicochemical, functional, and microbiological studies related to beetroot and its derived products.

9.2 Physicochemical Characteristics

9.2.1 Physical Characteristics

9.2.1.1 Average Weight of Beetroot

The average weight of beetroot varied from 298.5 to 395.5 g in Hungary (Székely et al., 2014). The average weight of beetroot in the Austrian investigations varied between 214 and 366 g (Wruss et al., 2015). Slosar et al. (2020) reported a broader range, with weights spanning from 187.8 g for the variety "Golden" to 412.1 g for "White Detroit." Mohammad et al. (2014) report that the lowest average weight of beetroot was 85.33 g. Comparative studies from Poland, Brazil, and Ukraine reported weight ranges of 149.4–202.4 g (Szopinska and Gawęda, 2013), 161.6–180.7 g (Da Silva et al., 2016), and 170.4–180.2 g (Vdovenko et al., 2018), respectively. Among these findings, the lowest reported beetroot weight in the literature was 85.33 g, while the highest reached 412 g.

9.2.1.2 Length, Width, and Diameter of Beetroot

The beetroot length ranged from 7.06 to 16.25 cm, and its diameter varied between 4.03 and 5.43 cm (Kale et al., 2018; Sakhare et al., 2019). Beetroot has a spherical or slightly flat root shape depending on the variety, growing conditions, soil type, and genetic variation (Ruboczki et al., 2015). Takácsné Hájos and Rubóczki (2012) grew twelve varieties of beetroot, having diameters ranging from 5.84 to 6.94 cm and lengths ranging from 6.35 to 7.07 cm. Mohammad et al. (2014) reported the length of the beetroot was 6.03 cm, with a width of 5.59 cm.

9.2.1.3 Firmness of Beetroot

The firmness of a beetroot is an important parameter of its freshness and quality of post-harvest fruit. Nayak et al. (2007) reported a beetroot hardness of 713 N using texture analysis, whereas Viskelis et al. (2019) observed firmness values ranging from 337 to 446 N/cm.

9.2.1.4 Edible Portion of Beetroot

The edible index represents the percentage of a beetroot that is suitable for consumption, calculated as a proportion of its total weight. On the other hand, the waste index indicates the percentage of the beetroot that is discarded as waste, including parts like peels, stems, or damaged sections unsuitable for consumption. Kale et al. (2018) and Sakhare et al. (2019) reported the edible index of beetroot was reported as 91.03% and 88.24%, while the waste index was 8.07% and 11.75%. These values reflect the proportion of the beetroot that can be consumed versus the portion discarded as waste.

9.2.1.5 Viscosity

Kale et al. (2018) determined the viscosity of the beetroot juice with the aid of a Lovibond Brookfield viscometer and reported a viscosity value of 0.72 Pa.s. Kruszewski et al. (2023) reported the viscosity of control beetroot juice as 5.51 mPa.s, while under the influence of high pressure processing treatment, the viscosity of beetroot juice varied from 3.76 to 4.01 mPa.s.

9.2.2 Chemical Properties of Beetroot

9.2.2.1 pH and Titratable Acidity

pH and titratable acidity (TA) of beetroot juice were 6.01 and 0.12 g citric acid/100 mL, while the pH and TA of high pressure processing (HPP) treated beetroot juice samples varied from 6.13 to 6.16, and TA was reduced to 0.11 g citric acid/100 mL (Kruszewski et al., 2023). Indirani et al. (2021) reported a pH of 5.94 for beetroot powder. On the other hand, Juszczak et al. (2010) reported a pH of 4.3 and a TA of 2.4 g malic acid/100 g for beetroot juice concentrate.

9.2.2.2 Moisture Content

The moisture content in the edible portion of beetroot was found to be 87.4% in Indian varieties of beetroot (Kale et al., 2018). Rodler and Tápanyagtáblázat (2005) reported that the moisture content in a flesh of beetroot was 90.9%. The Egyptian red beetroot variety contains 84.95% moisture, while the Iraqi cuisine variety contains 87.58%. These studies reported that the moisture content in beetroot is in the range of 84.95% to 90.9%. Sahni and Shere (2017) reported a lower moisture content of 6.82% for beetroot pomace powder. Similarly, Kerr and Varner (2019) also reported a lower moisture content of 1.74 to 1.93 g H_2O/100 g for beetroot powders dried by various drying methods like tray drying, freeze drying, drum drying, and vacuum belt drying.

9.2.2.3 Total Soluble Solids (TSS)

TSS represent the concentration of monosaccharides and disaccharides in a solution, which is an indicator of the sugar content in fruits and vegetables. Hegedusova et al. (2018) also define the dissolved sugar in fruits and vegetables exactly, measured in degrees Brix. Slosar et al. (2020) also worked on different flesh colors of beetroot and have different TSS due to their color. It reported that red-colored flesh gives 8.1–10.8°B, white color flesh gives 7.8 to 8.8°B, and red-white flesh gives 6.4–7.7°B. Székely et al. (2014) reported a comparable range of TSS in beetroot during a Hungarian experiment, measuring between 6.0 and 10.7°B. On the other hand, Stagnari et al. (2014) determined that a higher TSS content in beetroot is 13.9°B, and a higher content of TSS ranges from 14.2–21.5°B in beetroot was also shown in the Australian study (Irving, 2012). In India, Kale et al. (2020) found the TSS of an Indian beetroot variety as 9.0°B, while Sakhare et al. (2019) recorded a TSS value of 7.9°B. Moreover, factors such as cultivation systems (Szopinska and Gawęda, 2013) and water stress (Stagnari et al., 2014) have also been shown to influence TSS levels in beetroot.

9.2.2.4 *Carbohydrates*

Agarwal and Varma (2014) reported that the carbohydrate content in beetroot was 9.96 g/100 g. Similarly, Chauhan et al. (2020) recorded a carbohydrate content of 9.96 g/100 g and a sugar content of 7.96 g/100 g in the Indian beetroot variety. Kale et al. (2018) also analyzed the Indian beetroot variety and reported a carbohydrate content of 7.59%. In contrast, Shehatta et al. (2024) documented the carbohydrate content of Egyptian red beetroot at 7.94 g/100 g.

9.2.2.5 *Sugar Content*

Rodriguez-Sevilla et al. (1999) reported that sucrose is the main component in beetroot, accounting for 6.68%, while other components such as fructose and glucose are present at 0.127% and 0.488%, respectively. Kale et al. (2018) reported a reducing sugar content of 4.20% and total sugar content of 7.93% in the Indian variety of beetroot. Viskelis et al. (2019) recorded the total sugar content in beetroot grown at the Institute of Horticulture, Lithuania, ranging between 9.23% and 10.53%. Chauhan et al. (2020) reported the sugar content in beetroot to be 7.96%.

9.2.2.6 *Protein Content*

Costa et al. (2017) evaluated the protein content of chlorinated and non-chlorinated beetroot waste flour dried at various temperatures and reported that it varied from 12.63% to 12.69%. Similarly, Sahni and Shere (2017) reported a higher protein content of 15.85% for beetroot pomace powder. The protein content of vacuum-dried beetroots varied from 0.62 to 0.71 g/100 g (Mella et al., 2022). Mirmiran et al. (2020) reviewed the potential of beetroot and reported that raw beetroot, cooked and boiled beetroot, canned beetroot, and fresh beetroot juice have protein contents of 1.61, 1.68, 0.91, and 1.02 g/100 g, respectively. Agarwal and Varma (2014) reported a protein content of 1.68 g/100 g in beetroot. Similarly, Kale et al. (2018) recorded a protein content of 1.35%, while Chauhan et al. (2020) also observed 1.68 g/100 g in the Indian variety. Shehatta et al. (2024) reported the protein content in the Egyptian red beetroot of 1.49 g/100 g. Beetroot contains a considerable amount of both essential and non-essential amino acids. The amino acid profile per 100 g of the edible portion includes essential amino acids tryptophan (0.019 g), isoleucine (0.048 g), leucine (0.068 g), lysine (0.058 g), threonine (0.047 g), methionine (0.018 g), phenylalanine (0.046 g), and valine (0.056 g); non-essential amino acids tyrosine (0.038 g), cystine (0.019 g), arginine (0.042 g), histidine (0.021 g), alanine (0.060 g), glutamic acid (0.428 g), glycine (0.031 g), proline (0.042 g), aspartic acid (0.116 g), and serine (0.059 g), as reported by Nemzer et al. (2011).

9.2.2.7 *Fat Content*

The Indian variety of beetroot contains a low fat content, as Chauhan et al. (2020) recorded 0.18 g of fat, and Kale et al. (2018) reported a fat content of 0.3% in red beetroots. These values show that beetroot has a very low fat content. Moreover, beetroot contains 0.027 g of total saturated fatty acids, 0.032 g of total monounsaturated fatty acids, and 0.060 g of total polyunsaturated fatty acids per 100g of its edible portion (USDA-ARS, 2014). Further, the crude fat content of vacuum-dried beetroots varied from 0.27 to 0.21 g/100g (Mella et al., 2022). Mirmiran et al. (2020) reviewed the potential of beetroot and reported that raw beetroot, cooked and boiled beetroot, and canned beetroot have fat contents of 0.17, 0.18, and 0.14 g/100g, respectively. Costa et al. (2017) evaluated the fat content of chlorinated and non-chlorinated beetroot waste flour dried at various temperatures and reported that it varied from 1.31% to 1.33%. Further, Sahni and Shere (2017) reported a fat content of 1.44% for beetroot pomace powder.

9.2.2.8 Vitamins

Beetroot is a rich source of various vitamins. Neelwarne et al. (2013) reported that it contains ascorbic acid (4.9mg), thiamine (B_1) (31 μg), riboflavin (B_2) (0.057mg), pantothenic acid (0.155mg), vitamin E (0.04mg), vitamin A (33IU), and pyridoxine (B_6) (0.067mg). Yashwant et al. (2015) reported similarly the presence of vitamin A (2 μg), thiamine (B_1) (0.31mg), riboflavin (B_2) (0.27mg), and folic acid (80 μg). Chauhan et al. (2020) analyzed an Indian variety of beetroot and observed vitamins such as niacin (0.331 mg) and vitamin C (3.6mg).

9.2.2.9 Mineral Content

Beetroot is a rich source of essential minerals that play a significant role in reducing blood pressure, promoting cardiovascular health, and combating cancer. Consuming a variety of fruits and vegetables is essential to obtain sufficient amounts of all the different minerals. Kale et al. (2018) studies on an Indian variety of beetroot and reported the presence of seven minerals in beetroot and their content were as follows; sodium (Na) 72.58 mg/100g, iron (Fe) 0.75 mg/100g, potassium (K) 30.12 mg/100g, calcium (Ca) 12.20 mg/100g, manganese (Mn) 0.79 mg/100g, copper (Cu) 0.09 mg/100g and zinc (Zn) 0.21 mg/100g. Yashwant (2015) reported that the mineral content in beetroot is Na (77mg), Ca (16mg), Fe (0.79mg), phosphorus (P) (38mg), K (305mg), magnesium (Mg) (23mg), and Zn (0.35mg) per 100 grams of the edible portion. Similarly, Chauhan et al. (2020) identified mineral levels in beetroot, including Ca (16 mg/100g), Fe (0.79 mg/100g), Mg (23 mg/100g), P (38 mg/100g), K (305 mg/100g), Zn (0.35 mg/100g), and Na (77 mg/100g). Further, Moulick et al. (2023) reported four minerals, that is, Na (130.82 mg/100g), K (704.14 mg/100g), Ca (36.77 mg/100g), and Fe (0.36 mg/100g) in dried beetroot powder. Mirmiran et al. (2020) reviewed the potential of beetroot and reported that raw beetroot has mineral content as Ca (16 mg/100g), Fe (0.8 mg/100g), Mg (23 mg/100g), P (40 mg/100g), K (325 mg/100g), Na (78 mg/100g), and Zn (0.35 mg/100g). Cooked and boiled beetroot showed mineral content as Ca (16 mg/100g), Fe (0.79 mg/100g), Mg (23 mg/100g), P (38 mg/100g), K (305 mg/100g, Na (77 mg/100g), and Zn (0.35 mg/100g). Canned beetroot showed Ca (15 mg/100g), Fe (1.82 mg/100g), Mg (17 mg/100g), P (17 mg/100g), K (148 mg/100g), Na (194 mg/100g), and Zn (0.21 mg/100g)

9.2.2.10 Fiber Content

Fruits and vegetables are rich in dietary fiber, which is crucial for lowering the risk of various health issues (Moulick et al., 2023). Aulia and Sunarharum (2020) reported fiber content of 5.39% in fresh and 2.26% in powdered beetroot of variety rubra Mella et al. (2022) reported fiber content of vacuum-dried beetroots varied from 5.78 to 4.08 g/100g. Similarly, Sahni and Shere (2017) reported fiber content of 11.12% for beetroot pomace powder. Further,

Costa et al. (2017) evaluated the total dietary fiber (TDF), soluble dietary fiber (SDF), and insoluble dietary fiber (IDF) content of chlorinated and non-chlorinated beetroot waste flour dried at various temperatures and reported that TDF, SDF, and IDF ranged from 65.22% to 65.29%, 20.14% to 20.28%, and 45.01% to 45.13%, respectively. The fiber content in beetroot is influenced by factors such as variety, geographical location, and environmental conditions. Indian varieties generally show a fiber content of 1.9%, as reported by Kale et al. (2018), while Chauhan et al. (2020) recorded a fiber content of 2%. These values show that the Indian variety shows the fiber content in a range from 1.9% to 2%. The Egyptian red beetroot shows a slightly higher crude fiber content of 2.50 g/100 g (Shehatta et al., 2024).

9.2.2.11 Betalain Content

Betalains are the main natural pigments that are present in a huge amount (more than 99.0%) in beetroots (Nirmal et al., 2021). Betalains are typically categorized based on their distinctive structural features. Betalains are nitrogenous, water-soluble pigments found in higher plants, particularly in those belonging to the order Caryophyllales. Ten families within this order are known to produce betalains, including the Chenopodiaceae family, to which *Beta vulgaris* (commonly known as beetroot) belongs (Khan and Giridhar, 2015). Betalains are classified into two main subgroups: betacyanins, which produce reddish-violet hues, and betaxanthins, responsible for yellowish shades. Betacyanins commonly found in plants include betanin, isobetanin, protobetanin, and neobetanin, while betaxanthins include vulgaxanthin, miraxanthin, portulaxanthin, and indicaxanthin (Kanner et al., 2001). The average value of betalains in red beetroot was calculated to be around 1,000 mg/100 g of total solids, or 120 mg/100 g fresh weight (Marmion, 1991). Some of the edible and familiar sources of betalains are beetroot (B. vulgaris L. sp. vulgaris) and Swiss chard (B. vulgaris L. ssp. cicla). Commercially, the level of betacyanins among the beetroot varies from 0.44 to 0.62 g betalin/kg, and the yellow pigment ranges from 0.32 to 0.42 g vulgaxanthin/kg. Beetroot is a natural pigment that has antioxidant properties and is also used as a food colorant. The maximum portion, about 54% of betalains, is present in its peel, then 32% in the crown, and the remaining 14% in its flesh part (Kujala et al., 2000). Lee et al. (2014) reported that the yellow-colored beetroot variety *Burpee's Golden*, cultivated in the USA, contained a betacyanin concentration of 0.114 mg/g FW and a betaxanthin concentration of 0.187 mg/g FW in its flesh. Costa et al. (2017) evaluated the betalain content of chlorinated and non-chlorinated beetroot waste flour dried at various temperatures and reported that betalain content varied from 68.08 to 80.31 mg/g DM.

9.2.2.12 Antioxidant Activity

Beetroot is a rich source of antioxidants, similar to many other colorful vegetables (Guldiken et al., 2016). Beetroot juice (pH 5.45, 9°B) contains high levels of total phenolics (1169 mg GAE/L), flavonoids (925 mg catechin equivalents/L), and pigments (854 mg/L), contributing to its superior antioxidant properties (Fidelis et al., 2017). Betanin and its aglycone, betanidine, exhibit strong antioxidant activity and are effective in preventing lipid peroxidation (Wootton-Beard and Ryan, 2011; Kathiravan et al., 2014). Vasconcellos et al. (2016) further highlighted the antioxidant capacity of various beetroot products: beetroot chips (95.70%), beetroot powder (95.31%), cooked beetroot (85.79%), and beetroot juice (80.48%). Betalains are recognized as potent antioxidants due to their ability to scavenge reactive oxygen species and enhance antioxidant defenses (Zhang and Tsao, 2016). Betalains have demonstrated effectiveness in protecting low-density lipoprotein (LDL) molecules from oxidation, attributed to their strong radical-scavenging properties (Gengatharan et al., 2015; Ribas-Agustí et al., 2018).

Moreover, the protective effects of betalains on LDL molecules may also be linked to the activation of paraoxonase 1, an antioxidant enzyme produced by the liver, which plays a critical role in reducing oxidative damage (Delgado-Vargas et al., 2000). These findings underscore the potential of betalains as natural compounds for managing oxidative stress and promoting cardiovascular health. Slosar et al. (2020) reported that the values of antioxidant activity (AOA) varied from 314.73 mg to 972.50 mg TEAC $\cdot$ kg^{-1}DW. Red-fleshed beetroot cultivars were significantly higher AOA (645.88–972.50 mg TEAC $\cdot$ kg^{-1} DW.) than beetroot cultivars with non-typical flesh colour (314.73–513.60 mg TEAC $\cdot$ kg^{-1} DW. Kovarovic et al. (2017) tested the effect of cultivar on the AOA of beetroot, determined by the DPPH method. The AOA of red-fleshed cultivars (20.70%–21.83% inhibition of DPPH) was statistically significantly higher compared to the red-white cultivar of beetroot (8.37%). Mella et al. (2022) evaluated the AOA of red beetroots exposed to vacuum drying temperatures from 40°C to 80°C by determining DPPH activity and reported that DPPH varied from 2440 to 2869 µmol TE/100 g d.m. Costa et al. (2017) evaluated the DPPH (%inhibition) of chlorinated and non-chlorinated beetroot waste flour dried at various temperatures and reported that it varied from 61.14% to 71.34%.

9.2.2.13 Total Phenolic Content

Phenolic components are secondary metabolites of plants that influence the functional qualities of plant products, such as antioxidant activity, as well as their sensory qualities, such as flavor, taste, and color (Ribas-Agustí et al., 2018). Additionally, plants' growth, pigmentation, reproduction, and defense systems under various stress situations are all influenced by phenolic chemicals (Lattanzio et al., 2006). The significant antioxidant effect of phenolic compounds is a result of their ability to scavenge free radicals. Yasaminshirazi et al. (2020) studied the red-colored, yellow-colored, and white colored varieties of beetroot and recorded the total phenolic (TP) content in the range from 489.06 mg GAE 100 g^{-1} DW, 172.89 mg GAE 100 g^{-1} DW, and 216.09 mg GAE 100 g^{-1} DW, respectively. It indicates that beetroot with a red hue has greater phenolic content. Rubóczki et al. (2015) found that the total polyphenol content of ten various beetroots cultivated in Hungary ranged from 45.47 mg GAE/100 g to 83.37 mg GAE/100 g. Its phenolic concentration is the lowest, as compared to others. Kavalcová et al. (2015) examined the total phenolic compounds of four beetroot varieties grown in the Czech Republic. They found that the types Cylindra and Chioggia (white with red stripes) have phenolic compounds ranging from 368.75 to 887 mg/kg. Mella et al. (2022) evaluated the TP content of red beetroots exposed to varying vacuum-drying temperatures from 40°C to 80°C and reported that TP content varied from 348.5 mg GAE/100 g d.m.to 557.65 mg GAE/100 g d.m.

9.3 Functional Characteristics

Functional characteristics are the fundamental physicochemical properties that reflect the complex interaction between the composition, structure, molecular conformation, and physicochemical properties of food components together with the nature of the environment in which these are associated and measured (Chaudhary and Kumar, 2020). Figure 9.1 represents the various other functional characteristics of red beetroot.

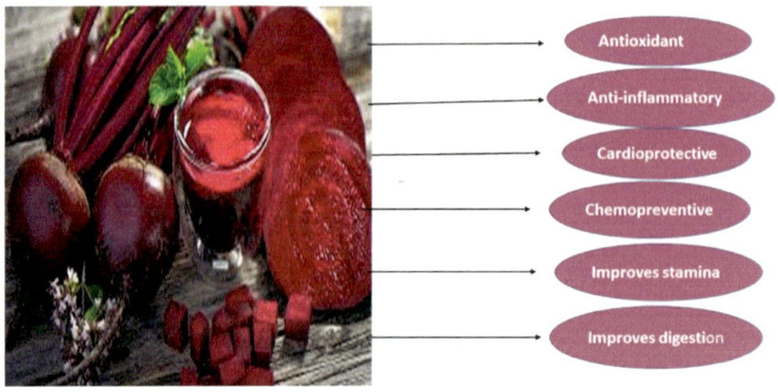

FIGURE 9.1
Image depicting the various other functional characteristics of red beetroot.

9.3.1 Hydration Properties

The hydration properties of fiber are important for its physiological role as well as for its interventions in techno-functional properties of the food. The hydration property of fiber governs its efficacy in stool bulking. Water holding capacity, water retention capacity, and swelling capacity provide information regarding the hydration capacity of fiber and give insights regarding its behavior during gut transit and food processing (Dhingra et al., 2012). Water holding capacity (WHC) is defined as the quantity of water that is bound to the fibers without the application of any external force. Water retention capacity (WRC) is defined as the quantity of water that remains bound to the hydrated fiber following the application of an external force. Water absorption capacity (WAC) refers to a product's ability to bind with water under conditions of limited water availability (Chandra and Samsher, 2013). Chaudhary and Kumar (2020) evaluated the WAC of tray-dried beetroot powder produced by applying pretreatments such as dipping in brine solution, sugar solution, blanching, and steam blanching, along with the untreated sample. From their studies, authors reported the highest WAC of 193% was observed for the sample obtained by steam blanching, while the untreated sample had the lowest WAC of 154%. Sahni and Shere (2017) reported functional properties of beetroot pomace powder, such as WHC=5.040 (g/g) and WRC=4.338 (g/g). Further, Costa et al. (2017) evaluated the WHC of beetroot waste flour, both chlorinated and non-chlorinated, dried at various temperatures, and reported that WHC ranged from 10.10 to 10.17 g water/g flour.

9.3.2 Oil Absorption Capacity (OAC)

OAC depends on the chemical structure of the plant polysaccharides, its surface properties, overall charge density, thickness, hydrophobic nature of the fiber particle, particle size, and drying (Sharoba et al., 2013). OAC enhances the powder's suitability for improving flavor and mouthfeel in food preparation. The ability of its proteins to bind with oil makes it valuable in food systems where optimal oil absorption is required (Chandra, 2013). Chaudhary and Kumar (2020) evaluated the OAC of tray-dried beetroot powder produced by applying pretreatments such as dipping in brine solution, sugar solution, blanching, and steam blanching, along with the untreated sample. From their studies, authors reported that OAC ranged from 4.4% to 21.8%, with the highest OAC of

21.8% being observed for the sample obtained by steam blanching, while the untreated sample had the lowest OAC of 4.4% Similarly, Sahni and Shere (2017) reported the OAC of 2.206 (g/g) for beetroot pomace powder. Further, Costa et al. (2017) evaluated the oil holding capacity (OHC) of beetroot waste flour, both chlorinated and non-chlorinated, dried at various temperatures, and reported that OHC ranged from 3.31 to 3.35 g oil/g flour.

9.3.3 Bulk Density

Bulk density (BD) is a critical parameter for the packaging and storage of powders. A lower bulk density indicates a larger volume required for packaging and may reduce shelf life, as the increased amount of occluded air raises the risk of oxidation (Martinez et al., 2015). Kerr and Varner (2019) reported BD ranging from 430 to 728 kg/m^3 for beetroot powders dried by various drying methods like tray drying, freeze drying, drum drying, and vacuum belt drying. Sahni and Shere (2017) reported BD of 0.631 g/cm^3 for beetroot pomace powder.

9.3.4 Foaming Properties

Foam is a colloidal suspension of many gas bubbles trapped in a liquid or solid. Small air bubbles are surrounded by thin liquid films. Chaudhary and Kumar (2020) evaluated the foam stability of tray-dried beetroot powder produced by applying pretreatments such as dipping in brine solution, sugar solution, blanching, and steaming, along with the untreated sample. From their studies, authors reported that the foam stability (FS) of beetroot powder ranged from 9.21% to 16.7%. The highest FS of 16.7% was observed for untreated and the lowest FS of 9.21% was observed for steam blanching. Ng and Sulaiman (2018) also evaluated functional properties of beetroot powder dried by foam mat drying and reported values of foam expansion and foam density for beetroot powder as 10.25% and 0.94 (g/cm^3), respectively.

9.3.5 Swelling Properties

Chaudhary and Kumar (2020) evaluated the swelling capacity of tray-dried beetroot powder produced by applying pretreatments such as dipping in brine solution, sugar solution, blanching, and steaming, along with the untreated sample, and reported that the swelling capacity of beetroot powder varied from 40.2 to 55.8 ml, with the highest swelling capacity (55.8 ml) observed for beetroot powder obtained by steam blanching and the lowest swelling capacity (40.2 ml) for untreated beetroot powder. Sahni and Shere (2017) reported a swelling capacity of 7.8 ml/g for beetroot pomace powder.

9.3.6 Water Activity and Hygroscopicity

Water activity ($a_{w)}$ is considered one of the important parameters in food preservation and processing. Hygroscopicity was related to the ability of water uptake of food, where high hygroscopicity means greater water uptake, which means more water physically bound by a solid matrix (Ng and Sulaiman, 2018). Fresh beetroots showed a_w of 0.9852, which was reduced to 0.2954 after microwave drying (Liu et al., 2021). Kerr and Varner

(2019) reported a_w and hygroscopicity values ranging from 0.159% to 0.303% and 16.22% to 18.69%, respectively, for beetroot powders dried by various drying methods like tray drying, freeze drying, drum drying, and vacuum belt drying. Kapoor et al. (2021) evaluated the spray-dried and freeze-dried beetroot powders for their functional properties and reported that the water activity of beetroot powder was 0.4041, while wettability and hygroscopicity ranged from 136 to 281 seconds and 20.15% to 23.18%, respectively. On the other hand, Ng and Sulaiman (2018) evaluated the functional properties of beetroot powder dried by foam mat drying and reported a_w and hygroscopicity values for beetroot powder as 0.3945 and 23.37 (g/100 g), respectively. Beetroot also has other functional properties, as mentioned below.

9.3.7 Beetroot as an Antioxidant

The demand for processed foods has never decreased since they were launched. Additionally, the addition of a particular quantity of antioxidants to processed meals has grown in popularity. Utilizing antioxidants to their full potential is especially important today to prolong shelf life and guarantee food quality for a variety of processed meals that are high in fat and protein (Ranawana et al., 2018). In addition, the antioxidant capacity of betanin, the major betalains in beetroot, is gradually receiving greater attention in response to public health demands. According to reports, betalains would not significantly lose their antioxidant qualities in gastrointestinal conditions, further preserving beetroot's effectiveness in treating human illnesses (Frank et al., 2005). Furthermore, the downregulation of superoxide dismutase brought on by the oxidative stress model was reversed by pretreatment with beetroot juice, which resulted in a threefold increase in superoxide dismutase activity. The authors' conclusion from this study is that beetroot does have promising antioxidant properties (Chen et al., 2021)

9.3.8 Beetroot as an Anti-Inflammatory Agent

Under normal conditions, inflammation is thought to be a healthy mechanism that controls our natural reaction to physical or biological stressors like infections, trauma, and other pathogens that could damage the body and upset homeostasis (Monteiro and Azevedo, 2010; Ricciotti and FitzGerald, 2011). Betalains and beetroot extracts have emerged as potent anti-inflammatory agents. Additionally, the beetroot-treated rats' kidney homogenates showed decreased levels of immune cells (TNF-α, IL-6, and MPO) and oxidative damage indicators (MDA), which may be directly linked to the NF-κB pathway being blunted (Clifford et al., 2015). The betalains found in beets are probably responsible for some of these effects; new research indicates that giving rats acute renal injury with 25 and 100 mg·kg·bm–1 of betanin for five days dramatically reduces their NF-κB DNA-binding activity (Tan et al., 2015). Betanins can modulate oxidative stress and reduce inflammation through multiple pathways, preventing the major events for the progression of CVD.

9.3.9 Beetroot as Cardio Protective

Clinical research is quickly increasing our understanding of how acute and long-term beetroot juice consumption affects blood pressure and vascular function. Consuming beetroot juice has also been demonstrated to lower blood pressure, enhance endothelial function, and significantly raise systemic NO production and plasma NO_2 levels (Kapil et al., 2010). The rate of erythrocyte xanthine oxidase expression or erythrocytic xanthine oxidoreductase, an enzyme involved in the reduction of NO_2 in active NO, in hypertensive states explains why beetroot juice administration was found to have a significantly stronger effect on blood pressure in hypertensive subjects than in normotensive subjects (Ghosh et al., 2013). The increase in free radicals leads to the peroxidation of fatty acids in LDL particles, increasing their negative charge, density, lysolecithin, cholesterol oxide, hydroxide, and hydroperoxide content, and cytotoxicity and decreasing polyunsaturated fatty acids, with consequent fragmentation of apolipoprotein B-100 (apo B-100) (Kovanen and Pentikainen, 2003). In an animal study to determine the protective effect of beetroot juice against doxorubicin-induced cardiotoxicity, the combination of beetroot juice with doxorubicin significantly reduced doxorubicin toxicity in rat cardiomyocytes by reducing the generation of ROS (Das et al., 2016). Therefore, beetroot juice supplement is postulated to be cardioprotective in patients treated with anthracycline chemotherapeutic drugs. The clinical application of beetroot juice as an adjunct therapy to combat cardiotoxicity caused by anthracycline drugs is promising and would have a beneficial public health impact (Kuriakose et al., 2016).

9.3.10 Beetroot as Chemopreventive

One prevalent side effect that lowers cancer patients' quality of life is fatigue. It is one of the most frequently reported complaints among cancer patients (Stasi et al., 2003). As a component of a balanced diet and lifestyle, functional foods are becoming more and more popular. Because of its chemopreventive and antioxidant properties, beetroot juice is proposed as a nitrate-dietary supplement for individuals receiving chemotherapy and those recovering from chemotherapy. Supplementing with beetroot may help reduce fatigue symptoms because of its high dietary nitrate content and consequent NO generation. NO is a common signaling molecule that plays crucial physiological roles in human tissues (Stamler and Meissner, 2001). In summary, beetroot is a valuable vegetable to be consumed for health maintenance and has great potential to be used for chemoprevention and in patients with cancer to manage fatigue and other side effects of chemotherapy.

9.3.11 Beetroot Improves Stamina and Digestion

The concentration of plasma nitrite was considerably raised by the beetroot juice. Consuming beetroot juice significantly lowered diastolic blood pressure both during the maximal cardiorespiratory exercise test and at rest. An alternate source of exogenous antioxidants that can strengthen the body's overall antioxidant defenses and delay the onset of diseases linked to oxidative stress is red beetroot (Wang et al., 2020). Scientific studies have confirmed that drinking beetroot juice can boost your stamina and athletic performance. One of the first such studies was conducted by researchers from the University of Exeter and published in the Journal of Applied Physiology in 2010. The men who drank beetroot juice had muscles that used less ATP to generate the same force, which meant that their bodies required less oxygen, according to the study. In contrast to males who drank the placebo, this enabled those men to engage in high-intensity activity for a longer period of time before feeling fatigued (Maheshwari et al., 2013).

9.4 Microbiological Studies on Beetroot

The promotion of healthier lifestyles has driven a rise in the consumption of fresh produce. However, these foods may pose a higher risk of food-borne illnesses, as they often bypass processing steps that ensure the effective removal or inactivation of pathogenic microorganisms before consumption. As a result, outbreaks of diseases linked to ready-to-eat fruits and vegetables have significantly increased in recent years. Identifying the sources and nature of microbial contamination in these products is crucial for developing effective mitigation strategies for food producers (Machado-Moreira et al., 2019). It is recognized that microbiological criteria are widely used in the food industry, but they have rarely been published. Traditionally, control of microorganisms in a food has been demonstrated by microbiological testing of samples at various stages of production and of the final product. Table 9.1 shows different microbiological studies on beetroot and its products.

Criteria have been developed to give some degree of assurance that food is safe and of suitable quality, and that it will remain so to the end of its shelf life, provided it is handled appropriately (Stannard, 1997). There are various parameters or microbiological tests available, which can be utilized for the evaluation of the quality of food materials. One of the commonly utilized tests is the total plate count (TPC) or viable count, which is used to quantify bacterial population of >250 CFU/mL for liquids and >2500 CFU/mL for solids, which were previously diluted ten-fold (Matthews et al., 2024). Czyzowska et al. (2020) evaluated the microbiological quality of fermented beetroot products, that is, grated beetroots and beetroot juice, during long-term cold storage. Authors evaluated fermented beetroot products for various microbial parameters, including total mesophilic count (TMC), lactic acid bacteria (LAB), Enterobacteriaceae, Yeast, and molds. Beetroot juice was evaluated by numerous researchers for various microbiological quality parameters, including total aerobic count (TAC), Aerobic mesophilic bacteria, aerobic plate count (APC), LAB, coliforms count, yeast and mould (YM) count, for understanding the effect of storage time, temperature, storage duration, and treatments on the quality of the product. In this context, Dusabumuremyi et al. (2021) evaluated the effect of temperature and storage duration on physical, chemical, and microbiological changes in beetroot juice during a storage period of 28 days at 4°C and at 24°C. Beetroot juice was evaluated for various physicochemical parameters, including pH, titratable acidity (TA), turbidity, color, TSS, and microbiological parameters such as TAC, total coliform count, and yeast. Further, Funmilayo and Ho (2024) evaluated the effect of storage time on the chemical and microbiological quality of pasteurized beetroot juice with enzyme extraction and clarification and stated that APC of beetroot juice had significantly increased from 5.35 log CFU/mL to 6.16 log CFU/mL for control 1 and 5.14 log CFU/mL to 5.94 log CFU/mL for control 2 during storage day of zero and 6th, respectively. Unpasteurized beetroot juice showed the YM count of 5.69 log CFU/mL and 5.06 log CFU/mL in control samples at 0 day. The pasteurized juice had the YM count of 1.98 log CFU/ mL.

On the other hand, Ubeira-Iglesias et al. (2019) evaluated the effect of high hydrostatic pressure (HPP) processing of beetroot juice and its effects on nutritional, sensory, and microbiological quality of beetroot juice and reported that initially, coliform and LAB counts were 2.62 CFU/mL and 3.71 log CFU/mL, respectively. The YM count and TAC varied between 1.16 and 3.66 log CFU/mL and 1.05 and 4.22 log CFU/mL. Results indicated that high pressure processing treatment significantly decreased the coliforms and LAB below their detection limit and resulted in a 2.5 and a 3.2 log reduction in YM count and aerobic bacteria, respectively.

TABLE 9.1

Different microbiological studies on beetroot and its products

Food product	Study	Parameters evaluated	Results	References
Grated beetroot	Evaluation of bioactive compounds and microbial quality of red beetroots and stored fermented red beetroot juice for a period of 7 to 10 months.	Total mesophilic count (TMC), lactic acid bacteria (LAB), Enterobacteriaceae, yeasts, and moulds	TMC varied from 1.8×10^5 to 8.0×10^5 (CFU/g) LAB varied from 4.5×10^4 to 5.6×10^4 (CFU/g) Yeasts and mold were 93 CFU/g in 7 month storage, while <10 in an 8 to 10 month storage period, while Enterobacteriaceae were <10 CFU/g.	Czyżowska et al. (2020)
Beetroot juice			TMC varied from 2.4×10^3 to 1.5×10^6 (CFU/mL) LAB varied from 2.2×10^5 to 1.4×10^6 (CFU/mL) Yeasts and mold were 1.2×10^3 to 4.0×10^3 (CFU/mL), while Enterobacteriaceae were <10 CFU/mL.	
Beetroot juice	Effect of storage temperature on physical-chemical and microbiological characteristics of beetroot juice	Total aerobic count (TAC), total coliforms, and yeasts	TAC and yeasts were absent till the lag phase and started to grow from 14 to 28 days. Coliforms were absent in juices stored at 4°C and 24°C.	Dusabumuremyi et al. (2021)
Freshly cut beetroots	Evaluation of physicochemical, microbial, and sensory quality of fresh-cut red beetroots in relation to sanitization method and storage duration	TAC, yeasts and mould (YM) count	TAC varied between 2.6 log CFU/g to 3.4 log CFU/g in various treated samples on washing day. The YM count varied between 1.1 log CFU/g to 2.2 log CFU/g in various treatments.	Chandra et al. (2015)
Beetroot juice	Evaluation of high hydrostatic pressure (HPP) processing of beetroot juice and its effects on nutritional, sensory, and microbiological quality	Aerobic mesophilic bacteria, LAB, and YM count along with coliforms.	Initially, coliform and LAB count was 2.62 CFU/mL and 3.71 log CFU/mL, respectively. The YM count and TAC varied between 1.16 and 3.66 log CFU/mL and 1.05 and 4.22 log CFU/mL. Results indicated that HPP treatment significantly decreased the coliforms and LAB below their detection limit and resulted in a 2.5 and a 3.2 log reduction in the YM count and aerobic bacteria, respectively.	Ubeira-Iglesias et al. (2019)
Beetroot juice	To study the effect of storage time on the chemical and microbiological quality of pasteurized beetroot juice with enzyme extraction and clarification.	APC and YM count	APC of beetroot juice had significantly increased from 5.35 log CFU/mL to 6.16 log CFU/mL for control 1 and from 5.14 log CFU/mL to 5.94 log CFU/mL for control 2 during storage days of zero and 6th, respectively. Unpasteurized beetroot juice showed YM count of 5.69 log CFU/mL and 5.06 log CFU/mL in control samples at 0 day. The pasteurized juice had a YM count of 1.98 log CFU/ mL.	Funmilayo and Ho (2024)

(Continued)

TABLE 9.1 (*Continued*)

Different microbiological studies on beetroot and its products

Food product	Study	Parameters evaluated	Results	References
Beetroot greens	Evaluation and characterization of nutritional, microbiological, and sensory properties of beet greens	Aerobic mesophilic bacteria, aerobic psychrophilic bacteria, LAB, total coliform bacteria, and YM count	The content of aerobic mesophilic and psychrophilic bacteria was 5.17 and 5.63 log CFU/g, respectively. The content of LAB was 2.86 log CFU/g, total coliform bacteria were 3.74 log CFU/g, and the YM count was 4.63 log CFU/g.	Fernández et al. (2017)
Ready to drink beetroot juice	To study the standardization of process conditions in batch thermal pasteurization and its effect on antioxidant, pigment, and microbial inactivation of Ready to Drink (RTD) beetroot (Beta vulgaris L.) juice	Samples were serially diluted, plated in total count agar for TAC, and in acidified Potato dextrose agar for YM count. Violet Red Bile Agar was used for coliform detection.	Results stated that the control sample had a microbial load of 6.53, 6.44, and 4.38 log CFU/mL for total plate count, coliforms, and YM count, respectively. Treatment of juice at 96°C for 540 seconds reduced the TPC to 1.30 log CFU/mL.	Kathiravan et al. (2014)
Red beetroot juice	Production of red beetroot juice by different methods: Kinetics of microbial growth, sugar consumption, and acid production	Total mesophilic and aerobic bacteria (TMAB), YM count, and LAB	Initially, the counts of YM and TMAB of beetroot juice samples produced by the spontaneous method were 4.85 log CFU/mL and 4.91 log CFU/mL, respectively and were found to be 8.56 log CFU/mL and 8.19 log CFU/mL, respectively, at the end of the fermentation process (96th hour). The average YM and TMAB counts of juice samples produced by adding *Lc. paracasei* were found to be 3.52 log CFU/mL and 8.59 log CFU/mL, and 6.62 log CFU/mL and 8.36 log CFU/mL, respectively, over the 0th and 44th hours of fermentation	Duyar et al. (2024)
Steamed cooked beetroots	Characterization of the physicochemical, phytochemical, and microbiological properties of steam-cooked beetroots during refrigerated storage.	Total viable count (TVC) at refrigerated storage for 9 days	Results showed that the shelf life of steam-cooked beetroots was 3 days (log CFU/g$<$7), whereas on the 6th day of storage, the TVC population reached 8.47 log CFU/g.	Intzirtzi et al. (2024)
Beetroot waste flour	Minimally processed beetroot waste as an alternative source to obtain functional ingredients	Total aerobic mesophilic count and total coliform test	Beetroot waste and its flour showed aerobic microorganism counts of 2.15×10^4 CFU/g and 1.96×10^2 CFU/g, respectively, with both materials containing less than 0.3 MPN/g of fecal coliforms at 45°C.	Costa et al. (2017)

9.5 Conclusion

In this article, an attempt is made to provide an insight into the various physicochemical and functional properties of beetroots and their related products. The article has been made even more comprehensive by providing explanations of various physical and chemical properties of beetroot in detail, along with various microbiological studies. These studies indicate that beetroots vary in length, diameter, and width, along with huge variation in their chemical properties like moisture, protein, fat, sugars, minerals, and vitamin content. Beetroots and their products can also be effectively utilized for the preparation of various other food products due to their wide range of functional properties, such as hydration properties, OAC, swelling properties, hydrophobicity, water activity, bulk density, and so forth. Moreover, beetroot's versatility allows it to be consumed in various forms, including raw, cooked, juiced, or as processed products like powders or supplements, although processing methods can influence its nutritional and antioxidant properties. Beetroot is a functional food with significant health benefits, making it an excellent choice for both dietary inclusion, and further research is needed in the context of disease prevention and health promotion.

Conflict of Interest

The authors declare no conflicts of interest

Abbreviations

AOA	Antioxidant activity
APC	Aerobic plate count
ATP	Adenosine triphosphate
aw	Water activity
°B	Degree brix
BD	Bulk density
Ca	Calcium
CFU	Colony-forming units
cm	Centimeters
Cu	Copper
CVD	Cardiovascular diseases
DM	Dry matter
DNA	2-deoxyribonucleic acid
DPPH	2, 2-diphenyl-1-picrylhydrazyl
Fe	Iron
FS	Foam stability
FW	Fresh weight
g	Grams
GAE	Gallic acid equivalent
IDF	Insoluble dietary fiber
IL-6	Interleukin-6

K	Potassium
kg	Kilograms
L	Liter
LAB	Lactic acid bacteria
LDL	Low density lipoproteins
LDL	Low density lipoproteins
m.Pa.s	Mili pascal seconds
MDA	Malondialdehyde
Mg	Magnesium
mg	Milligrams
mL	Milliliters
Mn	Manganese
N	Newton
Na	Sodium
NO	Nitric oxide
NO_2	Nitrogendioxide
OAC	Oil absorption capacity
OHC	Oil holding capacity
P	Phosphorus
ROS	Reactive oxygen species
SDF	Soluble dietary fiber
TA	Titrable acidity
TAC	Total aerobic count
TDF	Total dietary fiber
TE	Trolox equivalent
TEAC	Trolox equivalent antioxidant capacity
TMAB	Total mesophilic and aerobic bacteria
TMC	Total mesophilic count
TNF-α	Tumor necrosis factor-α
TP	Total phenolic
TPC	Total plate count
TSS	Total soluble solids
TVC	Total viable count
WAC	Water absorption capacity
WHC	Water holding capacity
WRC	Water retention capacity
YM	Yeasts and moulds
Zn	Zinc

References

Agarwal, K., & Varma, R. (2014). Biochemical screening of beetroot leaves. *International Journal of Pharmaceutical Sciences*, 1, 127–134.

Aulia, F., & Sunarharum, W. B. (2020, April). Beetroot (Beta vulgaris L. var. rubra L.) flour proportion and oven temperature affect the physicochemical characteristics of beetroot cookies. In *IOP Conference Series: Earth and Environmental Science* (Vol. 475, No. 1, p. 012040). IOP Publishing, Indonesia.

Azam, M., & Pandey, S. (2020). Efeect of hot-air drying on physical properties and colour characteristics of beetroot (Beta vulgaris L.). *Applied Biological Research*, 22(3), 215–225.

Baião, D. D. S., da Silva, D. V., Del Aguila, E. M., & Paschoalin, V. M. F. (2017). Nutritional, bioactive and physicochemical characteristics of different beetroot formulations. *Food Additives*, 6(6), 21–43.

Betoret, E., Betoret, N., Vidal, D., & Fito, P. (2011). Functional foods development: trends and technologies. *Trends in Food Science and Technology*, 22, 498–508.

Chandra, D., Choi, A. J., Kim, Y. P., & Kim, J. G. (2015). Physicochemical, microbial and sensory quality of fresh-cut red beetroots in relation to sanitization method and storage duration. *Italian Journal of Food Science/Rivista Italiana di Scienza degli Alimenti*, *27*(2), 208–220.

Chaudhary, V., & Kumar, V. (2020). Study on drying and rehydration characteristics of tray dried beetroot (Beta vulgaris L.) and functional properties of its powder. *Chemical Science Review and Letters*, 9(33), 98–108.

Chauhan, S., Chamoli, K., &Sharma, S. (2020). Beetroot-a review paper. *Journal of Pharmacognosy and* Phytochemistry, 9, 424–427.

Chen, L., Zhu, Y., Hu, Z., Wu, S., &Jin, C. (2021). Beetroot as a functional food with huge health benefits: Antioxidant, antitumor, physical function, and chronic metabolomics activity. *Food Science & Nutrition*, 9(11), 6406–6420.

Clifford, T., Howatson, G., West, D. J., & Stevenson, E. J. (2015). The potential benefits of red beetroot supplementation in health and disease. *Nutrients*, 7, 2801–2822.

Costa, A. P. D., Hermes, V. S., Rios, A. D. O., & Flôres, S. H. (2017). Minimally processed beetroot waste as an alternative source to obtain functional ingredients. *Journal of Food Science and Technology*, 54, 2050–2058.

Czyzowska, A., Siemianowska, K., Sniadowska, M., & Nowak, A. (2020). Bioactive compounds and microbial quality of stored fermented red beetroots and red beetroot juice. *Polish Journal of Food and Nutrition Sciences*, 70(1), 35–44.

Da Silva, A. O., De, F. E., Silva, E. F., & Klar, A. E. (2016). Yield of beet cultivars under fertigation management and salinity control in a protected environment. *Chilean Journal of Agricultural Research*, 76, 463–470.

Das, S., Filippone, S. M., Williams, D. S., Das, A., & Kukreja, R. C. (2016). Beet root juice protects against doxorubicin toxicity in cardiomyocytes while enhancing apoptosis in breast cancer cells. *Molecular and Cellular Biochemistry*, 421, 89–101.

Delgado-Vargas, F., Jimenez, A. R., & Paredes-Lopez, O. (2000). Natural pigments: carotenoids, anthocyanins, and betalains – Characteristics, biosynthesis, processing, and stability. *Critical Reviews in Food Science and Nutrition*, 40, 173–289.

Dhingra, D., Michael, M., Rajput, H., & Patil, R. T. (2012). Dietary fibre in foods: a review. *Journal of Food Science and Technology*, 49, 255–266.

Dusabumuremyi, J. C., Rubona, P., & Ndisanze, M. A. (2021). Effect of storage temperature on physical-chemical and microbiological characteristics of beetroot juice. *INES Scientific Journal*, 1(1), 43–52.

Duyar, S. M., Sari, F., & Karaoglan, H. A. (2024). Production of red beetroot juice by different methods: kinetics of microbial growth, sugar consumption, and acid production. *Heliyon*, *10*(9), e30448.

Fernández, M. V., Jagus, R. J., & Agüero, M. V. (2017). Evaluation and characterization of nutritional, microbiological and sensory properties of beet greens. *Acta Scientific Nutritional Health*, *1*(3), 37–45.

Fidelis, M., Santos, J. S., Coelho, A. L. K.,Rodionova, O. Y., Pomerantsev, A., & Granato, D. (2017). Authentication of juices from antioxidant and chemical perspectives: a feasibility quality control study using chemometrics. *Food Control*, 73, 796–805.

Frank, T., Stintzing, F. C., Carle, R., Bitsch, I., Quaas, D., Strass, G.,Bitsch, R., &Netzel, M. (2005). Urinary pharmacokinetics of beta-lains following consumption of red beet juice in healthy humans. *Pharmacological Research*, 52, 290–297.

Funmilayo, A. M. Z., & Ho, L. H. (2024). Effects of storage time on the chemical and microbiological quality of pasteurized beetroot (Beta vulgaris L.) juice with enzyme extraction and clarification. *Sarhad Journal of Agriculture*, 40, 69–80.

Gengatharan, A., Dykes, G. A., & Choo, W. S. (2015). Betalains: natural plant pigments with potential application in functional foods. *LWT- Food Science and Technology*, 64, 645–649.

Ghosh SM, Kapil V, Fuentes-Calvo I, Bubb KJ, Pearl V, Milsom AB, et al. (2013). Enhanced vasodilator activity of nitrite in hypertension: critical role for erythrocytic xanthine oxidoreductase and translational potential. *Hypertension*, 61(5):1091–102.

Guldiken, B., Toydemir, G., Nur Memis, K., Okur, S., Boyacioglu, D., & Capanoglu, E. (2016). Home-processed red beetroot (Beta vulgaris L.) products: changes in antioxidant properties and bioaccessibility. *International Journal of Molecular Sciences*, 17, 858.

Hegedusova, A., Slosar, M., Mezeyova, I., Hegedus, O., Andrejiova, A., & Szarka, K. (2018). *Methods for Estimation of Selected Biologically Active Substances*. Nitra, Slovak Republic: Slovak University of Agriculture.

Indirani, K. (2021). Evaluation of functional properties of beetroot powder (Beta vulgaris) and its suitability in developing a candy by incorporate. *International Journal of Science and Healthcare Research*, 6(1), 1–5.

Intzirtzi, E., Karabagias, V. K., Lazaridis, D. G., Karabagias, I. K., & Giannakas, A. E. (2024). Characterization of the physicochemical, phytochemical, and microbiological properties of steam cooked beetroots during refrigerated storage. *European Food Research and Technology*, 250(6), 1733–1743.

Irving, D. (2012). *Beetroot Stand Management*. Sydney, Australia: Horticulture Australia Ltd.

Juszczak, L., Witczak, M., Fortuna, T., & Solarz, B. (2010). Effect of temperature and soluble solids content on the viscosity of beetroot (Beta vulgaris) juice concentrate. *International Journal of Food Properties*, 13(6), 1364–1372.

Kale, R., Sawate, A., Kshirsagar, R., Patil, B., & Mane, R. (2018). Studies on evaluation of physical and chemicalcomposition of beetroot (*Beta Vulgaris* L.). *International Journal of Chemical Studies*, 6, 2977–2979.

Kanner, J., Harel, S., & Granit, R. (2001). Betalains: A new class of dietary cationized antioxidants. *Journal of Agricultural and Food* Chemistry, 49, 5178–5185.

Kapil, V., Milsom, A. B., Okorie, M., Maleki-Toyserkani, S., Akram, F., Rehman, F., ..., & Ahluwalia, A. (2010). Inorganic nitrate supplementation lowers blood pressure in humans: role for nitrite-derived NO. *Hypertension*, 56(2), 274–281.

Kapoor, N., Mohite, A. M., Sharma, N., & Sharma, D. (2021). Comparative analysis of freeze dried and spray dried beet-root powder according to physico-chemical, functional and color properties. *Bulletin of the Transilvania University of Brasov. Series II: Forestry Wood Industry• 02.Agricultural Food Engineering*, *14*(63), 153–162.

Kathiravan, T., Nadanasabapathi, S.& Kumar, R. (2014). Standardization of process conditionin batch thermal pasteurization and its effect on antioxidant, pigment and microbialinactivation of Ready to Drink (RTD) beetroot (Beta vulgaris L.) juice. *International Food Research Journal*, 21, 1305.

Kerr, W. L., & Varner, A. (2020). *Chemical and Physical Properties of Vacuum-Dried red Beetroot (Beta vulgaris) Powders Compared to Other Drying Methods*. Drying Technology.

Khan, M. I., & Giridhar, P. (2015). Plant betalains: chemistry and biochemistry. *Phytochemistry*, 117, 267–295.

Kovanen, P. T., & M. O. Pentikainen. 2003. Circulating lipoproteinsasproinflammatory and anti-inflammatory particles in atherogen-esis. *Current Opinion in Lipidology*,14 (5):411–9.

Kovarovic, J., Bystricka, J., Tomas, J., & Lenkova, M. (2017). The influence of variety of the content of bioactive compounds in Beetroot (*Beta vulgaris L.). Potravinarstvo Slovak Journal of Food* Sciences, 11, 106–112.

Kruszewski, B., Domian, E., & Nowacka, M. (2023). Influence of high-pressure homogenization on the physicochemical properties and betalain pigments of red beetroot (Beta vulgaris L.) juice. *Molecules*, 28(5), 2018.

Kujala, T. S., Loponen, J. M., Klika, K. D., & Pihlaja, K. (2000). Phenolics and betacyanins in red beetroot (Beta vulgaris) root: distribution and effect of cold storage on the content of total phenolics and three individual compounds. *Journal of Agricultural and Food Chemistry*, 48, 5338–5342.

Kuriakose, R. K., Kukreja, R. C., & Xi, L. (2016). Potential therapeutic strategies for hypertension-exacerbated cardiotoxicity of anticancer drugs. *Oxidative Medicine and Cellular Longevity, 2016*(1), 8139861.

Lattanzio, V., Lattanzio, V. M., & Cardinali, A. (2006). Role of phenolics in the resistance mechanisms of plants against fungal pathogens and insects. *Phytochemistry,* 661, 23–67.

Lee, E. J., Dami, A. N., Nguyen, C. T. T., Patil, B. S., Kim, J., & Yoo, K. S. (2014). Betalain and betaine composition of greenhouseorfield-produced beetroot (*Beta Vulgaris L.*) and inhibition of HepG2 cell proliferation. *Journal of Agricultural and Food Chemistry,* 62, 1324–1331.

Liu, Y., Sabadash, S., & Duan, Z. (2021). Effects of intermittent microwave drying conditions on characteristics and physical properties of beetroots. *Ukrainian Journal of Food Science*, 9(1).

Machado-Moreira, B., Richards, K., Brennan, F., Abram, F., & Burgess, C. M. (2019). Microbial contamination of fresh produce: what, where, and how?. *Comprehensive Reviews in Food Science and Food Safety*, 18(6), 1727–1750.

Maheshwari, R. K., Parmar, V., & Joseph, L. (2013). Latent therapeutic gains of beetroot juice. *World Journal of Pharmaceutical Research*, 2(4), 804–820.

Marmion, D. M. (1991). *Handbook of US Colorants: Foods, Drugs, Cosmetics and Medical Devices.* Hoboken, NJ: John Wiley & Sons.

Martinez, R. M., Longhi-Balbinot, D. T., Zarpelon, A. C., Staurengo-Ferrari, L., Baracat, M. M., Georgetti, S. R., …, & Casagrande, R. (2015). Anti-inflammatory activity of betalain-rich dye of Beta vulgaris: effect on edema, leukocyte recruitment, superoxide anion and cytokine production. *Archives of Pharmacal Research*, 38, 494–504.

Matthews, S., Siddiqui, Y., Supramaniam, C. V., & Ali, A. (2024). Boosting Capsicum annuum growth through non-native endophytic bacterial consortium. *Journal of Plant Growth Regulation*, *43*(8), 2739–2760.

Mella, C., Vega-Gálvez, A., Uribe, E., Pasten, A., Mejias, N., & Quispe-Fuentes, I. (2022). Impact of vacuum drying on drying characteristics and functional properties of beetroot (*Beta vulgaris*). *Applied Food Research*, *2*(1), 100120.

Mirmiran, P., Houshialsadat, Z., Gaeini, Z., Bahadoran, Z., & Azizi, F. (2020). Functional properties of beetroot (Beta vulgaris) in management of cardio-metabolic diseases. *Nutrition & Metabolism*, 17, 1–15.

Mohammad, A., Khamdelwal, B. M., Garg, S. K., & Pandey, S. (2014). Physical characteristics and mathematical modeling of drying of Beetroot (*Beta vulgaris L.*) JNKVV *Research Journal*, 48, 232–244.

Monteiro, R., & Azevedo, I. (2010). Chronic inflammation in obesity and the metabolic syndrome. *Mediators of Inflammation*, 2010(1), 289645.

Moulick, S. P., Jahan, F., Islam, M. B., Al Bashera, M., Hasan, M. S., Islam, M. J., …, & Bhuiyan, M. N. H. (2023). Nutritional characteristics and antiradical activity of turmeric (Curcuma longa L.), beetroot (Beta vulgaris L.), and carrot (Daucus carota L.) grown in Bangladesh. *Heliyon*, *9*(11),7.

Nayak, C. A., Suguna, K., Narasimhamurthy, K., &Rastogi, N. K. (2007). Effect of gamma irradiation on histological and textural properties of carrot, potato and beetroot. *Journal of Food Enginerring,* 79, 765–770.

Neelwarne, B. (2013). *Red Beet Biotechnology: Food and Pharmacentical Applications.* New York: Springer Science and Business, pp. 199–243.

Nemzer, B. Z., Pietrzkowski, A., Spoma, P. S., Thresher, T., Michalowski, S., &Wybranie, C. (2011). Betalainic and nutritionalprofiles of pigment-enriched red beetroot (*Beta vulgaris L.*) dried extracts. *Food Chemistry*, 127, 42–53.

Ng, M. L., & Sulaiman, R. (2018). Development of beetroot (Beta vulgaris) powder using foam mat drying. *LWT-Food Science & Technology*, 88, 80–86.

Nirmal, N. P., Mereddy, R., & Maqsood, S. (2021). Recent developments in emerging technologies for beetroot pigment extraction and its food applications. *Food Chemistry*, 356, 129611.

Panghal, A., Kumar, V., Dhull, S. B., Gat, Y., & Chhikara, N. (2017). Utilization of dairy industry waste-whey in formulation of papaya RTS beverage. *Current Research in Nutrition and Food Science*, 5, 168–174.

Paula, L., Ana, J. G., Hannu, P., Carlos, M. J., Eeva-Riikka, V., & Cristina, J. (2025). Role of red beetroot in bread for reducing mycotoxin risks: bioavailability of beetroot polyphenols and betalains with ochratoxin a, aflatoxin B1 and zearalenone in Caco-2 cells. *Food Chemistry*, 465, 142036.

Ramappa, D., GV, M. K., Kumargouda, V., Hallad, S. C., Aruna, T. N., Ravi, Y., & Murali, P. (2024). Standardization of vacuum frying techniques for the development of low-fat beetroot finger chips. *Food and Humanity*, 3, 100424.

Ranawana, V., Moynihan, E., Campbell, F., Duthie, G., & Raikos, V. (2018). Beetroot improves oxidative stability and functional properties of processed foods: singular and combined effects with chocolate. *Journal of Food Science and Technology*, 55, 2401–2409.

Ribas-Agustí, A., Martín-Belloso, O., Soliva-Fortuny, R.&Elez-Martínez, P. (2018). Food processing strategies toenhance phenolic compounds bioaccessibility and bioavailability in plant-based foods. *Critical Review Food Science and Nutrition*, 58, 2531–2548.

Ricciotti, E., & FitzGerald, G. A. (2011). Prostaglandins and inflammation. *Arteriosclerosis, Thrombosis, and Vascular Biology*, 31(5), 986–1000.

Rodler, I., & Tápanyagtáblázat, U. J. (2005). Budapest: Medicina Könyvkiadó Rt. pp. 308–309.

Rodriguez-Sevilla, M., Villanueva-Suárez, M. J., & Redondo-Cuenca, A. (1999). Effects of processing conditions on soluble sugars content of carrot, beetroot and turnip. *Food Chemistry*, 66:81–85.

Ruboczki, T., Raczko, V., & Takacsne Hajos, M. (2015). Evaluation of morphological parameters and bioactivecompounds in different varieties of beetroot (*Beta Vulgaris L.* Ssp. esculenta GURKE Var. Rubra L.). *International Journal of Horticultural Science*, 21, 31–35.

Sahni, P., & Shere, D. M. (2017). Comparative evaluation of physico-chemical and functional properties of apple, carrot and beetroot pomace powders. *International Journal of Food and Fermentation Technology*, 7(2), 317–323.

Sakhare, K. S., Sawate, A. R., Kshirsaga, R. B., & Taur, A. T. (2019). Studies on physical and chemical composition of beetroot (*Beta vulgaris* L.). *International Journal of Chemical* Studies, 7, 283–285.

Sharoba, A. M., Farrag, M. A., & El-Salam, A. (2013). Utilization of some fruits and vegetables wastes as a source of dietary fibers in cake making. *Journal of Food and Dairy Sciences*, 4(9), 433–453.

Shehatta, E. A., Abo-Raya, S. H., & Baioumy, A. A. (2024). Impact of different drying treatments on the biochemical and antioxidant activity properties of Egyptian red beetroot. *Food system*, 7, 151–156.

Slosar, M., Kopta, T., Hegedus, O., Hegedűsova, A., Mezeyova, I., Timoracka, M., & Mezey, J. (2020). Yield parameters, antioxidant activity, polyphenol and total soluble solids content of beetroot cultivars with different flesh colours. *Folia Horticulturae*, 32, 351–362.

Stagnari, F., Galieni, A., Speca, S., & Pisante, M. (2014). Water stress effects on growth, yield and quality traits of red beet. *Scientia Horticulturae*, 165, 13–22.

Stamler, J. S., & Meissner, G. (2001). Physiology of nitric oxide in skeletal muscle. *Physiological Reviews*, 81(1), 209–237.

Stannard, C. (1997). Development and use of microbiological criteria for foods. *Food Science and Technology Today*, 11(3), 137–177.

Stasi, R., Abriani, L., Beccaglia, P., Terzoli, E., & Amadori, S. (2003). Cancer-related fatigue: evolving concepts in evaluation and treatment. *Cancer*, 98(9), 1786–1801.

Suresh Chandra, S. C., & Samsher, S. (2013). Assessment of functional properties of different flours. *Journal of Agricultural Research*, *8*(48), 4849–4852.

Szekely, D., Illes, B., Steger-Mate, M., & Monspart-Senyi, J. (2016). Effect of drying methods for inner parameters of red beetroot *(Beta vulgaris L.). Acta University Sapientiae, Alimentaria*, 9, 60 –68.

Székely, D., Szalóki-dorkó, L., StéGer-Maté, M., SzaBó-nótin, B., Ivani, C. S, J., & MonSPart-Sényi, J. (2014). Distribution of antioxidant components in roots of different red beets *(Beta vulgaris L.)* cultivars. *Acta Alimentaria*, 43, 164–171.

Tan, D., Wang, Y., Bai, B., Yang, X., & Han, J. (2015). Betanin attenuates oxidative stress and inflammatory reaction in kidney of paraquat-treated rat. Food and Chemical Toxicology, *78*, 141–146.

Tulp, M., & Bohlin, L. (2004). Unconventional natural sources for future drug discovery. *Drug Discovery* Today, 9, 450–458.

Takácsné Hájos, M., & Rubóczki, T. (2012). Effects of environmental factors on morphological and quality parameters of table beet root. *International Journal of Horticultural Science, 18*(2), 1054.

Ubeira-Iglesias, M., Wilches-Pérez, D., Cavia, M. M., Alonso-Torre, S., & Carrillo, C. (2019). High hydrostatic pressure processing of beetroot juice: effects on nutritional, sensory and microbiological quality. *High Pressure Research*, 39(4), 691–706.

USDA Agricultural Research Service (USDA-ARS). (2014). *Composition of Foods Raw, Processed, Prepared USDA National Nutrient Database for Standard Reference, Release 28*. Nutrient Data Laboratory Home Page, USDA. Available at http://www.ars.usda.gov/nutrientdata.

Vasconcellos, J., Conte-Junior, C., Silva, D., Pierucci, A. P., Paschoalin, V., & Alvares, T. S. (2016). Comparison of total antioxidant potential, and total phenolic, nitrate, sugar, andorganic acid contents in beetroot juice, chips, powder, and cooked beetroot. *Food Science and Biotechnology*, 25, 79–84.

Vdovenko, S. A., Palamarchuk, I. I., Pantsyreva, H. V., Alexeyev, O. O., & Vdovenko, L. O. (2018). Energy efficient growing of red beet in the conditions of central forest steppe of Ukraine. *Ukrainian Journal of Ecology*, 8, 34–40.

Viskelis, J., Nevidomskis, S., Bobinas, C., Urbonaviciene, D. Bobinitel, R., Karkleliene, R., & Viskelis, P. (2019). Evaluation of beetroot quality during various storage conditions. *Food Balt*, 170–175.

Wang, T., Liu, L., Rakhmanova, A., Wang, X., Shan, Y., Yi, Y., …, &Lü, X. (2020). Stability of bioactive compounds and in vitro gastrointestinal digestion of red beetroot jam: effect of processing and storage. *Food Bioscience*, 38, 100788.

Wootton-Beard, P. C., Moran, A., & Ryan, L. (2011). Stability of the antioxidant capacity and total polyphenol content of commercially available vegetable juices before and after invitro digestion as measured by FRAP, DPPH, ABTS and Folin Ciocalteu methods. *Food Research International*, 44, 217–224.

Wruss, J., Waldenberger, G., Huemer, S., Uygun, P., Lanzerstorfer, P., Mullar, U., Hoglinger, O., & Weghuber, J. (2015). Compositional characteristics of commercial beetroot products and beetroot juice prepared from seven beetroot varieties grown in upper Austria. *Journal of Food Composition and Analysis*, 42, 46–55.

Yasaminshirazi, K., Hartung, J., Fleck, M., &Graeff-Hoenninger, S. (2020). Bioactive compounds and total sugar contents of different open-pollinated beetroot genotypes grown organically. *Molecules*, 25, 4884.

Yashwant, K. (2015). Beetroot: a super food. *International Journal of Engineering Studies and Technical Approach*, 1, 20–26.

Zhang, H., & Tsao, R. (2016). Dietary polyphenols, oxidative stress and antioxidant and anti-inflammatory effects. *Current Opinion Food Science*, 8, 33–42.

10

Current Scenario of Beetroot Processing

Sena Bakir and Sajad Ahmad Wani

10.1 Introduction

Beetroot (*Beta vulgaris*) is recognized as a highly valuable vegetable owing to its dense concentration of bioactive compounds, including betalains, ascorbic acid, carotenoids, polyphenols, flavonoids, and saponins (Chhikara et al., 2019). Betalains, the principal natural pigments of beetroot, are classified into red-violet betacyanins and yellow-orange betaxanthins, each characterized by distinct absorption peaks and important biological functions (Ravichandran et al., 2013). Compounds like betanin, indicaxanthin, and vulgaxanthin I and II have been linked to anti-inflammatory and antioxidant effects, highlighting their potential in managing inflammatory diseases (Wang et al., 2022).

The extensive health benefits associated with beetroot bioactives have positioned it as a promising ingredient in both the food and pharmaceutical sectors. Betalains, in particular, have been proposed as natural functional ingredients due to their antioxidative, anti-inflammatory, and potential anticancer properties (Fu et al., 2020). Various beetroot fractions, such as juice, powder, pomace, and extracts, have been utilized in developing functional foods like breads, snacks, dairy products, fermented beverages, and meat products (Punia Bangar et al., 2023). Beetroot powder, besides serving as a natural colorant (Ng & Sulaiman, 2018), has been explored as an alternative to nitrites in processed meat (Sucu & Turp, 2018).

The nitrate content of beetroot plays a significant role in promoting vascular health through the nitrate-nitrite-nitric oxide pathway, contributing to blood pressure regulation and improved exercise performance (Lidder & Webb, 2013; Mudgal et al., 2022). Experimental studies have also suggested that beetroot supplementation can enhance skeletal muscle function in heart failure models (Ferguson et al., 2016) and may help manage pregnancy-induced hypertension (Tropea et al., 2020).

However, the stability of beetroot's bioactive compounds is highly susceptible to environmental factors such as light, temperature, oxygen, and pH during processing and storage (Fu et al., 2020). Traditional processing methods like boiling, drying, and pickling have been reported to cause significant losses in betalain content and antioxidant activity (Guldiken et al., 2016; Czyżowska et al., 2020). Therefore, innovative processing approaches that minimize the degradation of bioactive compounds are essential to maximize beetroot's functional value.

This chapter explores the current advancements in beetroot processing and preservation technologies, highlighting both the challenges and future prospects in enhancing the stability and bioavailability of beetroot-derived bioactives.

DOI: 10.1201/9781003527084-10

10.2 Current Market Interest of Beetroot

Beetroot has experienced a surge in market interest in recent years, largely driven by its rich profile of bioactive compounds—phenolic acids, flavonoids, saponins, carotenoids, nitrates, and betalains—which contribute to its nutritional and functional benefits (Nistor et al., 2017). Consumers are increasingly seeking natural and health-promoting foods, and beetroot fits well within this trend due to its demonstrated antioxidant, anti-inflammatory, and cardiovascular properties.

One of the most significant trends is the expansion of beetroot-based functional foods and beverages. Beetroot is widely utilized in products like juices, smoothies, powders, nutritional supplements, and energy drinks, especially those marketed for cardiovascular health and athletic performance enhancement (Akan et al., 2021; Mudgal et al., 2022).

The affordability, ease of processing, and environmental friendliness of beetroot products further boost their appeal to both manufacturers and consumers. Red beetroot supplements, in particular, have become one of the fastest-growing segments in the functional food and beverage industries.

Beetroot is also favored for its natural colorant properties. The vibrant red betalains extracted from beetroot are increasingly used to replace synthetic dyes in foods, aligning with consumer preferences for "clean label" products free from artificial additives (Ng & Sulaiman, 2018).

Globally, countries like Hungary have integrated beetroot into traditional foods—serving it fresh, pickled, juiced, or dried—and the vegetable is valued both for its nutritional richness and its sensory appeal (Zin & Bánvölgyi, 2023).

Moreover, the use of beetroot by-products like peel extracts in bioactive food packaging has opened new markets, especially in the meat and processed food industries. Biofilms infused with beetroot extract have shown strong antioxidant and antimicrobial activities, adding functional value to packaging solutions (Chaari et al., 2022).

10.2.1 Market Future Potential

With the increasing emphasis on health, wellness, and sustainability, the beetroot market is expected to continue growing. Further research into beetroot's health benefits and improved processing and preservation techniques will likely expand its applications into new sectors, such as pharmaceuticals and functional cosmetics. Additionally, non-thermal processing and innovative packaging using beetroot derivatives offer exciting possibilities for future market diversification (Theba et al., 2024).

10.3 Modern Trends and Innovations in Beetroot Processing

In recent years, the growing consumer demand for natural, functional, and eco-friendly food products has driven significant innovation in beetroot processing. Beetroot's rich composition of bioactive compounds, particularly betalains and nitrates, has made it a target for developing new food products and sustainable technologies.

10.3.1 Non-Thermal and Mild Processing Technologies

Traditional thermal processing often leads to the degradation of sensitive compounds like betalains. Modern techniques such as pulsed electric field (PEF) pretreatment have emerged to minimize these losses. PEF pretreatment reduces the drying temperature by 20°C–25°C, maintaining the structural and color integrity of beetroot (Shynkaryk et al., 2008). This helps preserve antioxidant properties and enhances rehydration characteristics.

10.3.2 Vacuum-Steam Pulsed Blanching (VSPB)

VSPB represents a major innovation in beetroot preservation. Compared to conventional hot water or steam blanching, VSPB enhances uniform heat transfer, reduces nutrient loss, and minimizes pigment degradation. It creates a low-oxygen environment that protects delicate compounds from oxidation (Zhang et al., 2021).

10.3.3 Green Extraction Technologies

There is increasing interest in eco-friendly extraction methods to recover valuable beetroot bioactives without using harmful solvents. Techniques like ultrasound-assisted extraction and supercritical fluid extraction (CO_2 extraction) offer higher yields of betalains and polyphenols with minimal solvent usage (Fernando et al., 2021; Khaw et al., 2017). These methods improve purity and antioxidant activity, making beetroot extracts suitable for food and pharmaceutical applications.

10.3.4 Functional and Sustainable Food Applications

Beetroot fractions such as powders and extracts are incorporated into various functional foods—baked goods, dairy products, meat alternatives, and beverages—to enhance nutritional value and color naturally (Punia Bangar et al., 2023). The use of beetroot juice in sports drinks is particularly popular due to its nitrate content, which improves vascular function and exercise performance (Mudgal et al., 2022).

10.3.5 Bioactive and Intelligent Packaging

Innovations have extended beyond food into active and intelligent packaging. Beetroot peel extract has been used in biopolymer films (gelatin-sodium alginate) to create packaging that not only protects food but also offers antioxidant and antimicrobial benefits (Chaari et al., 2022). Other studies have developed freshness indicators using beetroot extract to monitor meat spoilage, further highlighting beetroot's potential in sustainable packaging (Guo et al., 2021).

10.3.6 Clinical and Nutraceutical Development

Highly purified beetroot extracts, such as betanin with enhanced antioxidant properties, are being studied for clinical use, particularly in cardiovascular health and inflammation management (Da Silva et al., 2023). The movement toward personalized nutrition is expected to drive the development of beetroot-based nutraceuticals targeting specific health conditions.

10.4 Current and Future Development in Beetroot Processing, Preservation, and Distribution

10.4.1 Current Developments

Beetroot (*Beta vulgaris*) has attracted substantial interest due to its bioactive compounds such as betalains, nitrates, phenolic acids, flavonoids, and amino acids (Thiruvengadam et al., 2024). These compounds not only offer nutritional benefits but also present opportunities for the food, nutraceutical, and pharmaceutical industries. Current developments focus on preserving beetroot's quality while enhancing its functional properties.

10.4.1.1 Key Techniques Currently in Use

Drying and Freezing: These methods are widely employed to preserve beetroot by reducing moisture content and extending shelf life. Drying methods such as hot air drying and solar drying are common but can affect the retention of sensitive compounds like betalains (Sharma et al., 2012; Essalhi et al., 2018).

PEF Pretreatment: PEF treatment improves drying efficiency by allowing lower drying temperatures and preserving pigment and nutrient content better than conventional methods (Shynkaryk et al., 2008).

Spray Drying and Foam Mat Drying: These techniques are used to produce beetroot powder, minimizing storage space and enhancing product stability (Carmo et al., 2018; Ng & Sulaiman, 2018).

Ohmic Heating and Microwave-Assisted Drying: These modern pre-treatments help retain polyphenols and antioxidant capacity better than conventional drying (Seremet (Ceclu) et al., 2020).

Challenges: They remain due to beetroot's high moisture content, susceptibility to microbial spoilage, and the instability of its natural pigments under heat, light, and oxygen exposure.

10.4.2 Future Developments

Future trends are aimed at more sustainable, efficient, and quality-preserving processing and distribution methods:

VSPB: This emerging technology provides uniform heating, reduces blanching time, minimizes nutrient loss, and is more eco-friendly than conventional blanching methods (Zhang et al., 2021).

Green Extraction Methods: Future extraction will increasingly rely on non-toxic, sustainable techniques such as ultrasound-assisted extraction, enzyme-assisted extraction, and supercritical CO_2 extraction to obtain high-purity bioactives with enhanced biological activity (Borjan et al., 2022; Fernando et al., 2021).

Bioactive Packaging Materials: The integration of beetroot by-products into biodegradable packaging films for active food preservation is gaining momentum, providing antioxidant and antimicrobial properties while reducing plastic waste (Chaari et al., 2022).

Intelligent Packaging Solutions: Innovations like color-changing films using beetroot extracts to monitor food spoilage are being developed, offering real-time quality indicators for consumers (Guo et al., 2021).

Expanded Distribution Strategies: As consumer interest in natural and functional foods rises, beetroot-based products are likely to be distributed through specialized health food markets, e-commerce platforms, and functional beverage sectors globally.

Customized Functional Foods and Clinical Applications: Research is moving toward developing beetroot products customized for specific health outcomes, such as cardiovascular health supplements or sports performance enhancers (Da Silva et al., 2023; Mudgal et al., 2022).

In the future, integrating precision agriculture, advanced processing technologies, and green logistics will further enhance the production, preservation, and distribution of high-quality beetroot products, making them more accessible and sustainable.

10.5 International Scenario of Production and Its Potential Role in Sustainable Agriculture

10.5.1 International Production Overview

Beetroot (*Beta vulgaris*) is cultivated globally, with major production concentrated in countries like Russia, the United States, Germany, France, Poland, and China. While beetroot is mainly grown for sugar production in the form of sugar beet, the demand for red beetroot used in food, beverages, and nutraceuticals has been rising steadily. Europe, especially Eastern Europe, remains a stronghold for beetroot cultivation and consumption, often incorporating it into traditional dishes and health-promoting products (Zin & Bánvölgyi, 2023).

In countries such as Hungary, beetroot is grown as a second crop with a growing duration of about 100–110 days and is utilized fresh, pickled, juiced, or processed into dried products (Zin & Bánvölgyi, 2023). In recent years, there has been a growing trend of beetroot-based functional products across North America and Asia, especially in the form of beetroot powders, juices, and supplements, responding to consumer demand for natural health-enhancing foods (Akan et al., 2021).

10.5.2 Potential Role in Sustainable Agriculture

Beetroot offers significant potential for promoting sustainable agriculture through multiple pathways:

Crop Diversification: Beetroot fits well into crop rotation systems, helping to improve soil structure, reduce soil-borne diseases, and enhance overall farm biodiversity. Its relatively short growing season allows farmers to optimize land use between major crop cycles.

Low Input Requirement: Beetroot generally requires fewer chemical fertilizers and pesticides compared to many other cash crops. Its natural hardiness makes it a suitable option for integrated pest management and organic farming systems.

Utilization of By-Products: Beetroot peel and pomace, often considered waste, are now being repurposed into value-added products such as natural colorants, antioxidants, and active packaging materials, supporting a circular economy model in agriculture (Chaari et al., 2022; Costa et al., 2017).

Climate Resilience: Beetroot exhibits adaptability to diverse climatic conditions, making it a promising crop in the context of climate change adaptation. Its moderate water requirements compared to other commercial crops also make it suitable for water-scarce regions.

Carbon Footprint Reduction: By promoting the use of natural food colorants like betalains from beetroot instead of synthetic dyes, agricultural systems can reduce their dependence on environmentally damaging chemical production processes (Fu et al., 2020).

Biodegradable Innovations: The development of beetroot-based biopolymers for packaging materials could significantly reduce plastic pollution, linking beetroot farming directly to environmental sustainability initiatives (Guo et al., 2021).

10.5.3 Future Outlook

To fully exploit beetroot's potential in sustainable agriculture, investment in varietal improvement, precision farming, and green processing technologies will be essential. Research into drought-tolerant and pest-resistant beetroot varieties could further enhance its viability as a sustainable crop. Additionally, policies promoting organic farming and sustainable supply chains will encourage broader adoption of beetroot in both local and global markets.

10.6 Conclusion and Future Perspective

Beetroot stands out as a multifunctional crop with substantial nutritional, therapeutic, and industrial value, owing to its rich composition of bioactive compounds such as betalains, nitrates, and polyphenols. Its application across diverse food products, functional beverages, and natural colorants continues to expand. However, the sensitivity of these compounds to processing and storage conditions necessitates the use of innovative and optimized techniques to retain their bioactivity and stability. Emerging technologies such as VSPB, PEF pretreatments, and green extraction methods offer promising solutions for preserving the quality of beetroot products. Moreover, the incorporation of beetroot by-products in sustainable packaging materials highlights its role in circular food systems. For the food industry to fully harness beetroot's potential, further research on advanced preservation methods, consumer preferences, and varietal improvement is essential.

With growing consumer demand for natural, functional, and sustainable food products, beetroot presents significant opportunities for innovation in both the food and pharmaceutical industries. Future research should focus on developing high-yield, nutrient-rich beetroot varieties tailored for specific industrial applications. Advanced, non-thermal processing technologies such as PEFs, vacuum-assisted drying, and green extraction methods hold promise for preserving beetroot's sensitive bioactive compounds while reducing environmental impact. Additionally, the exploration of beetroot-based active and intelligent packaging can offer eco-friendly alternatives to synthetic materials. Integrating beetroot into personalized nutrition and clinical formulations may further unlock its therapeutic potential. A multidisciplinary approach combining food science, material engineering, and biotechnology will be key to expanding the applications of beetroot in future health-oriented and sustainable food systems.

References

Akan, S., Tuna Gunes, N., & Erkan, M. (2021). Red beetroot: health benefits, production techniques, and quality maintaining for food industry. *Journal of Food Processing and Preservation*, 45(10). https://doi.org/10.1111/jfpp.15781

Borjan, D., Šeregelj, V., Andrejč, D. C., Pezo, L., Šaponjac, V. T., Knez, Ž., Vulić, J., & Marevci, M. K. (2022). Green techniques for preparation of red beetroot extracts with enhanced biological potential. *Antioxidants*, 11(5), 805. https://doi.org/10.3390/antiox11050805

Carmo, E. L. do, Teodoro, R. A. R., Félix, P. H. C., Fernandes, R. V. de B., Oliveira, É. R. de, Veiga, T. R. L. A., Borges, S. V., & Botrel, D. A. (2018). Stability of spray-dried beetroot extract using oligosaccharides and whey proteins. *Food Chemistry*, 249, 51–59. https://doi.org/10.1016/j.foodchem.2017.12.076

Chaari, M., Elhadef, K., Akermi, S., Ben Akacha, B., Fourati, M., Chakchouk Mtibaa, A., Ennouri, M., Sarkar, T., Shariati, M. A., Rebezov, M., Abdelkafi, S., Mellouli, L., & Smaoui, S. (2022). Novel active food packaging films based on gelatin-sodium alginate containing beetroot peel extract. Antioxidants, *11*(11), 2095. https://doi.org/10.3390/antiox11112095

Chhikara, N., Kushwaha, K., Sharma, P., Gat, Y., & Panghal, A. (2019). Bioactive compounds of beetroot and utilization in food processing industry: a critical review. *Food Chemistry*, 272, 192–200. https://doi.org/10.1016/j.foodchem.2018.08.022

Costa, A. P. D., Hermes, V. S., Rios, A. de O., & Flôres, S. H. (2017). Minimally processed beetroot waste as an alternative source to obtain functional ingredients. *Journal of Food Science and Technology*, 54(7), 2050–2058. https://doi.org/10.1007/s13197-017-2642-4

Czyżowska, A., Siemianowska, K., Śniadowska, M., & Nowak, A. (2020). Bioactive compounds and microbial quality of stored fermented red beetroots and red beetroot juice. *Polish Journal of Food and Nutrition Sciences*, 70(1), 35–44. https://doi.org/10.31883/pjfns/116611

Da Silva, D. V. T., Baião, D. dos S., Magalhães, A., Almeida, N. F., Conte, C. A., & Paschoalin, V. M. F. (2023). Combining conventional organic solvent extraction, ultrasound-assisted extraction, and chromatographic techniques to obtain pure betanin from beetroot for clinical purposes. Antioxidants, *12*(10), 1823. https://doi.org/10.3390/antiox12101823

Essalhi, H., Benchrifa, M., Tadili, R., & Bargach, M. N. (2018). Experimental and theoretical analysis of drying grapes under an indirect solar dryer and in open sun. *Innovative Food Science & Emerging Technologies*, 49, 58–64. https://doi.org/10.1016/j.ifset.2018.08.002

Ferguson, S. K., Holdsworth, C. T., Colburn, T. D., Wright, J. L., Craig, J. C., Fees, A., Jones, A. M., Allen, J. D., Musch, T. I., & Poole, D. C. (2016). Dietary nitrate supplementation: impact on skeletal muscle vascular control in exercising rats with chronic heart failure. *Journal of Applied Physiology*, 121(3), 661–669. https://doi.org/10.1152/japplphysiol.00014.2016

Fernando, G. S. N., Wood, K., Papaioannou, E. H., Marshall, L. J., Sergeeva, N. N., & Boesch, C. (2021). Application of an ultrasound-assisted extraction method to recover betalains and polyphenols from red beetroot waste. *ACS Sustainable Chemistry & Engineering*, 9(26), 8736–8747. https://doi.org/10.1021/acssuschemeng.1c01203

Fu, Y., Shi, J., Xie, S.-Y., Zhang, T.-Y., Soladoye, O. P., & Aluko, R. E. (2020). Red beetroot betalains: perspectives on extraction, processing, and potential health benefits. *Journal of Agricultural and Food Chemistry*, 68(42), 11595–11611. https://doi.org/10.1021/acs.jafc.0c04241

Guldiken, B., Toydemir, G., Nur Memis, K., Okur, S., Boyacioglu, D., & Capanoglu, E. (2016). Home-processed red beetroot (Beta vulgaris L.) products: changes in antioxidant properties and bioaccessibility. *International Journal of Molecular Sciences*, 17(6), 858. https://doi.org/10.3390/ijms17060858

Guo, Z., Ge, X., Li, W., Yang, L., Han, L., & Yu, Q. (2021). Active-intelligent film based on pectin from watermelon peel containing beetroot extract to monitor the freshness of packaged chilled beef. *Food Hydrocolloids*, 119, 106751. https://doi.org/10.1016/j.foodhyd.2021.106751

Khaw, K.-Y., Parat, M.-O., Shaw, P. N., & Falconer, J. R. (2017). Solvent supercritical fluid technologies to extract bioactive compounds from natural sources: a review. Molecules, *22*(7), 1186. https://doi.org/10.3390/molecules22071186

Lidder, S., & Webb, A. J. (2013). Vascular effects of dietary nitrate (as found in green leafy vegetables and beetroot) via the nitrate-nitrite-nitric oxide pathway. *British Journal of Clinical Pharmacology*, 75(3), 677–696. https://doi.org/10.1111/j.1365-2125.2012.04420.x

Mudgal, D., Singh, S., & Singh, B. R. (2022). Nutritional composition and value added products of beetroot: a review. *Journal of Current Research in Food Science*, 3(1), 1–9.

Ng, M. L., & Sulaiman, R. (2018). Development of beetroot (Beta vulgaris) powder using foam mat drying. *LWT*, 88, 80–86. https://doi.org/10.1016/j.lwt.2017.08.032

Nistor, O.-V., Seremet (Ceclu), L., Andronoiu, D. G., Rudi, L., & Botez, E. (2017). Influence of different drying methods on the physicochemical properties of red beetroot (Beta vulgaris L. var. Cylindra). *Food Chemistry*, 236, 59–67. https://doi.org/10.1016/j.foodchem.2017.04.129

Punia Bangar, S., Singh, A., Chaudhary, V., Sharma, N., & Lorenzo, J. M. (2023). Beetroot as a novel ingredient for its versatile food applications. *Critical Reviews in Food Science and Nutrition*, 63(26), 8403–8427. https://doi.org/10.1080/10408398.2022.2055529

Ravichandran, K., Saw, N. M. M. T., Mohdaly, A. A. A., Gabr, A. M. M., Kastell, A., Riedel, H., Cai, Z., Knorr, D., & Smetanska, I. (2013). Impact of processing of red beet on betalain content and antioxidant activity. *Food Research International*, 50(2), 670–675. https://doi.org/10.1016/j.foodres.2011.07.002

Seremet (Ceclu), L., Nistor, O.-V., Andronoiu, D. G., Mocanu, G. D., Barbu, V. V., Maidan, A., Rudi, L., & Botez, E. (2020). Development of several hybrid drying methods used to obtain red beetroot powder. *Food Chemistry*, 310, 125637. https://doi.org/10.1016/j.foodchem.2019.125637

Sharma, K. D., Karki, S., Thakur, N. S., & Attri, S. (2012). Chemical composition, functional properties and processing of carrot – a review. *Journal of Food Science and Technology*, 49(1), 22–32. https://doi.org/10.1007/s13197-011-0310-7

Shynkaryk, M. V., Lebovka, N. I., & Vorobiev, E. (2008). Pulsed electric fields and temperature effects on drying and rehydration of red beetroots. *Drying Technology*, 26(6), 695–704. https://doi.org/10.1080/07373930802046260

Sucu, C., & Turp, G. Y. (2018). The investigation of the use of beetroot powder in Turkish fermented beef sausage (sucuk) as nitrite alternative. *Meat Science*, 140, 158–166. https://doi.org/10.1016/j.meatsci.2018.03.012

Theba, T., Nayi, P., & Ravani, A. (2024). Beetroot-based blended juice: process development, physico-chemical analysis and optimization of novel health drink. *Food Chemistry Advances*, 4, 100607. https://doi.org/10.1016/j.focha.2024.100607

Thiruvengadam, M., Chung, I.-M., Samynathan, R., Chandar, S. R. H., Venkidasamy, B., Sarkar, T., Rebezov, M., Gorelik, O., Shariati, M. A., & Simal-Gandara, J. (2024). A comprehensive review of beetroot (*Beta vulgaris* L.) bioactive components in the food and pharmaceutical industries. *Critical Reviews in Food Science and Nutrition*, 64(3), 708–739. https://doi.org/10.1080/10408398.2022.2108367

Tropea, T., Renshall, L. J., Nihlen, C., Weitzberg, E., Lundberg, J. O., David, A. L., Tsatsaris, V., Stuckey, D. J., Wareing, M., Greenwood, S. L., Sibley, C. P., & Cottrell, E. C. (2020). Beetroot juice lowers blood pressure and improves endothelial function in pregnant eNOS$^{-/-}$ mice: importance of nitrate-independent effects. *The Journal of Physiology*, 598(18), 4079–4092. https://doi.org/10.1113/JP279655

Wang, Y., Fernando, G. S. N., Sergeeva, N. N., Vagkidis, N., Chechik, V., Do, T., Marshall, L. J., & Boesch, C. (2022). Uptake and immunomodulatory properties of betanin, vulgaxanthin I and indicaxanthin towards Caco-2 intestinal cells. Antioxidants, *11*(8), 1627. https://doi.org/10.3390/antiox11081627

Zhang, Y., Sun, B.-H., Pei, Y.-P., Vidyarthi, S. K., Zhang, W.-P., Zhang, W.-K., Ju, H.-Y., Gao, Z.-J., & Xiao, H.-W. (2021). Vacuum-steam pulsed blanching (VSPB): an emerging blanching technology for beetroot. *LWT*, 147, 111532. https://doi.org/10.1016/j.lwt.2021.111532

Zin, M. M., & Bánvölgyi, S. (2023). Emerging technology approach for extractability and stability of betalains from the peel of beetroot (Beta vulgaris L.). *Biomass Conversion and Biorefinery*, 13(12), 10759–10769. https://doi.org/10.1007/s13399-021-01975-z

Index

Note: **Bold** page numbers refer to tables; *italic* page numbers refer to figures.